WITHDRAWN FROM
TSC LIBRARY

Chernobyl – Catastrophe and Consequences

Jim T. Smith and Nicholas A. Beresford

Chernobyl – Catastrophe and Consequences

Published in association with
Praxis Publishing
Chichester, UK

Dr Jim T. Smith
Centre for Ecology and Hydrology
Winfrith Technology Centre
Dorchester
Dorset
UK

Dr Nicholas A. Beresford
Radioecology Group
Centre for Ecology and Hydrology
Lancaster Environment Centre
Cumbria
UK

SPRINGER–PRAXIS BOOKS IN ENVIRONMENTAL SCIENCES
SUBJECT *ADVISORY EDITOR*: John Mason B.Sc., M.Sc., Ph.D.

ISBN 3-540-23866-2 Springer-Verlag Berlin Heidelberg New York

Springer is part of Springer-Science + Business Media (springeronline.com)

Bibliographic information published by Die Deutsche Bibliothek

Die Deutsche Bibliothek lists this publication in the Deutsche Nationalbibliografie; detailed bibliographic data are available from the Internet at http://dnb.ddb.de

Library of Congress Control Number: 2005928344

Apart from any fair dealing for the purposes of research or private study, or criticism or review, as permitted under the Copyright, Designs and Patents Act 1988, this publication may only be reproduced, stored or transmitted, in any form or by any means, with the prior permission in writing of the publishers, or in the case of reprographic reproduction in accordance with the terms of licences issued by the Copyright Licensing Agency. Enquiries concerning reproduction outside those terms should be sent to the publishers.

© Praxis Publishing Ltd, Chichester, UK, 2005
Printed in Germany

The use of general descriptive names, registered names, trademarks, etc. in this publication does not imply, even in the absence of a specific statement, that such names are exempt from the relevant protective laws and regulations and therefore free for general use.

Cover design: Jim Wilkie
Project management: Originator Publishing Services, Gt Yarmouth, Norfolk, UK

Printed on acid-free paper

Contents

List of contributors xi

Editors and principal authors xiii

Contributing authors xv

List of figures xvii

List of tables xxi

List of abbreviations xxv

1   **Introduction** (*Jim T. Smith and Nick A. Beresford*) 1
   1.1   History of the accident 1
         1.1.1   Emergency response and early health effects 5
         1.1.2   Emergency clean up and waste disposal 7
         1.1.3   Radionuclides released and deposited 11
   1.2   Radiation exposures 16
         1.2.1   Health effects of radiation 16
         1.2.2   Exposure pathways and change of dose over time after Chernobyl 19
         1.2.3   Limiting the long-term dose to the population 23
         1.2.4   Unofficial resettlement of the abandoned areas 24
   1.3   Chernobyl in context 25
         1.3.1   Previous radioactive releases to the environment 25
         1.3.2   Natural radioactivity in the environment and medical radiation 27
   1.4   References 31

2 Radioactive fallout and environmental transfers (*Jim T. Smith and Nick A. Beresford*) ... 35

- 2.1 Pattern and form of radioactive depositions ... 35
  - 2.1.1 Element isotope ratios and 'hot' particles ... 37
  - 2.1.2 Break up of hot particles ... 40
- 2.2 Environmental transfers of radionuclides ... 41
  - 2.2.1 Migration of radionuclides in the soil ... 41
  - 2.2.2 Rates of vertical migration ... 44
  - 2.2.3 Change in external dose rate over time ... 46
  - 2.2.4 Resuspension of radioactivity ... 48
  - 2.2.5 Transport of radioactivity by rivers ... 53
- 2.3 Bioavailability, bioaccumulation and effective ecological half-lives ... 54
  - 2.3.1 Aggregated Transfer Factor and Concentration Ratio ... 54
  - 2.3.2 Physical, biological and ecological half-lives ... 55
  - 2.3.3 Changes in radiocaesium bioavailability over time ... 57
  - 2.3.4 Temporal changes in radiostrontium bioavailability ... 64
- 2.4 Characteristics of key Chernobyl radionuclides ... 65
  - 2.4.1 Radioiodine ... 65
  - 2.4.2 Radiostrontium ... 67
  - 2.4.3 Radiocaesium ... 69
  - 2.4.4 Plutonium and americium ... 71
- 2.5 References ... 73

3 Radioactivity in terrestrial ecosystems (*Jim T. Smith, Nick A. Beresford, G. George Shaw and Leif Moberg*) ... 81

- 3.1 Introduction ... 81
- 3.2 Agricultural ecosystems ... 85
  - 3.2.1 Interception of radioactive fallout by plants ... 85
  - 3.2.2 Transfer of radionuclides to crops and grazed vegetation ... 86
  - 3.2.3 Transfers to animal-derived food products ... 93
  - 3.2.4 Time changes in contamination of agricultural systems ... 103
  - 3.2.5 Very long-lived radionuclides in agricultural systems ... 107
- 3.3 Forest ecosystems ... 108
  - 3.3.1 Cycling of radioactivity in the forest ecosystem ... 108
  - 3.3.2 Transfer of radionuclides to fungi, berries and understorey vegetation ... 110
  - 3.3.3 Transfer of radionuclides to game and semi-domestic animals ... 116
  - 3.3.4 Radionuclides in trees ... 118
- 3.4 Radiation exposures from ingestion of terrestrial foods ... 119
  - 3.4.1 Reference levels of radioactivity in foodstuffs ... 119
  - 3.4.2 Radiation exposures from agricultural foodstuffs ... 121
  - 3.4.3 People now living in the abandoned areas ... 122
  - 3.4.4 Radiation exposures via the forest pathway ... 124
  - 3.4.5 Time dependence of exposures ... 126

		3.4.6	Comparison of radiocaesium transfers to various products	128

3.5 References 128

4 Radioactivity in aquatic systems (*Jim T. Smith, Oleg V. Voitsekhovitch, Alexei V. Konoplev and Anatoly V. Kudelsky*) 139

4.1 Introduction 139
    4.1.1 Distribution of radionuclides between dissolved and particulate phases 141
4.2 Radionuclides in rivers and streams 143
    4.2.1 Early phase 144
    4.2.2 Intermediate phase 149
    4.2.3 Long-term ^{137}Cs contamination of water 149
    4.2.4 Processes controlling declines in ^{90}Sr and ^{137}Cs in surface waters 150
    4.2.5 Influence of catchment characteristics on radionuclide runoff 152
4.3 Radioactivity in lakes and reservoirs 154
    4.3.1 Initial removal of radionuclides from the lake water ... 155
    4.3.2 The influence of lake water residence time 157
    4.3.3 The influence of lake mean depth d 159
    4.3.4 The influence of sediment–water distribution coefficient K_d 159
    4.3.5 Transport of ^{90}Sr in lakes 160
    4.3.6 Transport of ^{131}I in lakes 161
    4.3.7 Transport of Ruthenium in lakes 162
    4.3.8 Radionuclide balance in water of open lakes 162
    4.3.9 Closed lake systems 163
4.4 Radionuclides in sediments 165
4.5 Uptake of radionuclides to aquatic biota 168
    4.5.1 ^{137}Cs in freshwater fish 168
    4.5.2 Influence of trophic level on radiocaesium accumulation in fish 170
    4.5.3 Size and age effects on radiocaesium accumulation ... 170
    4.5.4 Influence of water chemistry on radiocaesium accumulation in fish 171
    4.5.5 ^{131}I in freshwater fish 173
    4.5.6 ^{90}Sr in freshwater fish 173
    4.5.7 Radiocaesium and radiostrontium in aquatic plants ... 174
    4.5.8 Bioaccumulation of various other radionuclides 174
4.6 Radioactivity in marine systems 175
    4.6.1 Riverine inputs to marine systems 177
    4.6.2 Transfers of radionuclides to marine biota 178
4.7 Radionuclides in groundwater and irrigation water 179
    4.7.1 Radionuclides in groundwater 179
    4.7.2 Irrigation water 180

viii Contents

	4.8	Radiation exposures via the aquatic pathway	180
	4.9	References	181

5 Application of countermeasures (*Nick A. Beresford and Jim T. Smith*).. 191
    5.1 Countermeasure techniques 191
        5.1.1 Methods of reducing uptake of radioiodine to the thyroid 192
        5.1.2 Methods of reducing the soil-to-plant transfer of radionuclides 192
        5.1.3 Methods of reducing the radionuclide content of animal-derived foodstuffs 193
        5.1.4 Countermeasures for freshwater systems 197
        5.1.5 Reduction of the external dose in residential areas 200
        5.1.6 Social countermeasures 201
    5.2 Countermeasures to reduce internal doses applied within the agricultural systems of the fSU 204
        5.2.1 Key foodstuffs contributing to ingestion doses 207
    5.3 Discussion 208
    5.4 References 209

6 Health consequences (*Jacov E. Kenigsberg and Elena E. Buglova*) 217
    6.1 Introduction 217
    6.2 Radiation-induced health effects 218
    6.3 Deterministic health effects after the Chernobyl accident 219
    6.4 Stochastic health effects after the Chernobyl accident 220
        6.4.1 Leukaemia 220
        6.4.2 Thyroid cancer 222
        6.4.3 Non-thyroid solid cancer 231
        6.4.4 Non-cancer diseases 232
    6.5 Discussion 232
    6.6 References 233

7 Social and economic effects (*Ingrid A. Bay and Deborah H. Oughton*).. 239
    7.1 Social and economic effects and their interactions 239
    7.2 Health detriments and associated harms due to radiation exposure 242
        7.2.1 Radiation exposure of the Chernobyl 'liquidators' 243
        7.2.2 Physiological health effects 244
        7.2.3 Psychological and social effects 245
    7.3 Economic impact 247
        7.3.1 Expenditures related to countermeasures 247
        7.3.2 Capital losses 249
        7.3.3 Rural breakdown 250
    7.4 Social costs of countermeasure implementation 251
        7.4.1 Evacuation and resettlement 252
        7.4.2 Countermeasures in agricultural food chains 254
        7.4.3 Compensation 254

		7.4.4	Communication and information	255
		7.4.5	The European response	257
		7.4.6	Risk perception	258
		7.4.7	Factors influencing risk perception	258
		7.4.8	Control	260
	7.5	Conclusions		261
	7.6	References		262
8	Effects on wildlife (*Ivan I. Kryshev, Tatiana G. Sazykina and Nick A. Beresford*)			267
	8.1	Terrestrial biota		268
		8.1.1	Radiation effects in forests	268
		8.1.2	Radiation effects in herbaceous vegetation	272
		8.1.3	Radiation effects in soil faunal communities and other insects	274
		8.1.4	Radiation effects in mammal populations	275
		8.1.5	Radiation effects in bird populations	276
	8.2	Freshwater biota		277
		8.2.1	Exposure of aquatic biota	277
		8.2.2	Radiation effects in aquatic biota	279
	8.3	The Chernobyl exclusion zone – a nature reserve?		280
	8.4	References		282
9	Conclusions (*Jim T. Smith and Nick A. Beresford*)			289
	9.1	Contamination of the environment		289
		9.1.1	Current radiation exposures in the Chernobyl affected areas	289
		9.1.2	Future environmental contamination by Chernobyl	290
		9.1.3	Countermeasures and emergency response	293
	9.2	Consequences of the accident		294
		9.2.1	Damage to the ecosystem	294
		9.2.2	Direct health effects of the accident	296
		9.2.3	Social and economic consequences	298
		9.2.4	Future management of the affected areas	300
		9.2.5	Chernobyl and the Nuclear Power Programme	301
	9.3	References		302
Index				307

Contributors

EDITORS AND PRINCIPAL AUTHORS

Jim Smith Centre for Ecology and Hydrology, Winfrith Technology Centre, Dorchester, Dorset, DT2 8ZD, UK.
E-mail: jts@ceh.ac.uk

Nick Beresford Centre for Ecology and Hydrology, Lancaster Environment Centre, Library Avenue, Bailrigg, Lancaster LA1 4AP, UK.
E-mail: nab@ceh.ac.uk

CONTRIBUTING AUTHORS

Ingrid Bay-Larsen Nordland Research Institute, Moerkvedtråkket 30, 8049 Bodø, Norway.
E-mail: ingrid.bay@umb.no

Elena Buglova Protection in Interventions Unit, Division of Radiation, Transport and Waste Safety, International Atomic Energy Agency, Wagramerstrasse 5, A-1400 Vienna, Austria.
E-mail: E.Buglova@iaea.org

Jacov Kenigsberg National Commission of Radiation Protection, Komchernobyl, 23 Masherov Ave., Minsk 220004, Belarus.
Email: jekenig@komchern.org.by

Alexei Konoplev Science & Production Association 'Typhoon', 82 Lenin Ave., Obninsk, Kaluga Region, 249020, Russia.
E-mail: konoplev@typhoon.obninsk.org

Ivan Kryshev Science & Production Association 'Typhoon', 82 Lenin Ave., Obninsk, Kaluga Region, 249020, Russia.
E-mail: ecomod@obninsk.com

Anatoly Kudelsky Institute of Geochemistry and Geophysics, 7 Kuprevich Str., Minsk 220141, Belarus.
E-mail: kudelsky@ns.igs.ac.by

Leif Moberg Swedish Radiation Protection Institute, S-171 Stockholm, Sweden.
E-mail: Leif.Moberg@ssi.se

Deborah Ougton The Ethics Programme, Forskningsparken, Gaustadallèen 21, 0349 Oslo, Norway.
Email: deborah.oughton@umb.no

George Shaw Division of Agricultural & Environmental Sciences, University of Nottingham, Sutton Bonington Campus, Loughborough, LE12 5RD, UK.

Tatiana Sazykina Science & Production Association 'Typhoon', 82 Lenin Ave., Obninsk, Kaluga Region, 249020, Russia.
E-mail: ecomod@obninsk.com

Oleg Voitsekhovitch Ukrainian Hydrometeorological Institue, 37 Nauka Ave., Kiev 252028, Ukraine.
Email: voitsekh@voi.vedos.kiev.ua

Editors and principal authors

Dr. Jim Smith is an expert in modelling radioactive pollution in terrestrial and freshwater ecosystems. He has coordinated three multi-national projects on the environmental consequences of Chernobyl and regularly works in the Chernobyl exclusion zone. He has worked with Ukrainian, Belarussian and Russian scientists on the Chernobyl accident for fourteen years and speaks Russian. Currently he is leading an EC INTAS project to evaluate the ecological effects of remediation of the Chernobyl Cooling Pond. Jim Smith has 53 papers on environmental radioactivity in the refereed scientific literature. Together with Nick Beresford and colleagues, he made the first long-term predictions of general ecosystem contamination by Chernobyl radiocaesium, for the first time quantitatively linking soil sorption kinetics to environmental mobility and uptake. He is a member of the International Atomic Energy Agency (IAEA) Expert Group on Chernobyl and Chairman of the UK Coordinating Group on Environmental Radioactivity.

Dr. Nick Beresford has 20 years experience as a radioecologist. His areas of expertise include the study of mechanisms controlling radionuclide and heavy metal transfer/ metabolism in (predominantly farm) animals and the development of countermeasures to reduce the entry of radioactivity into the human food chain. Recently he has worked, both nationally and internationally, on the development of guidance and methodologies to ensure protection of the environment from ionising radiation. In relation to the Chernobyl accident, he conducted extensive studies on upland sheep farms in the UK, tested and developed countermeasures and conducted experiments in the Chernobyl exclusion zone. He has also contributed to the development of approaches for sustainable management strategies for radioactively contaminated environments. He has published 62 papers in the refereed literature.

Contributing authors

Professor George Shaw holds the Chair in Environmental Science at Nottingham University. He has carried out research into the behaviour of radionuclides in agricultural and semi-natural ecosystems since 1987, and has studied semi-natural ecosystems in the Chernobyl 30-km exclusion zone. He has contributed to the IAEA's work on contaminated forests and timber products, as well as being a member of the IAEA/WHO Chernobyl Forum. **Dr. Leif Moberg** is the principal radioecologist at the Swedish Radiation Protection Authority (SSI) where he has worked for more than 20 years. He has worked on the consequences of the Chernobyl accident since 28 April, 1986 and for a number of years has been the spokesman for SSI on these issues. He is closely involved in the radiation-related work of a number of international organisations, including the EU, UNSCEAR, IAEA and OSPAR.

Dr. Oleg Voitsekhovitch is deputy director of the Ukrainian Hydrometeorological Institute and was responsible for monitoring radioactivity in aquatic systems after the Chernobyl accident and for advising on aquatic countermeasures. **Dr. Alexei Konoplev** is Head of the Centre for Environmental Chemistry at SPA 'Typhoon' in Russia. He is a leading expert on radionuclide interactions with soils and sediments, fuel particle disintegration and on radionuclide transfers to and in aquatic systems. He has worked on radioactivity in the Chernobyl area since 1986. **Professor Anatoly Kudelsky** is Head of the Hydrogeology Laboratory at the Institute of Geochemistry and Geophysics in Belarus. He is an expert in mobility of radioactivity in soils, groundwaters and surface water systems and has worked on Chernobyl problems since the accident.

Professor Jacov Kenigsberg is Chairman of the National Commission of Radiation Protection of the Republic of Belarus. He is a leading expert in radiation protection, radiation medicine and dosimetry. He has more than 200 scientific publications on the health consequences of the Chernobyl accident and radiation protection of the population. **Dr. Elena Buglova** is an expert on the health effects of ionising radiation. Formerly Head of the Laboratory of Radiation Hygiene and Risk Analysis at the Institute of Radiation Medicine and Endocrinology of Belarus, Dr. Buglova is currently working as a radiation protection specialist at the International Atomic Energy Agency.

Ingrid Bay has a Masters in radioecology and has worked on social and ethical issues of radioactive contamination in a number of international projects. She has held a scholarship in the Ethics Programme (University of Oslo) and is currently an associated researcher to Nordland Research Institute. **Professor Deborah Oughton** is a Research Fellow with the University of Oslo's Ethics Programme and Professor in Environmental Chemistry at the Norwegian University of Life Sciences. Her main scientific research interests lie within radioecology, particularly experimental studies on the transfer and biological effects of long-lived radionuclides. At the University of Oslo, her studies have focused on ethical issues within radiation protection, including the non-human environment, and the ethical, legal and social aspects of risk assessment and management in general.

Professor Ivan Kryshev is Head of Environmental Modelling and Risk Analysis at SPA 'Typhoon', Russia. He has 29 years experience in modelling the transfer of radionuclides in the environment, ecological dosimetry and analysis of countermeasures following the Chernobyl and Kyshtym accidents. He is an Academician of the Russian Academy of Natural Sciences, a Member of the Russian Scientific Commission of Radiation Protection and an Expert of the United Nations Scientific Committee on the Effects of Atomic Radiation. He is author of 280 scientific publications, including 18 books. **Dr. Tatiana Sazykina** is a leading research scientist at SPA 'Typhoon' in Obninsk, with 28 years experience in radioecological and ecosystem modelling. She is a key member of a number of international research projects into the effects of radiation on wildlife.

Figures

1.1	The destroyed Unit 4 reactor building at the Chernobyl Nuclear Power Plant	3
1.2	Aerial view of the destroyed reactor	3
1.3	The abandoned town of Pripyat with the Chernobyl Nuclear Power Station in the background	6
1.4	Satellite photo of the area around Chernobyl NPP	7
1.5	Sarcophagus construction, early September 1986	9
1.6	Sarcophagus photographed in 2003	9
1.7	Estimated daily releases of ^{131}I from the reactor for the period from the initial explosion to the extinction of the fire	12
1.8	(a) Percentage of the initial radioactivity remaining in the environment at different times after the Chernobyl accident, based on release data given in Table 1.2. (b) Changes in the amounts of some key radionuclides over time due to radioactive decay	14
1.9	^{137}Cs fallout in Ukraine, Belarus and Russia	15
1.10	^{137}Cs fallout in Europe	16
1.11	^{90}Sr and ^{239,240}Pu fallout in the Ukrainian part of the 30-km zone	17
1.12	(a) Fatal solid cancer; (b) leukaemia rates in the follow up group of people exposed to radiation from the Hiroshima and Nagasaki atomic bombs	18
1.13	Effective dose to the populations of Belarus, Russia and the Ukraine (excluding thyroid dose) during the period 1986–1995	20
1.14	Thyroid dose to children less than 18 years old in the two most affected regions of Belarus	21
1.15	Change in external gamma dose rate over time after the accident	22
1.16	Contrasting contributions of internal and external dose rates to overall dose in areas of different soil types, Bryansk Region, Russia	23
2.1	Electron micrograph of a uranium fuel particle from Chernobyl	36
2.2	Concentration of different radionuclides in the air and deposition as a function of distance from Chernobyl, expressed as a ratio of radionuclide : ^{137}Cs	38
2.3	Half-time of dissolution of fuel particles as a function of soil pH	41

2.4	Examples of activity–depth profiles of various radionuclides in soils	43
2.5	Decline in the observed dispersion coefficient as a function of time in a grassland soil at Veprin, Belarus	44
2.6	External dose rate 0.05 m above the ground as a function of ^{137}Cs inventory in the soil	47
2.7	Change in annual effective external dose for rural indoor workers living in wood-framed houses	48
2.8	Change in resuspension factor as a function of time after fallout at a number of sites around Europe	49
2.9	Annual mean ^{137}Cs resuspension factors measured at 20 different sites around Europe at large distances from Chernobyl for two different time periods	50
2.10	Illustration of Cs sorption to specific FES on illitic clay minerals and competition for sorption sites by ions of similar hydrated radius (K^+, NH_4^+) but not ions with much larger hydrated radius such as Ca^{2+}, Mg^{2+}	57
2.11	Illustration of the dynamic model for radiocaesium sorption to illitic clay minerals showing rapid uptake to 'exchangeable' sites and slower 'fixation' in the mineral lattice	57
2.12	Schematic diagram indicating timescales of release of radiocaesium from soils to terrestrial and aquatic ecosystems during the years after a fallout event	58
2.13	Illustration of changes in radiocaesium in milk in a system with declining activity concentrations in vegetation and relatively rapid rates of uptake and removal from milk	59
2.14	(a) Examples of changes in ^{137}Cs activity concentration in different ecosystem components after Chernobyl. (b) Frequency distribution of effective ecological half lives in different ecological components during the first five years after Chernobyl. (c) Long-term changes in ^{137}Cs in brown trout, Norway, and perch, terrestrial vegetation and water, UK	60
2.15	Change in CR of wheat on soddy–podzolic soil	65
3.1	Ranges in ^{137}Cs activity concentration in various products from the Luginsk district, Zhitomir region, Ukraine in 1995	84
3.2	Plot of T_{ag} vs. organic matter content in soils in 5 catchments in Cumbria, UK	88
3.3	CR of ^{90}Sr and ^{137}Cs in various vegetables in Finland, clay and silt soils, 1987	93
3.4	Comparison between calcium intake and F_m for strontium with additional recent data for cattle	96
3.5	Variation in ^{137}Cs activity concentration in 1,144 sheep, Cumbria, UK	99
3.6	Increase in feed–milk transfer coefficient over time at a farm in Bavaria	101
3.7	Time changes in (a) ^{131}I and (b) ^{137}Cs in air, grass and milk in north-western Italy during the first month after the accident	104
3.8	Time changes in the aggregated transfer factor of ^{137}Cs in the decade after the accident	106
3.9	Seasonal trends in the ^{137}Cs activity concentrations of study ewes at one of the farms of Beresford et al. (1996)	106
3.10	Major storages and fluxes in radionuclides in contaminated forest ecosystems	110
3.11	Radiocaesium profiles in forest soils in (a) the Chernobyl 30-km zone at two different times after the accident and (b) in a forest soil in Germany (in 1996) contaminated by Chernobyl and weapons test fallout	111
3.12	Aggregated transfer factor for ^{137}Cs in a very highly accumulating mushroom species	113

3.13	Change in ^{137}Cs activity concentration in roe deer meat in a spruce forest, Ochsenhausen, Germany	117
3.14	A summary of the ^{137}Cs activity concentration measured in the milk of cattle owned by people living within the 30-km exclusion zone	124
3.15	A comparison of the consumption rate of fungi and the whole body ^{137}Cs burden determined in people living in an urban area of Russia	125
4.1	Pripyat–Dnieper River–Reservoir system showing Chernobyl and Kiev with the Kiev Reservoir in between	140
4.2	Fraction of a radionuclide absorbed to particulates as a function of suspended solids concentration in water for different values of K_d	143
4.3	The change in activity concentration of ^{137}Cs and ^{90}Sr in the Pripyat River over time after the accident	145
4.4	The initial activity concentrations of radionuclides in various rivers vs. the total amount released from the reactor	147
4.5	(a) Normalised activity concentration of ^{137}Cs in the dissolved phase of different rivers after Chernobyl. (b) Correlation between the normalised ^{137}Cs activity concentration and the percentage catchment coverage of organic, boggy soils in six different catchments	153
4.6	Radionuclide transfers in a catchment–lake system	154
4.7	Comparison of initial ^{137}Cs activity concentration in 15 lakes determined from measurements with that estimated from a simple dilution model	155
4.8	Change in the ^{137}Cs activity concentration in water and fish of: (a) a small shallow lake in Germany, Lake Vorsee and; (b) the large, deep Lake Constance	156
4.9	(a) The relationship between ^{137}Cs removal rate from 14 lakes and the removal rate of water through the outflow. (b) The relationship between the fraction of the total ^{137}Cs transferred to the outflow and the lake water residence time. (c) The relationship between ^{137}Cs removal rate and the lake mean depth	158
4.10	Changes in average annual content of ^{137}Cs and ^{90}Sr in the water of the first (Vishgorod, Kiev Reservoir) and last (Novaya Kahovka, Kahovka Reservoir) reservoirs of the Dnieper cascade	161
4.11	Graphs of ^{137}Cs activity–depth profiles in sediments in (a) Baltic Sea, muddy and sandy sediments; (b) Lake Constance; (c) Lake Svyatoe, Kostikovichy, Belarus.	167
4.12	Illustration of a simple model for uptake in fish via the food chain	169
4.13	Radiocaesium in fish in the Kiev Reservoir after Chernobyl, illustrating the 'size effect' in predatory perch, but not in the non-predatory roach	171
4.14	Relationship between ^{137}Cs concentration factor in fish and the potassium concentration in 17 lakes around Europe	172
4.15	Radiocaesium in the Baltic and Black Seas	177
5.1	Nick Beresford live-monitoring a sheep in upland west Cumbria in 1993 to determine Cs-137 activity concentration in muscle	196
5.2	Decrease in ^{137}Cs activity concentrations in perch in Lake Svyatoe over a 15-year period after a potassium countermeasure was applied	199
5.3	Variability within the ^{137}Cs activity concentration of private milk within the Belarussian village	208

Figures

6.1	Increase in thyroid cancer in children (aged 0–18 years at the time of the Chernobyl accident) in Belarus during the period 1986–2002	229
6.2	Increase in excess thyroid cancer risk in the period 1991–1995 in children born between 1971 and 1986	230
7.1	Interaction between health, social and economic effects	241
8.1	Area of Red Forest where coniferous trees were killed as a consequence of acute irradiation but deciduous trees continued to grow	268
8.2	Pripyat sports stadium	281
8.3	Kestrels nesting on the roof of a tower block, Pripyat	281
9.1	Rise in world primary energy consumption from 1970–2025	301
9.2	World consumption of nuclear energy from 1970 and projected future use	302

Tables

1.1	Confirmed cases of acute radiation sickness in emergency workers.	5
1.2	Physical half-lives and amounts of radionuclides released from Chernobyl . . .	12
1.3	Estimates of releases of some additional radionuclides compared with ^{137}Cs. .	13
1.4	Main pathways and nuclides contributing to the population exposure after the Chernobyl accident	20
1.5	Radiation exposures of different groups after Chernobyl.	21
1.6	Population dynamics within abandoned settlements of Belarus in selected years after Chernobyl.	25
1.7	Summary of previous major releases of radioactive material to the environment	26
1.8	Examples of some measurements of ^{137}Cs in the environment before the Chernobyl accident	27
1.9	Population average doses from natural radiation sources and average dose in various European and North American countries from medical diagnostic procedures	28
1.10	Doses from various X-ray medical diagnostic procedures	28
1.11	Primordial radionuclides and some of their decay products.	29
1.12	Concentrations of natural radioactive potassium in various foodstuffs	30
2.1	Radionuclide resuspension factors from agricultural activity, traffic and forest fires compared with natural wind resuspension.	52
2.2	Summary of mean values of rate of decline in ^{137}Cs activity concentrations in different environmental compartments, and comparison with rate of diffusion of ^{40}K into the illite lattice.	61

Radioiodine Isotope data.	65
Examples of stable iodine concentrations in the environment	67
Radiostrontium isotope data	67
Examples of stable strontium concentrations in the environment	68
Radiocaesium isotope data.	69
Examples of stable caesium concentrations in the environment	70
Plutonium and americum isotope data	71

Tables

3.1	Average ratio of fresh weight:dry weight of various products	83
3.2	Illustrative productivities and radiocaesium transfer factors of food products derived from different ecosystem types	84
3.3	Percentage of the total plant contamination from different contamination pathways	87
3.4	Soil–grass aggregated transfer factor for radiocaesium	89
3.5	^{137}Cs and ^{90}Sr soil-to-grass aggregated transfer coefficient for different soil groups, Bragin, Belarus, 1994–1995	90
3.6	Aggregated transfer factors of ^{90}Sr and ^{137}Cs to various crops	92
3.7	Recommended factors for radiocaesium to convert CR or T_{ag} values for cereals to values for other crops	93
3.8	Recommended factors for radiostrontium to convert CR or T_{ag} values for cereals to values for other crops	94
3.9	Recommended transfer coefficients for radiocaesium and dry matter feed intake rates	95
3.10	Radiocaesium transfer coefficients to various organs of cows, goats and sheep	97
3.11	Examples of radioactivity concentrations in milk and meat of domestic animals in various parts of Europe contaminated by the Chernobyl accident	98
3.12	Feed–milk transfer coefficient following intake of contaminated herbage by cows	100
3.13	Ratios of activity concentrations of ^{90}Sr and ^{137}Cs in milk products to those in milk	103
3.14	Estimated activity concentrations in milk and beef from a hypothetical pasture located in an area very highly contaminated by transuranium elements	108
3.15	Estimated activity concentrations of cereals and potatoes grown on hypothetical agricultural land in an area very highly contaminated by transuranium elements	108
3.16	^{137}Cs in various components of a pine forest, Bourakovka, Chernobyl in 1990. ^{90}Sr (from weapons tests) in different components of a pine forest in Sweden in 1990	109
3.17	Aggregated transfer factors of ^{137}Cs in various species of edible fungi collected in Belarus	112
3.18	Comparison of mean T_{ag} values for ^{137}Cs and ^{90}Sr in fungi in the Bragin district of Belarus, 1994–1995	113
3.19	Comparison of concentration ratios of ^{137}Cs with other radionuclides in understorey vegetation	114
3.20	Range in transfer factors and effective ecological half-lives observed in game during the first few years after Chernobyl	117
3.21	^{137}Cs and ^{90}Sr in game animals in the Bragin district of Belarus, 1994–1995	118
3.22	Radiocaesium transfer factors in different parts of trees at Dityatki, 28 km south of Chernobyl during 1987	118
3.23	Agreed CFILs of radionuclides in foods in place in the EC	120
3.24	Intervention limits for the ^{137}Cs activity concentration in foodstuffs within Belarus, Russia and the Ukraine as in place in 1999	121
3.25	Average annual consumption of foodstuffs by the population of a village in Bryansk, Russia, before and after the Chernobyl accident	122
3.26	Example of consumption rates of different foodstuffs and the contribution of each foodstuff to the daily ^{137}Cs intake, as determined during June/July 1997 in Milyach, the Ukraine	123

3.27	Mean effective dose in 15 forest units in the Novozybokov district, Bryansk region, Russia.	125
3.28	^{137}Cs transfer factors and illustrations of activity concentrations of different foodstuffs from measurements made in the early 1990s.	127
4.1	K_d values for radiostrontium, radioiodine, radiocaesium and plutonium in freshwaters.	142
4.2	Radionuclide levels in the Pripyat River at Chernobyl.	146
4.3	Temporary allowable levels of radionuclides in drinking water in the Ukraine at different times after Chernobyl.	147
4.4	Estimates of the initial rate of decline of radionuclides in river water after Chernobyl.	148
4.5	Rates of change in ^{137}Cs and ^{90}Sr activity concentrations in different rivers in the medium to long term (1987–2001) after Chernobyl.	150
4.6	Comparison of radiocaesium K_d determined from removal rate measurements with K_d measured in the field or laboratory.	160
4.7	Mean ^{137}Cs and ^{90}Sr activity concentration in inflow streams compared with concentrations in the lake water/outlet of different lakes.	163
4.8	Normalised water concentrations of ^{137}Cs and ^{90}Sr in various water bodies 4–10 years after fallout.	164
4.9	Radionuclides in Chernobyl Cooling Pond bed sediments approximately one month after the accident, expressed as a percentage of the total amount in both sediments and water.	165
4.10	Typical radionuclide activity concentrations in the most contaminated silty sediments of the Cooling Pond.	166
4.11	^{90}Sr concentration factors in freshwater fish after Chernobyl.	173
4.12	Mean CF of radiocaesium in aquatic plants.	175
4.13	Radionuclide CFs in biota of the Dnieper River in June 1986.	176
4.14	Radionuclides in marine macroalgae and fallout compared to ^{137}Cs in July 1986 and August–September 1987.	178
5.1	The reduction achieved in the radiocaesium content of fungi following commonly used cooking procedures.	203
5.2	Summary of the effectiveness of different agricultural countermeasures to reduce ^{137}Cs activity concentrations employed within the fSU.	205
5.3	Suggested feeding regime for beef cattle at various times prior to slaughter and the effect on the activity concentration in meat.	205
5.4	Changes in the amount of meat and milk produced by collective farms with ^{137}Cs activity concentrations in excess of intervention limits.	206
6.1	The most critical radiation-induced health effects resulting from a radiation exposure.	218
6.2	Examples of stochastic health effects from exposure to radiation.	219
6.3	Emergency workers with ARS.	220
6.4	Distribution of external doses to emergency workers as recorded in the registry of emergency workers.	222
6.5	Results of studies of the risk of thyroid cancer development following acute	

	external radiation from atomic bombs and from radiation therapy at an age of <20 years.	224
6.6	Results of studies of the risk of thyroid cancer development following exposure to ^{131}I at an age of <20 years.	225
6.7	Risk of radiation-induced thyroid cancer following radiation exposure at an age of <20 years.	227
6.8	Thyroid dose distribution for various age groups in Belarus.	228
6.9	Risk of thyroid cancer for children and adolescents of Belarus considering gender and age at the time of the accident.	230
6.10	Thyroid cancer risk for the exposed adult population of Belarus.	231
7.1	Estimated number of people affected by the accident in terms of evacuation, resettlement, people living in contaminated areas, liquidators and invalids.	239
7.2	Registered cases of 'class 16 illnesses' in the Ukraine (1982–1992).	246
7.3	Estimates of total expenditures, proportion of national budgets, numbers of newly built houses, schools and hospitals as a part of remediation and relocation actions in the three most affected countries; Belarus, Russia and the Ukraine.	248
7.4	Chernobyl budget expenditures for the Ukraine (US$ million in 2000).	249
7.5	Selected losses related to agricultural/forest land, and economic units.	250
7.6	Benefits and costs of remediation efforts.	252
7.7	Factors commonly used to explain the perception of risk.	259
8.1	Distribution of radiation damage in forests around the Chernobyl NPP.	269
8.2	The dynamics of external irradiation dose from soil at study plots with coniferous trees close to the Chernobyl NPP.	270
8.3	Radionuclide activity concentration in pine needles and estimated internal dose rate in October 1987.	271
8.4	Frequency of meiotic chromosomal aberrations in pine microsporocytes.	272
8.5	Temporal dynamics of conditions in study pine stands.	273
8.6	Frequency of mutation in the stamen filament hair of spiderwort at different dose rates.	274
8.7	Effects of chronic radiation exposure on reproduction and off-spring of silver carp in the Chernobyl Cooling Pond.	278
9.1	Estimated activity concentrations of long-lived radionuclides at Kopachi, 6 km south-east of Chernobyl.	291
9.2	Tentative prediction of future contamination by Chernobyl.	292

Abbreviations

ACF	aggregated concentration factor
ARS	acute radiation syndrome
BNFL	British Nuclear Fuels Plc.
BAF	bioaccumulation factor
CF	concentration factor
CR	concentration ratio
CEC	cation exchange capacity
CFILs	Council Food Intervention Limits
CLL	Chronic Lymphocytic Leukaemia
CI	confidence interval
d.w.	dry weight
EC	electrical conductivity
ERR	excess relative risk
EAR	excess absolute risk
fSU	former Soviet Union
f.w.	fresh weight
FES	'Frayed Edge Sites'
GDL	Generalised Derived Limit
IAEA	International Atomic Energy Agency
ICRP	International Commission on Radiation Protection
IRG	inert radioactive gases
NPP	Nuclear Power Plant
NWT	nuclear weapons test
RBE	relative biological effectiveness
RN	radionuclide
SEER	US Program Surveillance Epidemiology and Results
TUE	transuranium elements

UNSCEAR	United Nations Scientific Committee on the Effects of Atomic Radiation
UNDP	United Nations Development Programme
UNICEF	United Nations Children's Fund
UN-OCHA	UN Office for the coordination of Humanitarian Affairs
WHO	World Health Organisation

1

Introduction

Jim T. Smith and Nick A. Beresford

The explosion at Unit 4 of the Chernobyl nuclear power station was the worst nuclear accident in history. Radioactive fallout from the accident (directly or indirectly) affected the lives of hundreds of thousands of people in the former Soviet Union and contamination spread throughout Europe. In the 19 years since the accident, thousands of scientific papers have been published on Chernobyl and its consequences. In this book we have tried to summarise this vast literature, focusing particularly on the long-term consequences of the accident to people and the environment.

There are many historical accounts of the Chernobyl accident (e.g., Shcherbak, 1989; IAEA, 1991; UNSCEAR, 2000; Mould, 2000; Kryshev and Ryazantsev, 2000; OECD/NEA, 2002). Whilst this book focuses primarily on the longer term impacts of the accident, here we briefly summarise the history of the accident and its immediate consequences. In particular, we aim to put Chernobyl within the context of other (natural and man-made) sources of radioactivity in the environment. This chapter also introduces some key concepts and units of radiation measurement and risk assessment. Many of these will be familiar to some readers, and are therefore where possible included in boxes separate from the main text.

1.1 HISTORY OF THE ACCIDENT

At the time of the accident in 1986, Chernobyl was one of four nuclear power stations in the Ukraine and was part of a rapid expansion in nuclear generating capacity. The Chernobyl power station consisted of four 'RBMK-1000'-type reactors, the first of which, Unit 1, began electricity generation in 1977. Electricity generation at Unit 4 (the reactor at which the accident occurred) was begun in 1983 and in 1986 two other Units (5 and 6) were being built. Construction of these last units stopped after the accident.

The accounts of the accident and its immediate aftermath (Shcherbak, 1989; IAEA, 1991; Mould, 2000) make truly chilling reading. There are still some uncertainties regarding the exact causes and events leading to the accident, though the key factors are now known. The accident occurred during an experiment to test the behaviour of an electrical system which powered the station in the event of a failure of the main electricity supply. In order to conduct the experiment, the reactor thermal power output had to be reduced to 700–1,000 MegaWatts (MW), about 25% of its maximum power output.

At 13:00 on 25 April, 1986, the plant operators began reduction of the reactor power in preparation for the experiment. At 14:00, however, the operators received a request from Kiev to continue supplying electricity until 23:10 that evening, so the experiment was postponed. At 23:10 reduction of the reactor power output began again and at just after midnight on 26 April, reactor power was 720 MW. Approximately 30 minutes later, however, power output had fallen to just 30 MW. This unexpected fall in the power output is believed to have been due to a problem in the operation of the automatic control rods (which were designed to control the reactor power under low-power conditions).

At 01:00, the operators had stabilised reactor power at 200 MW by removing some of the control rods. During the next 20 minutes the operators varied the flow rate of water in the coolant circuit, leading to a significant variation in temperature of the inlet water. The reactor has been described as being in an unstable condition during this period (UNSCEAR, 2000): the coolant flow was almost completely liquid water with no stream entrained. At 01:22:30 the operator received an automatic printout which indicated that the reactor should be shut down immediately (IAEA, 1986). This warning was ignored. At 01:23 the experiment began, despite the fact that:

- the reactor power output was well below that required by the experimental procedure;
- certain reactor safety systems had been deliberately disabled in order to carry out the experiment; and
- the number of control rods in the reactor was only half the minimum required for its safe operation.

Thirty seconds after the experiment began, the reactor power began to increase rapidly and ten seconds later the operators attempted a full emergency shut down by re-inserting the control rods. The reactor power was now increasing exponentially leading to a failure in the pressurised cooling water system. Eight seconds later, the reactor exploded (an explosion of steam, not a nuclear explosion) scattering burning core debris over the surrounding area. The ruined reactor is shown in Figures 1.1 and 1.2.

> **Box 1.1. Design flaw in the RBMK reactor.**
>
> The RBMK nuclear reactor used at Chernobyl, in contrast to most nuclear reactors, had what is known in the nuclear industry as a 'positive void coefficient'. In an accident situation, should cooling water be lost or turned to steam, most reactors (with 'negative void coefficient') naturally reduce their power output. In the RBMK reactor, loss of cooling water results in an increase in power output and consequent temperature rise in the reactor core. This in turn causes more of the coolant water to turn to steam, leading, potentially, to an uncontrolled rise in power output.

Figure 1.1. The destroyed Unit 4 reactor building at the Chernobyl Nuclear Power Plant (NPP). The edge of the Cooling Pond can be seen top left.

Figure 1.2. Aerial view of the destroyed reactor.

Over 100 firemen were called to the scene and they worked with plant personnel to put out many small fires in the reactor building and on the roofs of Unit 4 and the adjacent Unit 3 building. This work exposed the emergency workers to extremely high doses of radiation. The report of the IAEA International Chernobyl Project (IAEA, 1991) describes the scene:

> By dawn on the Saturday [26 April], more than 100 firemen had succeeded in putting out the roof fires, and by about 05:00 all but the graphite fire in the [reactor] core had been extinguished. These courageous actions by the early firefighters and plant personnel resulted in many injuries, but they were essential to preventing the spread of the fire to the other units and to preventing a hydrogen explosion or fire that might have ignited the oil in the turbines. Many firemen stayed on the alert on the premises for several hours after the fire was out, which resulted in a number of radiation exposures.
>
> Radiation levels were so high in the damaged part of the plant and just outside it that monitoring equipment in the plant could not measure them. Available portable radiation meters went off-scale and systematic monitoring became impossible. It seems that many of those who entered the buildings to rescue others, fight fires, perform critical operations or assess damage did not appreciate the radiation risk.

Although the initial fires had been put out, the destroyed reactor core continued to burn. During the days after the explosion, helicopters were used to dump thousands of tonnes of various materials onto the exposed reactor core. These materials included boron, lead, sand and clay to smother the fire, absorb radiation and reduce nuclear reactions in the molten core material. In total, 1,800 helicopter flights were made at great risk to the pilots (UNSCEAR, 2000). Despite the heroic efforts of firemen, helicopter pilots and many other emergency workers to put out the fire, the reactor continued to burn for 10 days.

Box 1.2. Myths and revelations.

Soon after the accident, an article appeared in the *New York Times* claiming that the Ukrainian word 'Чорнобиль' (Chernobyl) translates to English as 'Wormwood' (a bitter herb) and quoting a verse from the Book of Revelations:

> The third angel sounded and there fell a great star from heaven, burning, as it were a lamp, and it fell on the third part of the rivers and upon the fountains of water; And the name of the star is Wormwood; and the third part of the waters became Wormwood, and many men died of the waters because they were made bitter.

This has been interpreted by some as giving an apocalyptic dimension to the tragedy, particularly since radioactivity polluted rivers and reservoirs in the Ukraine. In fact, the herb named Chernobyl ('Чернобыльник' – Russian, Чорнобиль – Ukrainian) is the Mugwort (*Artemesia vulgaris*). The Wormwood (Полынь горькая, polyn gorkaya – Russian, Полин гіркий' polyn girkiy – Ukrainian) is a related, but different species, *Artemesia absinthum*.

1.1.1 Emergency response and early health effects

In the early stages of the accident, many power plant operators and emergency workers were exposed to very high doses of radiation. This was a result of external gamma radiation from the exposed reactor core and core debris, as well as exposure to beta radiation from contamination of their skin and clothes (see Box 1.3 for a description of different radiation types). One hundred and thirty four emergency workers were confirmed as suffering from acute radiation sickness (UNSCEAR, 2000), 28 of whom died during the months after the accident (Table 1.1). Internal radiation exposure of these people (mainly from inhalation of radioiodine and radiocaesium) was in general much lower than the external exposure. Chapter 6 presents a fuller discussion of the health consequences of the accident.

Box 1.3. Charateristics of some radioactive emissions.

Radiation type	Description	Stopped by:	Approximate relative biological effectiveness*
Alpha particle	Helium nucleus	Air or outer layers of skin	20
Beta particle	Electron	Few mm of aluminium	1†
Gamma ray	Electromagnetic wave	Few cm of lead	1

* Relative biological effectiveness is used to convert radiation energy absorbed by the human body into a radiation dose: for a given absorbed energy alpha radiation is estimated to be approximately 20 times more biologically damaging than high-energy beta or gamma.
† For beta energies <10 keV a value of 3 is often used.

Table 1.1. Confirmed cases of acute radiation sickness in emergency workers. Adapted from UNSCEAR (2000).

Degree of acute radiation sickness	Range of external radiation dose*	Number of people affected	Number of deaths
Mild	0.8–2.1 gray	41	0
Moderate	2.2–4.1 gray	50	1
Severe	4.2–6.4 gray	22	7
Very severe	6.5–16 gray	21	20
Total		*134*	*28*

* See Box 1.4 for a definition of the gray.

Measures to protect both the people on the site, and the population of the surrounding areas were, in the very early stages of the accident, inadequate. Firemen had not been trained in radiation protection and had no dosimeters to control their radiation exposure. Although potassium iodide tablets (to block radioiodine uptake by the

Figure 1.3. The abandoned town of Pripyat with the Chernobyl Nuclear Power Station in the background.

thyroid) were distributed to power plant workers within half an hour of the accident (UNSCEAR, 2000), there was 'no systematic distribution' (IAEA, 1991) of tablets to the population of Pripyat, a town approximately 3 km from the plant (Figures 1.3 and 1.4). Face masks to protect from inhalation of radioactivity were not available to the population and there were no official warnings for people to stay indoors, out of the contaminated air. Many children in Pripyat were playing outdoors on 26 April (the accident occurred in the early hours of 26 April), unaware of the potential danger.

At 14:00 on Sunday 27 April, the 44,000 population of Pripyat were evacuated in 1,200 buses. On 2 May it was decided to evacuate people and cattle from an area of approximately 30 km radius around the plant (the '30-km zone'), the boundary being based on a map of radiation dose rate. By 6 May, the entire 30-km zone had been evacuated. Subsequent mapping of contamination later led to more evacuations, including areas in Belarus and the Bryansk region of Russia around 150 km to the northwest of the reactor. In total, approximately 116,000 people (Belyaev *et al.*, 1996) and 60,000 cattle (UNSCEAR, 2000) were initially evacuated from an area of approximately 3,500 km^2. In subsequent years many more people were evacuated, reaching approximately 350,000 (UNDP/UNICEF, 2002). At present, many of the evacuated areas remain uninhabited, though some small areas have been re-settled.

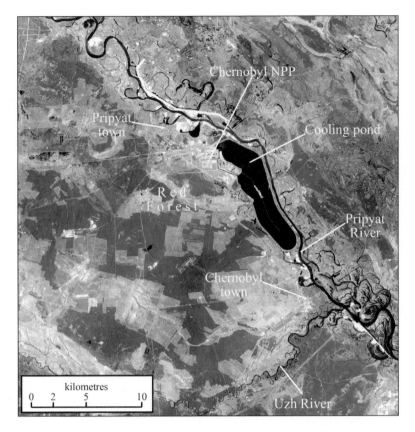

Figure 1.4. Satellite photo of the area around Chernobyl NPP. Note that the town of Chernobyl is about 15 km south of the plant: it had much lower radioactive fallout than many other areas. In this book 'Chernobyl' refers to the nuclear reactor rather than the town unless otherwise stated.
Photo adapted by Simon Wright from the original with the kind permission of Valery Kashparov of the Ukrainian Institute of Agricultural Radiology (UIAR, 2001).

1.1.2 Emergency clean up and waste disposal

A concrete structure (the 'shelter' or 'sarcophagus') was built around the destroyed reactor building in order to prevent further releases of radioactive material (Figures 1.5 and 1.6). The sarcophagus was built rapidly under extremely difficult conditions; work was completed in November 1986 (Belyaev *et al.*, 1996). Since its construction, there have been concerns about the structural integrity of this temporary building. The sarcophagus was (necessarily) built using existing parts of the reactor building as support and the stability of these existing structures is not precisely known.

There has been particular concern that the sarcophagus could collapse in the event of an earthquake, for example (though seismic activity in this area is not high).

> **Box 1.4. Some units and definitions**
>
> The amount of radioactive material is measured in **becquerels** (Bq). An amount of radioactive material equal to one becquerel will, on average, undergo radioactive decay at a rate of one decay per second. For example, in one becquerel of radiocaesium (^{137}Cs), one radiocaesium atom will, on average, decay every second, emitting a beta particle and a gamma ray. In terms of the mass of substance, one becquerel of ^{137}Cs is a very small amount: 2.3×10^{-15} moles or 3.1×10^{-13} grammes.
>
> Because of radioactive decay, the amount of a radioactive substance is continually decreasing. This decrease is measured by the **radioactive half-life**, defined as the time taken for a given amount of radioactive substance to decline by one-half. For radiocaesium, for example, this half-life is 30.1 years, so by 2016, 30.1 years after the accident, the amount of Chernobyl-derived radiocaesium in the environment will have halved. Most of the radionuclides released had a much shorter half-life than radiocaesium, leading to rapid declines in the radiation exposures during the first days and weeks after the accident.
>
> The amount of radiation absorbed in the human body is measured in **grays** (Gy) – one gray equals one joule of radiation energy absorbed per kilogram. This absorbed energy can be estimated quite precisely using models for the transfer of ionising radiation through tissues. The damage that the absorbed radiation does to people is much more difficult to quantify, particularly at the relatively low doses associated with environmental contamination. There are, however, a number of studies relating absorbed dose to risk of adverse health effects (primarily cancer risk).
>
> Because different types of radiation damage cells in different ways, radiation risk is measured in *dose equivalents*, **sieverts** (Sv) rather than in total energy absorbed. For high-energy beta and gamma radiation, one gray of absorbed energy results in a dose equivalent of approximately one sievert. For the (potentially) more damaging alpha particles, one gray of absorbed energy results in a dose equivalent of approximately 20 sieverts. The term *effective dose* is also often used: this represents the weighted dose to the whole body.

Before 1996, the sarcophagus had withstood an earthquake of Richter Scale 4 intensity (Borovoi, 1996). In 1996, this author reported that 'up to this time, no hazardous deteriorations in the upper structures [of the sarcophagus] had been reported'. Nearly 10 years later, however, there is serious concern that deterioration of the structure could result in a potential failure to withstand a major earthquake, or a collapse of damaged structures within the sarcophagus. Such a collapse could release a proportion of the large amount of radioactive dust which has accumulated in the sarcophagus since its construction. It is estimated that there is approximately 2 tonnes of dust which could potentially (under extreme conditions) be resuspended

Figure 1.5. Sarcophagus construction, early September 1986: construction of the northern cascade wall.
Photo: The Kurchatov Institute (Russia) and the ISTC-Shelter Project (Ukraine).

Figure 1.6. Sarcophagus photographed in 2003.
Photo: Dave Timms (Portsmouth University, UK).

(Borovoi et al., 1999). In order to suppress resuspension of radioactive dust, a system spraying dust-suppressing chemicals has been installed in the sarcophagus.

Water from rainfall can penetrate the sarcophagus, even though work has been carried out to seal the many small gaps and cracks in the original building. It has been estimated that approximately $1,000\,m^3$ of water (including $140\,m^3$ of water sprayed as part of the dust suppression system) is accumulated in the building per year (Krinitsyn et al., 1998). This water can potentially damage building structures, and leach radionuclides from the building. Potential groundwater contamination around the Chernobyl site is discussed further in Chapter 4.

Release of radioactive dust from a partial or complete collapse of the sarcophagus would be a risk to people working on the Chernobyl site, though there would be no significant transport of material outside the 30-km exclusion zone (IAEA, 1991; OECD/NEA, 2002). In 1999, work was completed to strengthen the sarcophagus roof and stabilise the ventilation stack, which was in danger of collapse (EBRD, 2000). Recently a 20,000-tonne steel shelter has been planned as a protective cover for the sarcophagus.

Decontamination of the Chernobyl site

The other power plant units and the surrounding site were decontaminated using various methods including (IAEA, 1991; Hubert et al., 1996):

- liquid sprays, sand blasting and steam cleaning surfaces and equipment;
- removal and burial of contaminated topsoil;
- polymer coatings applied to surfaces and then stripped off to remove embedded radioactivity, especially dust particles; and
- resurfacing roads and other tarmac areas.

Decontamination measures, involving tens of thousands of workers succeeded in significantly reducing radiation exposures in and around the power plant. In October 1986 Unit 1 of the plant restarted, and in November Unit 2 re-started (Belyaev et al., 1996). Unit 3 (the reactor closest to the destroyed Unit 4) re-started in 1987. A report by the IAEA (1991) described the situation inside the plant several years after the accident: 'The floors inside the Chernobyl plant building are covered in thick plastic. The control room of Unit 4 is dark and instrument panels are covered with plastic sheets. The room itself does not show any signs of damage due to the accident'. All of the reactors on the site are now closed: the last operating reactor, Unit 3 closed in the year 2000.

Decontamination of the (much larger) urban and rural areas in the 30-km zone was largely unsuccessful. Removal of contaminated topsoil, road resurfacing, demolition of some buildings and other decontamination techniques succeeded in reducing radiation exposure in localised areas. 'However, the efforts made eliminated only a small fraction of the deposited radionuclides and succeeded only in redistributing the rest. In many cases the surfaces quickly became re-contaminated by radioactive materials migrating from trees in highly contaminated forests' (IAEA, 1991).

Decontamination efforts created an enormous amount of radioactive waste.

Approximately 14 PBq (1.4×10^{16} Bq) of radioactive wastes (approximate volume: 1×10^6 m^3) were disposed of by burial (Kholosha et al., 1996). These wastes included reactor debris, including graphite and other core material dispersed by the explosion, as well as contaminated building materials, vegetation, soil and waste generated during the clean up operation (Kholosha et al., 1996; Bugai et al., 1996). Such waste includes demolished houses and household goods. Several square kilometres of pine forest which died from excess radiation exposure (the so-called 'Red Forest') were cut down and buried.

There are approximately 800 waste burial sites in the 30-km zone (Kholosha et al., 1996; OECD/NEA, 2002). Many of the lower level disposal sites were hastily constructed in sandy soil in '2 to 3.5 m deep trenches with no isolating covers or liners' (Bugai et al., 1996). Other waste is buried in thirty large trenches 'each the size of a football field and about 10 m deep. These trenches are layered with clay and sand' (IAEA, 1991). Higher level waste is stored in containers with 1 m thick concrete walls. The potential groundwater contamination from waste disposal sites is discussed in Chapter 4.

Several hundred thousand civilian and military personnel were involved in the 'clean up' and recovery operation within the 30-km evacuated zone during the years after the accident (UNSCEAR, 2000). These people are termed 'liquidators' and include all workers in the zone from 1986 to 1989. Approximately 292,000 people worked in the zone in the period 1986–1987. During 1988–1989, when exposures were much lower, approximately 566,000 people worked in the exclusion zone (UNDP/UNICEF, 2002). The average exposure received by liquidators working in 1986–1987 was 100 mGy, but approximately 4% of the group received more than 250 mGy (Boice, 1997). In compensation for their work, liquidators were issued with a card entitling them to some special privileges from the state. These privileges were in the form of improved medical care, financial compensation (such as higher pensions) or other benefits such as vacations in sanatoria (health resorts) and free passes for public transport. The radiation doses received, and health effects on these workers are discussed further in Chapters 6 and 7.

1.1.3 Radionuclides released and deposited

Releases of radionuclides to the atmosphere occurred during the initial reactor explosion and continued over an approximately 10-day period. Figure 1.7 (from data in UNSCEAR, 2000) shows the estimated daily releases of ^{131}I during the period 26 April–5 May, 1986. Of the 190.3 tonnes of radioactive material in the reactor core, approximately 6.7 tonnes were released to the environment surrounding the reactor and further afield (UNSCEAR, 2000). A large number of different radioactive elements were released (Table 1.2). Very rough estimates of releases of some additional radionuclides not included in Table 1.2 were made based on measurements made in soil profiles in the Ukraine and Russia (Murumatsu et al., 2000 Carbol et al., 2003), as shown in Table 1.3.

It should be noted that the release estimates given in Table 1.2 may be an overestimate for some radionuclides. The estimates in Table 1.2 imply a release of

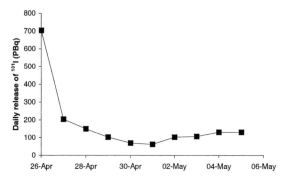

Figure 1.7. Estimated daily releases of ^{131}I (in PBq: 1 PBq = 10^{15} Bq) from the reactor for the period from the initial explosion to the extinction of the fire. (Releases of ^{137}Cs followed a similar temporal pattern.)
From data in UNSCEAR (2000).

Table 1.2. Physical half-lives and amounts of radionuclides released from Chernobyl. Radionuclides (RNs) of primary radiological concern (over different timescales) after the accident are in bold.

Radionuclide	Physical half-life[a]	% of amount in core released[b]	Amount released (Bq)[c]	Amount of RN released, as ratio RN : ^{137}Cs
Sr-89	50.5 days	2.9	1.15×10^{17}	1.35
Sr-90	**28.8 years**	**4.3**	**1.0×10^{16}**	**0.12**
Zr-95	64.0 days	3.4	1.96×10^{17}	2.31
Mo-99	2.75 days	>2.7**	>1.68×10^{17}**	1.98
Tc-99	2.11×10^5 years	2–3[d]	9.7×10^{11}[d]	0.000 011
Ru-103	39.3 days	>4.5**	>1.68×10^{17}**	1.98
Ru-106	372.6 days	>8.5**	>7.3×10^{16}**	0.86
I-131	**8.04 days**	**57**	**1.76×10^{18}**	**20.7**
I-133	20.8 hours	19	9.1×10^{17}[e]	10.7
Te-132	3.26 days	26	1.15×10^{18}	13.5
Xe-133	5.25 days	100	6.5×10^{18}	76.5
Cs-134	**2.07 years**	**32**	**5.4×10^{16}**	**0.64**
Cs-137	**30.1 years**	**33**	**8.5×10^{16}**	**1.00**
Ba-140	12.8 days	4.0	2.4×10^{17}	2.82
Ce-141	32.5 days	3.5	1.96×10^{17}	2.31
Ce-144	284 days	3.0	1.16×10^{17}	1.36
Np-239	2.35 days	1.6	9.45×10^{17}	11.1
Pu-238	**87.7 years**	**2.7**	**3.5×10^{13}**	**0.000 41**
Pu-239	**2.41×10^4 years**	**3.2**	**3.0×10^{13}**	**0.000 35**
Pu-240	**6.54×10^3 years**	**2.8**	**4.2×10^{13}**	**0.000 49**
Pu-241	**14.4 years**	**3.3**	**6.0×10^{15}**	**0.071**
Am-241[f]	**432.2 years**		**4.2×10^{12}**[f]	**0.000 05**[f]
Cm-242	163 days	2.1	9.0×10^{14}	0.011

[a] Half-life data from CRC (1988); [b] from inventory data in Sich *et al.* (1994), quoted in UNSCEAR (2000); [c] from Devell *et al.* (1996) and Dreicer *et al.* (1996), quoted in UNSCEAR (2000); [d] Uchida *et al.* (1999); [e] Khrouch *et al.* (2000), quoted in UNSCEAR (2000); [f] estimates of ^{241}Am release from Ivanov *et al.* (1994) (relatively little ^{241}Am was released from the reactor, but environmental concentrations will become more significant since ^{241}Pu decays to ^{241}Am).
** Other estimates given in UNSCEAR (2000) suggest that releases were not significantly greater than these values.

Table 1.3. Estimates of releases of some additional radionuclides (not given in Table 1.2) compared with ^{137}Cs.

Radionuclide	Half-life[a]	Total deposition* in soil profile (kBq m^{-2})		Ratio RN : ^{137}Cs in soil profile		Ratio RN : ^{137}Cs in reactor core[d]
		6 km[c]	140 km[b]	6 km[c]	140 km[b]	
^{60}Co	5.27 years	7.35	1.7	0.0024	0.001	No data
^{125}Sb	2.76 years	235	60	0.076	0.04	0.06
^{154}Eu	8.5 years	–	0.64	0.017[†]	0.0004	0.05
^{137}Cs	30.1 years	3,095	1,600	1.0	1.0	1.0

[a] Half-life data from CRC (1988); [b] from soil profile data at a site in Bryansk Region, Russia, Carbol et al. (2003); [c] from soil profile data at a site in Kopachy, Ukraine, Murumatsu et al. (2000); [d] from inventory data in Sich et al. (1994), quoted in UNSCEAR (2000).
* Decay corrected to 26 April, 1986.
[†] From ratio of activity concentrations in the Of-horizon of a forest soil, Lux et al. (1995).

approximately 3–4% of the total inventory of ^{90}Sr and isotopes of Pu, for example. Recent mapping of the inventory of these radionuclides in soils (Kashparov et al., 2003) found that the release was lower, being $1.5 \pm 0.5\%$ of the reactor inventory at the time of the accident.

The fuel used in the RBMK reactor was uranium dioxide slightly enriched to 2% ^{235}U (natural uranium contains about 0.7% ^{235}U and 99.3% ^{238}U). A large proportion of the material released from Chernobyl were particles of this uranium dioxide fuel. The uranium release from Chernobyl is generally not considered to be a significant health or environmental risk. Because of its low specific activity (i.e., low bequerels per gram of radioisotope), the total activity of U isotopes released was much lower than, for example, Pu isotopes, despite the large mass of uranium fuel released.

Although a large mass of uranium fuel was released, concentrations of U in soils around Chernobyl were only slightly elevated above natural levels. Typically, concentrations of natural uranium in soils range from 0.5–5.0 mg kg^{-1}. In soils in the Chernobyl area, Mironov et al. (2002) observed concentrations of natural U in the range 0.2–3.4 mg kg^{-1}. This compared with U of Chernobyl origin in the range 5×10^{-6}–2.0 mg kg^{-1}; the highest values at the sites studied were found at Chistogolavka and Pripyat, within a few km of the reactor.

The vast majority of radionuclides released by the accident had very short halflives. Thus, the amount of radioactivity in the environment (and the radiation exposure) declined very rapidly in the weeks to months after the accident. One year after the accident, only approximately 2% of the total activity originally released was remaining in the global environment. Ten years after the accident, less than 1% remained. This rapid decline is illustrated in Figure 1.8(a). This is based on data in Table 1.2 and should be considered illustrative only as some (relatively, not radiologically important) radionuclides are not given in Table 1.2.

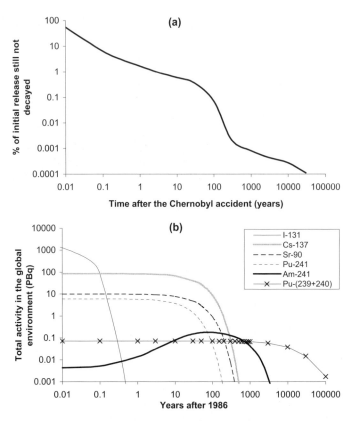

Figure 1.8. (a) Percentage of the initial radioactivity remaining in the environment at different times after the Chernobyl accident, based on release data given in Table 1.2. Note that ingrowth of ^{241}Am is accounted for: ingrowth of other radionuclides is not, however the influence of this is expected to be minor. (b) Changes in the amounts (in PBq: $1\,\text{PBq} = 10^{15}\,\text{Bq}$) of some key radionuclides over time due to radioactive decay. The increase in ^{241}Am activity concentration is due to ingrowth from ^{241}Pu.

A large proportion of the radioactivity released was in the form of small particles which deposited mainly within 30 km of the reactor (see Chapter 2). More volatile radionuclides were emitted in gaseous form, or as extremely small particles. Because of the heat of the explosion and subsequent fire, the cloud of emissions (the 'release plume') rose to an altitude of 1–2 km (Smith and Clark, 1989). This plume was carried by the wind, spreading radioactivity over large parts of Europe. Weather conditions varied during the 10 days of emissions, so the transport of the plume, and subsequent deposition of radioactivity, was complex. The plume initially travelled to the north of the plant, contaminating large areas of Belarus (the Ukrainian–Belarussian border is just a few km north of Chernobyl). Slightly more radiocaesium was deposited on the territory of Belarus than in Ukraine. Deposition of ^{137}Cs was 15 PBq in Belarus, 13 PBq in Ukraine

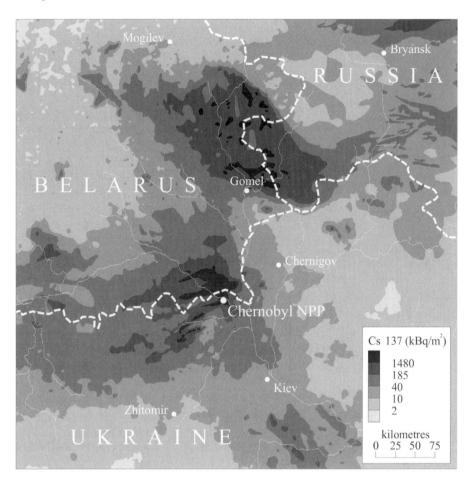

Figure 1.9. ^{137}Cs fallout in Ukraine, Belarus and Russia.
Produced by Simon Wright from the original data with the kind permission of Valery Kashparov of the Ukrainian Institute of Agricultural Radiology (UIAR, 2001; Kashparov et al., 2003).

and 29 PBq in Russia: a further 27 PBq was deposited in other European countries (De Cort et al., 1998). Radioactivity from Chernobyl was first detected in western Europe by monitoring equipment at a Swedish nuclear power station.

In places outside the 30-km zone, depositions of radiocaesium from the plume to the ground surface mainly occurred by washout of radioactivity from the plume by rainfall. Hence the pattern of radiocaesium deposition (Figures 1.9 and 1.10) was largely influenced by rainfall patterns during the period of releases. Deposition of ^{131}I was mainly 'dry' deposition, so the fallout pattern was somewhat different to that for ^{137}Cs. Maps of the fallout of various long-lived radionuclides in Ukraine, Belarus and Russia and over Europe are shown in Figures 1.9–1.11. The mechanisms and rates of deposition of different radionuclides are discussed further in Chapter 2.

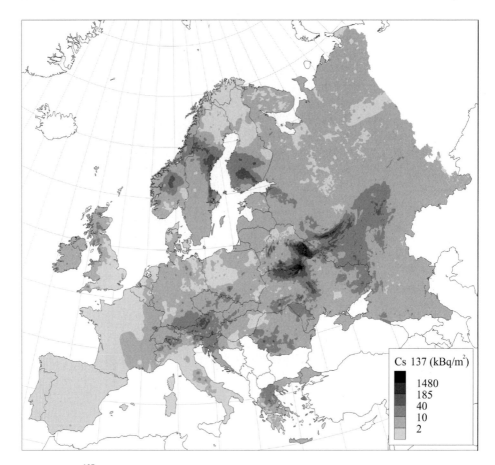

Figure 1.10. ^{137}Cs fallout in Europe.
^{137}Cs deposition was interpolated by Simon Wright (CEH Lancaster UK) from contours presented by De Cort et al. (1998), with the kind permission of Marc De Cort, European Commission.

1.2 RADIATION EXPOSURES

1.2.1 Health effects of radiation

Radiation exposure can lead to the death of cells in the body or to damage to DNA which can, over time, lead to cancer. Extremely high radiation doses (such as those received by firemen in the early stages of the accident) lead to acute radiation syndrome with symptoms which include nausea, skin burns, destruction of white blood cells and temporary or permanent sterility. These symptoms can to a certain extent be treated, but in serious cases often result in death. The effects of extremely high radiation doses are termed *deterministic effects* – above a certain exposure level, they will always occur.

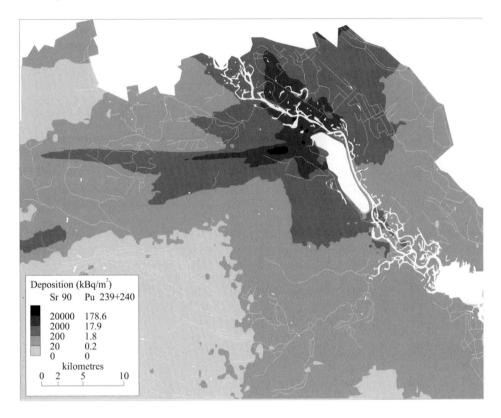

Figure 1.11. ^{90}Sr and ^{239,240}Pu fallout in the Ukrainian part of the 30-km zone.
Produced by Simon Wright from the original data with the kind permission of Valery Kashparov of the Ukrainian Institute of Agricultural Radiology (UIAR, 2001; Kashparov *et al.*, 2003).

All but a few of those affected by Chernobyl were exposed to much lower levels of radiation. These low-level exposures can result in what are termed *stochastic effects* (i.e., probabilistic effects). There is a certain probability that a cancer will result from a low-level exposure, but in most cases there will be no effect on the individual. The probability of a health effect in an individual increases in proportion to the radiation dose received.

Using epidemiological studies, primarily of survivors of the Hiroshima and Nagasaki atomic bombs (Figure 1.12), radiation protection agencies have estimated the lifetime fatal cancer risk to people from exposure to ionising radiation. Risk estimates (ICRP, 1991) predict a 5% lifetime risk of fatal cancer per sievert of dose to the general population. This estimate is for exposures at a low dose rate rather than the high-dose rate exposures in the atomic bomb survivors.

The International Commission on Radiation Protection (ICRP) risk estimate implies that if a population is exposed to low dose rate radiation leading to an average dose equivalent of 0.1 Sv (100 mSv) to each person, an additional 0.5%

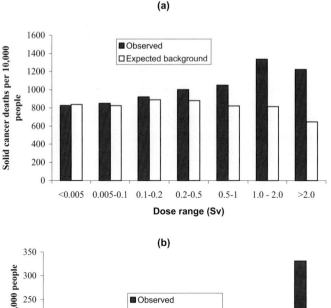
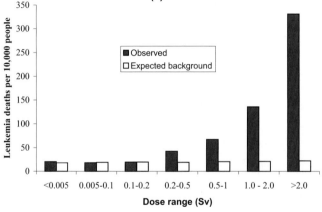

Figure 1.12. (a) Fatal solid cancer; (b) leukaemia rates in the follow up group of people exposed to radiation from the Hiroshima and Nagasaki atomic bombs using data presented in Pierce et al. (1996). The study group includes about half of all survivors who were within 2.5 km of the explosions (Pierce et al., 1996). Data is for the period 1950–1990 and is presented for solid cancers and leukaemia separately. The expected background values are 'estimates of what would be expected for this cohort had there been no radiation exposure' (Pierce et al., 1996). Note that the data here is presented as a number per 10,000 people exposed, though much fewer than 10,000 people were exposed to very high doses (greater than 0.5 Sv). Of the 4,741 cancer deaths observed in all persons exposed to more than 0.005 Sv in the study group, 454 were attributable to radiation exposure.

risk of fatal cancers in that population can be predicted. These cancers do not occur immediately, but may arise many years after exposure. Since dose rates to most individuals after Chernobyl were of order 0.1 Sv or less (see Table 1.5, p. 21), the expected increase in fatal cancers of <1% can be anticipated to be difficult to detect against a 'background' cancer mortality rate of around 20–25% in populations in industrialised countries.

It should be noted that incidence of non-fatal cancer also increases as a function of radiation exposure. Risk estimates (ICRP, 1991) predict a 1% lifetime risk of non-fatal cancer per sievert of dose to the general population.

In radiation risk assessments it is current practice to assume that even very low dose radiation carries with it an associated cancer risk, though this is a matter of some debate within (and outside) the radiation protection community. Epidemiological studies have not shown clear evidence of increased cancer risk at very low doses and low dose rates. The large-scale studies required to show the very small increase in cancers expected have in many cases proved inconclusive. Thus the risks from very low doses (and low dose rates) of radiation could perhaps be described as 'theoretical', but for radiation protection purposes, it is conservatively assumed that every exposure to radiation carries a potential cancer risk.

Exposure to low-level radiation can potentially result in *hereditary effects* on subsequent generations. Evidence of effects on offspring has been observed in studies on laboratory animals (UNSCEAR, 2001). Studies on the children of the survivors of the Hiroshima and Nagasaki bombs have, however, shown no evidence of hereditary effects of radiation (Neel *et al.*, 1990). As stated in an UNSCEAR report (2001) 'Radiation exposure has never been demonstrated to cause hereditary effects in human populations'.

1.2.2 Exposure pathways and change of dose over time after Chernobyl

Radiation exposures from radioactivity released changed significantly over time both in terms of the level of exposures and the pathways and radionuclides contributing to the dose. The dose pathways and major radionuclides contributing to the dose have been summarised by Balonov and co-workers (Balonov *et al.*, 1996), as shown in Table 1.4. Exposures decreased significantly over time as a result of radioactive decay of short-lived isotopes. In the short-term, doses to the population were dominated by ingestion (primarily from consumption of milk) and inhalation of short-lived radioiodine isotopes, particularly ^{131}I, and their rapid transfer to the thyroid (Pröhl *et al.*, 2002). After the ^{131}I and other short-lived isotopes had decayed, exposures were dominated by ingestion of, and external exposure to, isotopes of radiocaesium.

Because of the importance of doses to the thyroid, these are reported separately to effective doses to the whole body. The range of doses received by the affected populations is illustrated in Figures 1.13 and 1.14. Figure 1.13 shows the effective doses received by the populations of the Ukraine, Belarus and Russia during the period 1986–1995, excluding doses to the thyroid (from data in UNSCEAR, 2000). Figure 1.14 shows the dose to the thyroid (measured as absorbed energy per kg, grays) to children in the two most affected regions of Belarus (from data in UNSCEAR, 2000). Thyroid doses and their effects are discussed further in Chapter 6.

The average radiation exposures to the various groups affected by the Chernobyl accident have been summarised by Cardis *et al.* (2001) and are presented here in Table 1.5. Note that this does not include specific doses to the thyroid

Table 1.4. Main pathways and nuclides contributing to the population exposure after the Chernobyl accident.
Balonov et al. (1996).

Exposed population	Time after accident	External exposure		Internal exposure	
		β	γ	Inhalation	Ingestion
Evacuated population	1–11 days	^{106}Ru/Rh ^{144}Ce/Pr ^{132}Te/I	^{132}Te/I ^{131}I IRG*	^{131,133}I ^{132}Te/I TUE**	^{131}I ^{132}Te/I ^{134,137}Cs
Population of the contaminated area	<100 days	^{106}Ru/Rh ^{132}Te/I	^{132}Te/I ^{131}I ^{134,137}Cs	^{131}I TUE**	^{131}I ^{134,137}Cs ^{89}Sr
	>100 days	–	^{134,137}Cs ^{106}Ru/Rh	TUE** ^{106}Ru/Rh ^{144}Ce/Pr	^{134,137}Cs ^{90}Sr/Y

* IRG denotes inert radioactive gases (Kr, Xe); ** TUE denotes radionuclides from transuranium elements (Pu, Am, Cm).
Note that radioactive decay products are named: for example, ^{132}Te/I denotes ^{132}Te and its decay product ^{132}I.

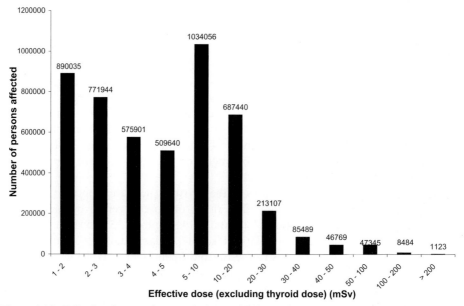

Figure 1.13. Effective dose to the populations of Belarus, Russia and the Ukraine (excluding thyroid dose) during the period 1986–1995. Note that data for doses of 1–2 mSv is only for the affected regions of Belarus and Russia.
From data in UNSCEAR (2000).

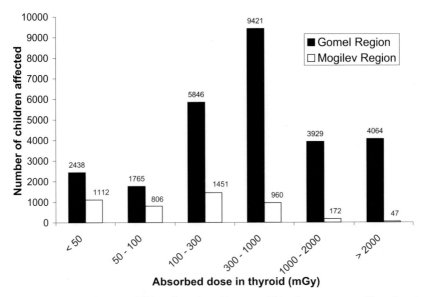

Figure 1.14. Thyroid dose to children less than 18 years old in the two most affected regions of Belarus.
From data in UNSCEAR (2000).

Table 1.5. Radiation exposures of different groups after Chernobyl.
From information presented in Cardis et al. (2001).

Population group	Approximate number of people	Average radiation dose (mSv)
'Liquidators' – accident recovery workers working in the 30-km zone during 1986–1987	200,000*	100
'Evacuees' – people evacuated from Pripyat and other parts of the 30-km zone and parts of Belarus, in April and May, 1986	134,000	10
'Residents of the strict control zones' – people who continued to live in areas of ^{137}Cs deposition >555 kBq m^{-2} (15 Ci km^{-2})‡	270,000†	50
'General population of the contaminated territories' – people living in regions of ^{137}Cs contamination >37.5 kBq m^{-2} (1 Ci km^{-2})‡	6,400,000	5–16**

* Other studies give a higher number, but the figure given here is for those working in and around the reactor during 1986–1987; † population estimate at the time of the accident; ** range in estimates of the average dose. ‡ The authors (Cardis et al., 2001) do not specify the time interval over which these exposures apply. Our interpretation of their source (Balonov et al., 1996) suggests that they are for the period 1986–1995

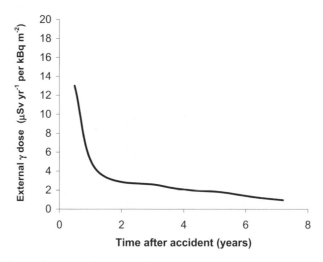

Figure 1.15. Change in external gamma dose rate over time after the accident. This data accounts for occupancy of different locations during the day: the dose rate during time spent in the home, for example, is much lower than when working in a contaminated field. Adapted from Balonov et al. (1996).

(see Figure 1.14), nor does it include doses to the emergency workers exposed during the accident (see Table 1.1).

External gamma-ray exposures declined significantly over time (Figure 1.15): it was estimated (Balonov et al., 1996) that approximately 60% of the external gamma dose was received during the first 10 years after the accident, with 40% of the dose remaining in the years 1996–2056. This decline in exposure over time is due to radioactive decay and migration of radionuclides into deeper layers of the soil.

Doses from ingestion of contaminated food also declined significantly over time as a result of radioactive decay, migration and 'fixation' of radiocaesium to the soil, and the implementation of countermeasures (see Chapters 2–5). The relative importance of external and internal doses is strongly dependent on dietary habits and on the soil type on which food products are grown. Using data from Balonov et al. (1996), Figure 1.16 illustrates that on black earth soils (which have the ability to strongly fix radiocaesium) average doses were dominated by external exposures. In contrast, on turf–podzol soil (where radiocaesium transfer to grass for cattle grazing and agricultural products is high), ingestion doses dominate the average exposures. Note also that high consumption of certain 'wild' foods (mushrooms, berries, freshwater fish) leads to relatively high ingestion doses (see Chapter 3 for more details).

From measurements of the resuspension of contaminated dust (see Chapter 2 for a discussion of resuspension rates), Jacob et al. (1996) estimated inhalation doses from Pu isotopes and ^{137}Cs. The study focused particularly on tractor drivers who may be exposed to high concentrations of contaminated dust during agricultural work, and concluded that, although there are uncertainties, 'with sufficient care it can be concluded that even at sites inside the 30-km zone, lifetime doses per year of

Radiation exposures

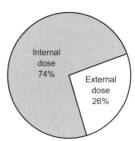

Figure 1.16. Contrasting contributions of internal and external dose rates (in the absence of countermeasures) to overall dose in areas of different soil types, Bryansk Region, Russia. From data in Balonov *et al.* (1996).

work will hardly exceed ... 35 µSv effective dose for agricultural workers' (Jacob *et al.*, 1996). This implies that doses from inhalation of resuspended material make a negligible contribution to radiation exposures in comparison with external and ingestion doses.

1.2.3 Limiting the long-term dose to the population

In 1988 a plan known as the '350 mSv Concept' was developed to limit the lifetime radiation dose to the populations in the Chernobyl affected areas (Belyaev *et al.*, 1996). The aim of this concept was to better define the evacuated areas using more comprehensive information than was available in 1986. The proposal defined 350 mSv as the safe lifetime dose to the most exposed people in a particular area. Previously evacuated areas which would give rise to lifetime doses less than 350 mSv could be resettled, while those areas not previously evacuated which could exceed the limit would be evacuated. Although the 350 mSv Concept was not formally adopted by the USSR, it illustrates the thinking behind dose control measures at the time.

Following the break-up of the Soviet Union, control of dose limitation fell to the governments of the separate affected countries – the Ukraine, Belarus and Russia. The situation therefore became more complex with, in some cases, different regulatory limits being adopted in the different countries. The general approach, however, was the same for all three countries and was based on the 'Chernobyl 1991 Concept' (Belyaev *et al.*, 1996). This concept was based on the following principles (Belyaev *et al.*, 1996):

- No restrictions or countermeasures need be applied in areas where the annual effective dose in 1991 was less than $1\,\text{mSv}\,\text{yr}^{-1}$.
- Where doses were above $1\,\text{mSv}\,\text{yr}^{-1}$, measures should be put in place to reduce contamination of foodstuffs in order to ensure that the individual effective dose was less than $5\,\text{mSv}\,\text{yr}^{-1}$ and that, over time, maximum doses reduced to $1\,\text{mSv}\,\text{yr}^{-1}$ or less.

- If necessary more people would be relocated: 'Each person living in a contaminated territory shall have the right to make their own decision about continuing to live in the given territory ... based on unbiased information about the radiation situation, socio-economic and other aspects of life' (Belyaev et al., 1996).

In practice, relocations, compensation for affected people and countermeasure applications were largely based on zones of different surface contamination by ^{137}Cs. These zones, based on curies (Ci) per square kilometre (1 Ci $= 3.7 \times 10^{10}$ Bq) were defined as follows (UNDP/UNICEF, 2002):

- <1 Ci km^{-2} (<37 kBq m^{-2}): area designated as not significantly contaminated by Chernobyl.
- 1–5 Ci km^{-2} (37–185 kBq m^{-2}): *zone of enhanced radiation control*. Periodic monitoring is carried out and some compensation is given to the population.
- 5–15 Ci km^{-2} (185–555 kBq m^{-2}): *zone with the right to resettle*. Resettlement is voluntary.
- 15–40 Ci km^{-2} (555–1480 kBq m^{-2}): *zone of strict radiation control*. Resettlement is obligatory in Ukraine, in Russia it is voluntary if the dose is less than 5 mSv yr^{-1}.
- >40 Ci km^{-2} (>1480 kBq m^{-2}): *zone of obligatory resettlement*.
- Territories adjacent to Chernobyl which were evacuated in 1986–1987. *Zone of evacuation* or '30-km exclusion zone'.

The zones in which strict radiation controls are in place (>15 Ci km^{-2}) cover an area of approximately 7,000 km^2 of Belarus, 2,660 km^2 of Russia and 1,290 km^2 of the Ukraine (De Cort et al., 1998). In 1995, approximately 150,000 people lived in these areas (UNSCEAR, 2000).

Currently, it is the policy within the three most affected countries of the former Soviet Union (fSU) to implement protective measures where the total (external + internal) dose exceeds 1 mSv y^{-1} (Balonov et al., 1999; Kenik et al., 1999).

1.2.4 Unofficial resettlement of the abandoned areas

There are people who are living within settlements which are officially abandoned in all three affected fSU countries. Information concerning the population dynamics and radiation exposures of these people was summarised in Beresford and Wright (1999). Some of the resettlers have returned to where they used to live, whereas, others who did not previously live in the area have decided to settle. In Russia, a large proportion of people unofficially living within abandoned settlements (in 1999) were ethnic Russians originating from Asian republics of the fSU. Changes in the populations of abandoned settlements within Belarus are presented in Table 1.6. Following the initial reduction in population there is no apparent trend. The population of abandoned areas of the Ovruch, Narodichi and Polesskoje regions of the Ukraine in 1998 was 5,534, approximately 13% of whom were children below the age of 14 (Ukrainian Institute of Agricultural Radiology, unpublished). There are

Table 1.6. Population dynamics within abandoned settlements of Belarus* in selected years after Chernobyl.
From Beresford and Wright (1999).

Region	1986	1987	1988	1989	1990	1991	1992	1994	1996	1998
Gomel	44,457	19,614	17,923	18,084	17,627	7,076	2,836	1,947	1,703	1,769
Mogilev	9,578	n/a	n/a	8,567	n/a	3,112	1,133	31	15	n/a

*Collated from the annual census data of the State Committee on Analysis and Statistics of Republic of Belarus, Minsk.
n/a – not available.

also a number of settlements unofficially inhabited within the 30-km exclusion zone (in addition to the town of Chernobyl which provides temporary accommodation for people officially working within the zone). The inhabited settlements within the 30-km zone are in the relatively less contaminated southern area.

1.3 CHERNOBYL IN CONTEXT

1.3.1 Previous radioactive releases to the environment

The Chernobyl accident was on a much greater scale than previous accidental releases of radioactivity to the environment. The largest nuclear accident prior to Chernobyl was the explosion in 1957 of a high-level waste tank at the Mayak plutonium production and reprocessing facility in Siberia. Releases of a mixture of radionuclides, including long-lived ^{90}Sr (with a physical half-life of 28 yr) resulted in evacuation and removal from agricultural production of approximately 1,000 km^2 of land. By 1997, 82% of this land had been reclaimed (Joint Norwegian–Russian Expert Group, 1997). Releases from Mayak, and other previous releases of radioactivity to the environment, are summarised in Table 1.7.

Following the 1957 fire at the Windscale nuclear reactor in the UK, a ban on the consumption of milk because of high ^{131}I activity concentrations was implemented over an area extending to a maximum of 518 km^2 (Jackson and Jones, 1991). It is probable that, at present-day intervention levels, temporary precautionary bans on foodstuffs, including meat and milk, would also have been implemented as a consequence of radiocaesium contamination (Wright et al., 2003). This may also have been the case for some food products in some areas as a consequence of fallout from the atmospheric nuclear weapons testing era (predominantly 1952–1963). The accident at Three Mile Island in the USA did not result in significant contamination of the environment and food chain; the highest activity concentration in a food product determined in a sample of goats milk was only 1.5 Bq l^{-1} of ^{131}I, collected 2 km from the site (Katherine, 1984).

Nuclear weapons explosions have also released radioactivity into the environment. Long-term environmental contamination from the Hiroshima and Nagasaki bombs was not significant: the high-radiation doses to survivors came primarily from

Table 1.7. Summary of previous major releases of radioactive material to the environment. Note that the summary is not comprehensive, and only data for three radionuclides are presented. These release data should not be interpreted in terms of significance of the releases to environmental or human health: the impact of radionuclide releases is not solely determined by the amount of radioactivity released.

Release event	Area*	Release of some key radionuclides to the environment (PBq)		
		^{137}Cs	^{90}Sr	^{131}I
Chernobyl, 1986[a]	Significant part of Europe	85	10	1,760
Hiroshima atomic bomb, 1945[†b]	Few km radius around epicentre	0.1	0.085	52
Atmospheric nuclear weapons testing, 1952–1981**[c]	Global, primarily northern hemisphere	949	578	**
USA atmospheric weapons tests, Nevada Test Site**[d]	USA states, particularly Nevada	**	**	5,550 released to atmosphere, 1,390 deposited to ground
Three Mile Island, USA[e]	No significant environmental contamination	–	–	Negligible
Mayak, discharges to the Techa River, 1949–1956[f]	Techa and Ob rivers	13	12	–
Mayak accident, 1957[f]	Approximately 300 km × 50 km area of Siberia	0.3	40	–
Waste discharges from Sellafield, 1964–1992[g]	Irish Sea	41	6	–
Windscale accident, 1957[g]	518 km^2 area of northern England	0.022	7.4×10^{-5}	0.74

† Note that the radiation health effects of the Hiroshima and Nagasaki bombs resulted primarily from gamma and neutron radiation from the initial explosion. Radioactive fallout to the environment (detailed here for Hiroshima) was minor in comparison.
* Indicative area only – the contaminated area depends on how you define 'contaminated'.
** ^{131}I data is given for the US atmospheric weapons tests only: ^{137}Cs and ^{90}Sr data are global totals for the period 1952–1981.
[a] From UNSCEAR (2000), see Table 1.2; [b] Gudiksen *et al.* (1989); [c] Cambray *et al.* (1989); [d] NCI (1997); [e] Katherine (1984); [f] NATO (1998); [g] Gray *et al.* (1995).

Table 1.8. Examples of some measurements of ^{137}Cs in the environment before the Chernobyl accident.

Product	^{137}Cs activity concentration (Bq kg^{-1})	Notes
Root crops	0.5–2 d.w.*	West Cumbria, England, 1984–1985 (1)
Barley	<1 d.w.	
Sheep meat	220 (max. f.w.*)†	West Cumbria, England, 1965–1966 (2)
Sheep meat	<10 (d.w.)	Yorkshire, England, 1982 (3)
Sheep meat	<2 (d.w.)	South coast of England, 1984 (3)
Reindeer	300 (f.w.)†	Finland, 1985–1986 (pre-accident) (4)
Freshwater fish (trout)	31 (f.w.)	Norway, 1985 (4)

† Note that ^{137}Cs in upland sheep in West Cumbria is likely to have been influenced by the 1957 Windscale accident and that animals in some semi-natural arctic and upland environments showed significantly higher bio-accumulation of ^{137}Cs from nuclear weapons testing (NWT) than intensively farmed animals.
* d.w. – dry weight; f.w.– fresh weight (meat is circa 75% water).
(1) Horrill and Howard (1985); (2) ARCL (1966); (3) Howard (1987); (4) Eisler (2003).

radiation during or shortly after the explosions. Subsequent atmospheric (i.e., above ground) testing of nuclear weapons caused fallout of relatively low-level radioactivity (particularly ^{90}Sr and ^{137}Cs) globally, mainly in the northern hemisphere. Several hundred atmospheric nuclear weapons tests were carried out by the USA, USSR and the UK until a test ban treaty was signed in 1963. Limited atmospheric nuclear weapons tests were carried out by France and China in the early 1970's.

Environmental levels of radiocaesium before Chernobyl

Radiocaesium and radiostrontium were still present in the global environment (at relatively low concentrations) prior to the Chernobyl accident as a result of atmospheric nuclear weapons testing and (to a lesser extent) discharges from nuclear facilities. Table 1.8 gives examples of some measurements of ^{137}Cs in the environment before the Chernobyl accident.

1.3.2 Natural radioactivity in the environment and medical radiation

People are continually exposed to ionising radiation from natural sources. This natural background radiation provides an important context within which to consider radioactive contamination and exposures from the Chernobyl accident. Natural background radiation doses have recently been reviewed by Thorne (2003). Table 1.9 summarises average background doses and includes averaged doses from diagnostic medical exposures (e.g., X-rays). Table 1.10 shows doses from some individual diagnostic medical procedures. Natural background doses to

Table 1.9. Population average doses from natural radiation sources (Thorne, 2003) and average dose in various European and North American countries from medical diagnostic procedures (estimated from data in Hart and Wall, 2002). Doses from therapeutic medical procedures are not shown.

Background radiation source	Mean annual dose (mSv yr^{-1})
Cosmic rays	0.38
Cosmogenic radionuclides (mainly ^{14}C)	0.012
Primordial radionuclides: external dose	0.48
Primordial potassium-40: internal dose	0.165
Primordial uranium, thorium series: internal dose	0.12
Radon-220, 222 (mainly lung irradiation)	1.2
Total (mean) natural background radiation	**2.4**
Medical diagnostic procedures (X-rays)	0.72
Total (mean) from natural and medical sources	**3.1**

Table 1.10. Doses from various X-ray medical diagnostic procedures.
Adapted from European Commission (2000).

X-ray examination	Effective dose (mSv)
Limbs and joints (excluding hip)	<0.01
Chest	0.02
Skull	0.07
Lumbar spine	1.3
Hip	0.3
Abdomen	1.0
Barium enema	7.0
Computed Tomography (CT), chest	8.0

individuals vary considerably around these averages as a result of, for example, variation in geological conditions and dietary habits.

Cosmic rays

Particles originating from the sun and sources outside the Solar System continually impact on the Earth's atmosphere. Interactions of these particles with the upper atmosphere produce secondary radiations, most of which are absorbed in the atmosphere, but some reach the Earth's surface. The shielding effect of the atmosphere means that cosmic radiation doses at the Earth's surface are relatively low – approximately 0.04 μSv hr^{-1}. This atmospheric shielding effect, however, reduces rapidly with altitude, hence the cosmic radiation dose rate increases rapidly with altitude.

At a height of 12 km (a typical cruising altitude of intercontinental aircraft), the cosmic radiation dose rate is approximately $8\,\mu\text{Sv}\,\text{hr}^{-1}$, around 200 times higher than at ground level. Radiation doses from cosmic rays are around $5\,\text{mSv}\,\text{yr}^{-1}$ for airline crew regularly flying intercontinental routes.

Interactions of cosmic rays with the atmosphere can produce tiny amounts of radioactive particles such as ^{14}C, ^{7}Be and ^{3}H, however the dose rates arising from these 'cosmogenic' particles are very small (Table 1.9).

Primordial radionuclides

Primordial radionuclides (most importantly, ^{238}U, ^{235}U, ^{232}Th and ^{40}K) were formed by the same stellar processes which formed the other heavy elements of the Earth. Their half-lives are of the same order as the age of the Earth (4.5 billion years) and hence are still present in the Earth's crust today (Table 1.11). The radionuclides ^{238}U, ^{232}Th (and their decay products) and ^{40}K form the major part of the natural radiation dose. Due to its low natural abundance, ^{235}U and its decay products do not form a significant part of natural radiation exposure.

Radiation doses from primordial radionuclides are primarily from external gamma radiation, ingestion and inhalation (Table 1.9). Both external and internal doses can vary significantly according to differences in the geology of a region. Different dietary habits also influence internal doses. As stated in Thorne (2003), 'consumption of shellfish, offal and meat from animals such as reindeer and caribou … can lead to enhanced dietary intakes of uranium and thorium series radionuclides. Intakes of ^{210}Pb and ^{210}Po are of particular importance, and these are often dominated by consumption of seafood'. Brazil nuts are one of the most radioactive foods, containing up to $250\,\text{Bq}\,\text{kg}^{-1}$ of natural radium isotopes. Concentrations of natural ^{210}Po in marine organisms in the Baltic Sea are in the range 0.8–$30\,\text{Bq}\,\text{kg}^{-1}$ (Nielsen *et al.*, 1999).

Potassium-40 is found in foodstuffs (Table 1.12). The internal dose from ^{40}K is an important component of the background dose, however, this does not vary much between individuals. The potassium concentration of the human body is maintained at an approximately constant level, regardless of the intake of potassium in the diet. Therefore different levels of dietary intake of stable potassium (and hence ^{40}K) have little effect on internal radiation dose (since the ratio of stable K to ^{40}K is similar in most environmental components).

Table 1.11. Primordial radionuclides and some of their decay products.

Primordial radionuclide	Half-life (years)	Abundance	Radioactive decay products (not a complete list)
^{238}U	4.46×10^{9}	99.27% of natural U	^{234}U, ^{230}Th, ^{226}Ra, ^{222}Rn, ^{210}Pb, ^{210}Po
^{235}U	7.04×10^{8}	0.72% of natural U	^{231}Th, ^{223}Ra, ^{219}Rn
^{232}Th	1.4×10^{10}	100% of natural Th	^{228}Ra, ^{228}Th, ^{220}Rn
^{40}K	1.25×10^{9}	0.012% of natural K	No radioactive decay products

Table 1.12. Concentrations of natural radioactive potassium (^{40}K) in various foodstuffs.
From data in IAEA (1991).

Food product	Natural ^{40}K activity concentration* (Bq kg^{-1})	Stable potassium concentration*† (mg kg^{-1})
Whole milk	44	1,460
Dried milk	300	9,930
Beef, lamb, poultry	100	3,310
Eggs	44	1,460
Fish	90	2,980
Potatoes	170	5,630
Soya	440	14,600
Green vegetables	150	4,970

*Fresh weight except where specified; †calculated using specific activity of natural potassium (30.2 Bq g^{-1}).

Naturally occurring radon gas is the largest contributor to exposures to natural radiation. Radon gas is produced in soils and rocks as a decay product of uranium (almost all ^{222}Rn, half-life 3.8 days) and thorium (^{220}Rn, half-life 55.6 s). Radon gas decays in the atmosphere, and its (non-gaseous) decay products may become attached to small particulates in the air. Inhalation of radon and its decay products leads to radiation exposure to the lungs.

Concentrations of the two radon isotopes, ^{222}Rn and ^{220}Rn, are roughly similar to each other in soils, rocks and the (ground-level) atmosphere (UNSCEAR, 2000). Their concentrations can vary considerably according to the bedrock and soil properties. Granitic rocks often have high concentrations of uranium and thorium and hence produce significant amounts of radon.

Inhalation doses from ^{222}Rn and its decay products are approximately ten times higher than those from ^{220}Rn and its decay products (Thorne, 2003). Radon concentrations in air inside buildings are significantly higher than outside and the major part of exposure to radon and its decay products is received indoors.

Variation in natural background dose rates

Dose rates from natural radioactivity vary considerably around the average values given in Table 1.9. This variation can be due to different occupational exposure: for example, long-haul airline crews can receive doses from cosmic radiation of around 5 mSv yr^{-1} above the average. Underground (non-coal) miners in the UK, for example, received mean occupational doses of 11.7 mSv yr^{-1} during 1997 (HSE, 1998). There is also a significant variation in radiation exposures which is not linked to occupation. Using data on population exposure to natural background radiation from 15 countries worldwide, UNSCEAR (2000) estimate that the range of exposures to people is approximately 1–10 mSv yr^{-1}, the average being 2.4 mSv yr^{-1}. In some cases, natural background radiation can be much higher than 10 mSv yr^{-1}. For example in some parts of the USA, exposures from outdoor and indoor radon can be up to 60 mSv yr^{-1} (Steck et al., 1999). In Finland, more than 100,000 people are exposed to $>10\,\mathrm{mSv\,yr^{-1}}$ natural background radiation (UNSCEAR, 2000). The large range in natural background exposures is to a large extent due to variations in indoor radon concentrations which are dependent in a complex way on geology, soil characteristics, air flows and building construction.

High levels of natural radiation exposures such as this are comparable to the dose rates (excluding thyroid doses from short lived radioiodine) observed in the Chernobyl affected populations during the period 1986–1995 (Figure 1.13). A natural radiation dose of $5\,\mathrm{mSv\,yr^{-1}}$ gives an accumulated dose of 50 mSv over a 10-year period, for example. This is greater than the dose accumulated by most of the populations exposed by Chernobyl.

1.4 REFERENCES

ARCL (1966) *Agricultural Research Council Radiobiological Laboratory Annual Report 1965–1966*. Agricultural Research Council ARCL-16, London.

Balonov, M., Jacob, P., Likhtarev, I. and Minenko, V. (1996) Pathways, levels and trends of population exposure after the Chernobyl accident. In: Karaoglou, A., Desmet, G., Kelly, G.N. and Menzel, H.G. (eds), *The Radiological Consequences of the Chernobyl Accident*, pp. 235–249. European Commission, Brussels.

Balonov, M., Anisimova, L.I. and Perminova, G.S. (1999) Strategy of population protection and area rehabilitation in Russia in the remote period after the Chernobyl accident. *Journal of Radiological Protection*, **19**, 261–269.

Belyaev, S.T., Demin, V.F., Kutkov, V.A., Bariakhtar, V.G. and Petriaev, E.P. (1996) Characteristics of the development of the radiological situation resulting from the accident, intervention levels and countermeasures. In: Karaoglou, A., Desmet, G., Kelly, G.N. and Menzel, H.G. (eds), *The Radiological Consequences of the Chernobyl Accident*, pp. 19–28. European Commission, Brussels.

Beresford, N.A. and Wright S.M. (eds) (1999) *Self-help Countermeasure Strategies for Populations Living within Contaminated Areas of the Former Soviet Union and an Assessment of Land Currently Removed from Agricultural Usage*. Institute of Terrestrial Ecology: Grange-over-Sands, U.K., 82 pp.

Boice, J.D. (1997) Leukaemia, Chernobyl and epidemiology. *Journal of Radiological Protection*, **17**, 129–133.

Borovoi, A.A. (1996) The sarcophagus: What do we know, what should we do? *Nuclear Engineering International*, **41**, 28–30.

Borovoi, A.A., Bogatov, S.A. and Pazukhin, E.M. (1999) Sarcophagus: Current state and environmental impact. *Radiochemistry*, **41**, 390–401.

Bugai, D.A., Waters, R.D., Dzhepo, S.P. and Skal'skij, A.S. (1996) Risks from radionuclide migration to groundwater in the Chernobyl 30-km zone. *Health Physics*, **71**, 9–18.

Cambray, R.S., Playford, K., Lewis, G.N.J. and Carpenter, R.C. (1989) *Radioactive fallout in air & rain, results to the end of 1988*. Atomic Energy Authority Report AERE R 13575, HMSO Publications, London.

Carbol, P., Solatie, D., Erdmann, N., Nylén, T. and Betti, M. (2003) Deposition and distribution of Chernobyl fallout fission products and actinides in a Russian soil profile. *Journal of Environmental Radioactivity*, **68**, 27–46.

Cardis, E., Richardson, D. and Kesminiene, A. (2001) Radiation risk estimates in the beginning of the 21st century. *Health Physics*, **80**, 349–361.

CRC (1988) *Handbook of Chemistry and Physics* (68th Edition). CRC Press Inc., Boca Raton, Florida.

De Cort, M., Fridman, Sh.D., Izrael, Yu.A., Jones, A.R., Kelly, G.N., Kvasnikova, E.V., Matveenko, I.I., Nazarov, I.M., Stukin, E.D., Tabachny, L.Ya. and Tsaturov, Yu.S. (1998) *Atlas of Caesium deposition on Europe after the Chernobyl accident*. EUR 16733, EC, Office for Official Publications of the European Communities, Luxembourg.

Devell, L., Güntay, S. and Powers, D.A. (1996) *The Chernobyl reactor accident source term. Development of a consensus view*. NEA/CSNI/R(95)24, OECD Nuclear Energy Agency, Paris.

Dreicer, M., Aarkrog, A., Alexakhin, R., Anspaugh, L., Arkhipov, N.P. and Johansson, K.-J. (1996) Consequences of the Chernobyl accident for the natural and human environments. *Proceedings of the Conference: One decade after Chernobyl: summing up the consequences of the accident*. STI/PUB/1001 IAEA, Vienna.

EBRD (2000) Shelter implementation plan. European Bank for Reconstruction and Development, London, 24 pp (available online: http://www.iaea.or.at/worldatom/Press/Focus/Chernobyl-15/shelter-fund.pdf).

Eisler, R. (2003) The Chernobyl Nuclear Power Plant reactor accident: Ecotoxicological update. In: Hoffman, D.J., Rattner, B.A., Burton, G.A.Jr. and Cairns, J.Jr. (eds), *Handbook of Ecotoxicology* (2nd edition), pp. 702–736. CRC Press, Boca Raton, Florida.

European Commission (2000) *Referral guidelines for imaging*. Radiation Protection Report 118. European Commission, Luxembourg.

Gray, J., Jones, S.R. and Smith, A.D. (1995) Discharges to the environment from the Sellafield Site, 1951–1992. *Journal of Radiological Protection*, **15**, 99–131.

Gudiksen, P.H., Harvey, T.F. and Lange, R. (1989) Chernobyl source term, atmospheric dispersion and dose estimation. *Health Physics*, **57**, 697–705.

Hart, D. and Wall, B.F. (2002) *Radiation exposure of the UK population from medical and dental X-ray examinations*. Report of the UK National Radiological Protection Board, NRPB-W4, NRPB, Chilton, 41 pp.

Horrill, A.D. and Howard, B.J. (1985) *The distribution and dynamics of radionuclides in the terrestrial environment*. Progress report. 50 pp. ITE Project 873. Department of the Environment, London.

Howard, B.J. (1987) Cs uptake by sheep grazing tidally-inundated and inland pastures near the Sellafield reprocessing plant. In: Coughtrey, P.J., Martin, M.H. and Unsworth, M.H. (eds), *Pollutant Transport and Fate in Ecosystems*, pp. 371–383. British Ecological Society special publication no. 6, Blackwell Scientific, Oxford.

HSE (1998) *Central index of Dose Information. Summary of Statistics for 1997*. HSE Books, Health and Safety Executive, London.

Hubert, P., Ramzaev, V., Antsypov, G., Sobotovich, E. and Anisimova, L. (1996) Local strategies for decontamination. In: Karaoglou, A., Desmet, G., Kelly, G.N. and Menzel, H.G. (eds), *The Radiological Consequences of the Chernobyl Accident*, pp. 411–424. European Commission, Brussels.

IAEA (1986) *Summary report on the post-accident review meeting on the Chernobyl accident*. Safety series No. 75-INSAG-1. IAEA, Vienna.

IAEA (1991) *The International Chernobyl Project Technical Report*. International Atomic Energy Agency, Vienna, 640 pp.

ICRP (1991) *Recommendations of the International Commission on Radiological Protection*. International Commission on Radiation Protection Publication 60, Pergamon Press, Oxford.

Ivanov, E.A., Ramzina, T.V., Khamyanov, L.P., Vasilchenko, V.N., Korotkov, V.T., Nosovskii, A.V. and Oskolkov, B.Y. (1994) Radioactive contamination of the environment with ^{241}Am as a result of the Chernobyl accident. *Atomic Energy*, 77, 629–633.

Jackson, D. and Jones S.R. (1991) Reappraisal of environmental countermeasures to protect members of the public following the Windscale nuclear reactor accident 1957. In: *Comparative Assessment of the Environmental Impact of Radionuclides Released During Three Major Nuclear Accidents: Kyshtym, Windscale and Chernobyl*. EUR 13574, 1015–1055, CEC, Luxembourg.

Jacob, P., Roth, P., Golikov V., Balonov, M., Erkin, V., Likhtariov, I., Garger, E. and Kashparov, V. (1996) Exposures from external radiation and from inhalation of resuspended material. In: Karaoglou, A., Desmet, G., Kelly, G.N. and Menzel, H.G. (eds), *The Radiological Consequences of the Chernobyl Accident*, pp. 251–260. European Commission, Brussels.

Joint Norwegian–Russian Expert Group (1997) *Sources contributing to radioactive contamination of the Techa River and areas surrounding the MAYAK production association, Urals, Russia*. Norwegian Radiation Protection Authority, Oslo (ISBN 82-993079-6-1).

Kashparov, V.A., Lundin, S.M., Zvarych, S.I., Yoshchenko, V.I., Levchuk, S.E., Khomutinin, Y.V., Maloshtan, I.M. and Protsak, V.P. (2003) Territory contamination with the radionuclides representing the fuel component of Chernobyl fallout. *Science of the Total Environment*, 317, 105–119.

Katherine, R.L. (1984) *Radioactivity in the Environment: Sources, Distribution and Surveillance*. Harwood Academic Publishers, New York (ISBN 3-7186-0203-2).

Kenik I., Rolevich I., Ageets V., Gurachesky, V. and Poplyko, I. (1999) Long-term strategy of rehabilitation of Belarusian territories contaminated by radionuclides. *Radioprotection*, 34, 13–24.

Kholosha, V., Sobotovitch, E., Proscura, N., Kozakov, S. and Korchagin, P. (1996) Management problems of the restricted zone around Chernobyl. In: Karaoglou, A., Desmet, G., Kelly, G.N. and Menzel, H.G. (eds), *The Radiological Consequences of the Chernobyl Accident*, pp. 339–343. European Commission, Brussels.

Khrouch, V., Gavrilin, Yu., Shinkarev, S., et al. (2000) *Case-control study of Chernobyl-related thyroid cancer among children of Belarus: Estimation of individual doses. Part I*. Submitted for publication.

Krinitsyn, A.P., Simanovskaya, I.Ya. and Strikhar, O.L. (1998) Action of water on construction and fuel-containing materials in the facilities of the Chernobyl Sarcophagus. *Radiochemistry*, 40, 287–297.

Kryshev, I.I. and Ryazantsev, E.P. (2000) *Ecological Safety of Nuclear Energy Complexes of Russia*. Izdat, Moscow, 383 pp (in Russian).

Lux, D., Kammerer, L., Rühm, W. and Wirth, E. (1995) Cycling of Pu, Sr, Cs, and other long-living radionuclides in forest ecosystems of the 30 km zone around Chernobyl. *Science of the Total Environment* 173/4, 375–384.

Mironov, V.P., Matusevich, J.L., Kudrjashov, V.P., Boulyaga, S.F. and Becker, J.S. (2002) Determination of irradiated reactor uranium in soil samples in Belarus using ^{236}U as irradiated uranium tracer. *Journal of Environmental Monitoring*, 4, 997–1002.

Mould, R.F. (2000) *Chernobyl Record*. Institute of Physics Publishing, Bristol.

Muramatsu Y., Rühm, W., Yoshida, S., Tagami, K., Uchida, S. and Wirth, E. (2000) Concentrations of ^{239}Pu and ^{240}Pu and their isotopic ratios determined by ICP-MS in soils collected from the Chernobyl 30 km zone. *Environmental Science and Technology*, 34, 2913–2917.

NATO (1998) *Radioactive contamination of rivers and transport through rivers, deltas and estuaries to the sea.* NATO report No. 225, 136 pp.

NCI (1997) *Estimated Exposures and Thyroid Doses Received by the American People from Iodine-131 in Fallout Following Nevada Atmospheric Nuclear Bomb Tests.* US National Cancer Institute, Bethesda, USA (available online at: https://cissecure.nci.nih.gov/ncipubs/).

Neel, J.V., Schull, W.J., Awa, A.A., Satoh, C., Kato, H., Otake, M. and Yoshimoto, Y. (1990) The children of parents exposed to atomic bombs: estimates of the genetic doubling dose of radiation for humans. *American Journal of Human Genetics*, **46**, 1053–1072.

Nielsen, S.P., Bengtson, P., Bojanowsky, R., Hagel, P., Herrmann, J., Ilus, E., Jakobson, E., Motiejunas, S., Panteleev, Y., Skujina, A. and Suplinska, M. (1999) The radiological exposure of man from radioactivity in the Baltic Sea. *Science of the Total Environment*, **237/238**, 133–141.

OECD/NEA (2002) *Update of Chernobyl: Ten Years On* (H. Métivier, ed.). OECD, Paris, 155 pp (available online at http://www.nea.fr/html/rp/chernobyl/welcome.html).

Pierce, D.A., Shimizu, Y., Preston, D.L., Vaeth, M. and Mabuchi, K. (1996) Studies of the mortality of atomic bomb survivors. Report 12, Part 1. Cancer: 1950–1990. *Radiation Research*, **146**, 1–27.

Pröhl, G., Mück, K., Likhtarev, I., Kovgan, L. and Golikov, V. (2002) Reconstruction of the ingestion doses received by the population evacuated from the settlements in the 30-km Zone around the Chernobyl reactor. *Health Physics*, **82**, 173–181.

Shcherbak, I. (1989) *Chernobyl: A Documentary Story.* Macmillan Press, London.

Sich, A.R., Borovoi, A.A. and Rasmussen, N.C. (1994) *The Chernobyl accident revisited: source term analysis and reconstruction of events during the active phase.* MITNE-306, Massachusetts Institute of Technology, Department of Nuclear Engineering Report.

Smith, F.B. and Clark, M.J. (1989) *The transport and deposition of airborne debris from the Chernobyl nuclear power plant accident.* Meteorological Office Scientific Paper No. 42, HMSO, London.

Steck, D.J., Field, R.W. and Lynch, C.F. (1999) Exposure to atmospheric radon. *Environmental Health Perspectives*, **107**, 123–127.

Thorne, M. C. (2003) Background radiation: Natural and man-made. *Journal of Radiological Protection*, **23**, 29–42.

Uchida. S., Tagami, K., Rühm, W. and Wirth, E. (1999) Determination of ^{99}Tc deposited on the ground within the 30-km zone around the Chernobyl reactor and estimation of ^{99}Tc released into the atmosphere by the accident. *Chemosphere*, **39**, 2757–2766.

UIAR (2001) *Contamination of the ChNPP 30-km Zone version 2.0.* CD-ROM: Ukrainian Institute of Agricultural Radiology, Kiev.

UNDP/UNICEF (2002) *The human consequences of the Chernobyl nuclear accident: A strategy for recovery.* UNDP Report 240102 (available online: http://www.undp.org/dpa/publications/chernobyl.pdf).

UNSCEAR (2000) *Report to the General Assembly: Sources and effects of ionizing radiation. Volume II, Annex J.* United Nations, New York, pp. 453–551 (available online at: http://www.unscear.org).

UNSCEAR (2001) *Hereditary effects of radiation.* United Nations Scientific Committee on the Effects of Atomic Radiation: Report to the General Assembly with Scientific Annexe. United Nations, New York, 156 pp (available online at: http://www.unscear.org).

Wright, S.M., Smith, J.T., Beresford, N.A. and Scott, W.A. (2003) Monte-Carlo prediction of changes in areas of west Cumbria requiring restrictions on sheep following the Chernobyl accident. *Radiation and Environmental Biophysics*, **42**, 41–47.

2

Radioactive fallout and environmental transfers

Jim T. Smith and Nick A. Beresford

2.1 PATTERN AND FORM OF RADIOACTIVE DEPOSITIONS

The pattern of deposition of radioactivity from Chernobyl was complex (see Figures 1.9–1.11). The release of radionuclides from the reactor occurred over a period of 10 days and the radionuclide composition of the release varied during this time. The initial explosion and subsequent fire deposited fuel particles ('hot' particles, see Figure 2.1) principally within an area of radius of 30 km around the reactor. Within this area, the majority of fallout was in the form of these hot particles (Bobovnikova *et al.*, 1991). Kashparov *et al.* (1999) distinguish two forms of particles released at different stages of the accident: 'non-oxidised' particles of uranium dioxide fuel released during the initial explosion and 'oxidised' particles released during the subsequent reactor fire. More than 90% of the release of ^{90}Sr, ^{141,144}Ce, Pu isotopes and ^{241}Am was in the form of fuel particles of average diameter around 10 µm (Kashparov *et al.*, 1999; Mück *et al.*, 2002), and within 30 km of the plant most (approximately 50–75%) of the ^{137}Cs was in fuel particles (Krouglov *et al.*, 1998; Kashparov *et al.*, 1999), though the (relatively volatile) caesium isotopes were also dispersed much farther afield.

Volatile radionuclides such as ^{134,137}Cs, ^{132}Te, ^{131,133}I and ^{103,106}Ru, were also released from the core by evaporation and dispersed around Europe. These radionuclides attached to small dust particles ('aerosols') in the atmosphere, though a large part of the radioiodine remained in gaseous form. For example, approximately 75% of the radioiodine in air in the UK and 68–76% in Italy was in gaseous form (Cambray *et al.*, 1987; Spezzano and Giacomelli, 1991). Radionuclides can be deposited on the Earth's surface either in dry conditions ('dry' deposition) or in precipitation ('wet' deposition). At an upland site in the UK, most (80%) of the radioiodine was deposited by dry deposition, whereas almost all of the caesium and ruthenium isotopes were deposited as wet deposition (Livens *et al.*, 1992).

Figure 2.1. Electron micrograph of a uranium fuel particle from Chernobyl. The large aggregate consists of small uranium oxide crystallites.
Courtesy of B. Salbu and H. C. Lind, NLH, Norway.

The dry deposition of radionuclides occurs in three ways: the fallout of heavier particles near to the source; the direct adsorption of gaseous elements onto the earth's surface; and the impaction of airborne particles with the surface. Small aerosol particles and gaseous elements are assumed to be deposited at a rate proportional to their concentration in air (Smith and Clark, 1989):

$$D = v_{part}C_{part} + v_{gas}C_{gas} \qquad (2.1)$$

where D is the amount of radionuclide deposited (Bq m^{-2}), C_{part} is the time-integrated concentration in air in the form of particulates (Bq m^{-3} s) and v_{part} is an empirically estimated dry deposition velocity (m s^{-1}) for particulates. C_{gas} is the time-integrated concentration in air in gaseous form (Bq m^{-3} s) and v_{gas} is an empirical dry deposition velocity (m s^{-1}) for gaseous elements. During the Chernobyl release, dry deposition velocites were estimated by measuring radionuclide depositions and time-integrated air concentrations in areas far from Chernobyl which were without rainfall during the passage of the release plume. Dry deposition velocities in the UK were estimated to be: $v_{part} = 4.6 \times 10^{-4}$ m s^{-1} and $v_{gas} = 3.6 \times 10^{-3}$ m s^{-1} (Smith and Clark, 1989).

Wet deposition of radionuclides occurs through washout by rain or snow, or through deposition when there is fog or ground-level cloud ('occult deposition'). Wet deposition is generally more efficient than dry. The 'hot spot' of radioactive fallout 150–250 km to the northwest of Chernobyl (Figure 1.9) and areas of comparatively high deposition throughout Europe (Figure 1.10) were caused by rainfall in this area 'washing out' radioactivity during the passage of the plume. Despite the complexity of washout processes, wet deposition can be estimated (Chamberlain, 1970; Smith and Clark, 1989) by assuming that washout by rainfall is proportional to the radionuclide concentration in air:

$$D = w_r C R \quad (2.2)$$

where C is the average concentration of radionuclide in air ($Bq\,m^{-3}$) during the rainfall event, R (m) is the total amount of rainfall and w_r is an empirically determined washout ratio. Note that w_r may vary with the intensity of rainfall. Values of w_r ([dimensionless]) measured in the UK during the passage of the Chernobyl plume were in the range 5.8–7.7×10^5: similar values were observed for weapons test fallout during 1958–1959 (Smith and Clark, 1989; Chamberlain, 1970).

Interception of the fallout by vegetation

The extent and type of vegetation cover can influence the dry deposition of radionuclides. In particular, trees efficiently intercept radioactivity and increased deposition of radioactivity has been observed as a plume of radioactivity passes over the edge of a forest. Interception of both wet and dry deposited radionuclides is important in determining initial activity concentrations in foodstuffs after a deposition event. Chapter 3 gives more information on interception of fallout by vegetation.

2.1.1 Element isotope ratios and 'hot' particles

The distribution of ^{137}Cs around Europe has been determined and mapped (Figure 1.10), but depositions of other radionuclides must be estimated from (often limited) measurements of their relative concentrations compared to ^{137}Cs. A comprehensive review and analysis of radionuclide deposition relative to ^{137}Cs has been carried out by Mück et al. (2002). Measurements were reviewed of the ratio of radionuclide : ^{137}Cs both in air samples and of radioactivity deposited on the ground. Empirical formulae were determined for estimating these ratios as a function of distance from Chernobyl. Mück et al. (2002) found that different isotopes of the same element were transported in the same way, so their relative concentrations in air and fallout even at large distances from the plant were similar to those in the reactor at the time of release. For example, the mean ratio of ^{134}Cs : ^{137}Cs from calculations of the reactor core inventory was 0.59 ± 0.06, whilst measurements in the release plume and from deposition gave a very similar mean ratio of 0.55 ± 0.02 (Mück et al., 2002). Note, however, that there is evidence that the isopic composition of the release varied during the 10-day duration of the accident (IAEA, 1986).

Though different isotopes of the same element behaved similarly, different elements showed very different fallout patterns as a function of distance from the

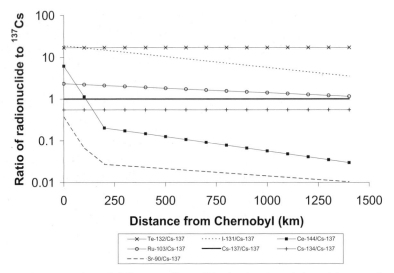

Figure 2.2. Concentration of different radionuclides in the air and deposition as a function of distance from Chernobyl, expressed as a ratio of radionuclide : ^{137}Cs. These relationships were derived by Mück et al. (2002) from estimations of the composition of the reactor core at the time of the accident, and a review of air concentration measurements and radionuclide deposition data.

reactor. Radionuclide : ^{137}Cs ratios as a function of distance from Chernobyl estimated by Mück et al. (2002) are illustrated in Figure 2.2. The (fuel particle associated) ^{90}Sr and ^{144}Ce show rapid declines with distance (relative to ^{137}Cs) whilst the more volatile ^{103}Ru and ^{132}Te show an almost constant deposition compared to ^{137}Cs. The (also volatile) ^{131}I declines in relation to ^{137}Cs due to its higher dry deposition velocity.

Even at large distances from Chernobyl, radionuclides could be deposited in particulate form. Fuel particles of <~1 μm diameter were found as far away as Norway and Sweden (Devell et al., 1986; Salbu et al., 1994a). Approximately 75% of the ^{137}Cs in Swedish rainwater was associated with high molecular weight particles (>10,000 Da), though this may largely have been a result of attachment to and subsequent washout of dust particles. Nine fuel particles found in north-eastern Poland contained from 4,990–139,000 Bq of ^{103}Ru and 1,050–28,000 Bq of ^{106}Ru in each particle (decay corrected to 26 April, 1986) (Schubert and Behrend, 1987). Concentrations of ^{137}Cs in these particles were much lower.

In general, most fallout at large distances was not in the form of fuel particles. Of the radiocaesium fallout in the UK, for example, approximately 70% was water soluble (Hilton et al., 1992). Close to the reactor (generally within the 30-km zone), radionuclides mainly associated with hot particles were much less bioavailable. Chemical extractions carried out on soils taken from various locations in the 30-km zone during the summer of 1987 showed that 80–90% of ^{90}Sr and ^{137}Cs were in non-exchangeable forms (Bobovnikova et al., 1991). The ^{137}Cs was assumed to be partly in fuel particles and partly 'fixed' to clay minerals in the soil.

^{90}Sr is, however, not believed to be strongly 'fixed' to mineral components of soils: ^{90}Sr deposited to the soil in solution is expected to remain in available form for long periods of time (e.g., Tikhomirov and Sanzharova, 1978, quoted in Coughtrey and Thorne, 1983). The fact that 80–90% of ^{90}Sr was in non-exchangeable forms therefore implied (Bobovnikova et al., 1991) that 80–90% of the ^{90}Sr was in hot particles at that time.

The radioactivity concentration of hot particles varied considerably with particle size and declined rapidly over time after the accident as short-lived radionuclides decayed away. A study in 1997 (Zheltonovsky et al., 2001) of ten hot particles found within 5 km of Chernobyl observed between 90 and 650 Bq of ^{137}Cs per particle (^{90}Sr was not measured). Measurements of α-activity in hot particles (Zheltonovsky et al., 2001; Kashkarov et al., 2003) gave activities of up to approximately 10 Bq per particle.

Box 2.1 Cancer risk from hot particles.

Ingestion or inhalation of hot particles leads to a highly localised exposure of the tissue in which the hot particle is deposited. It has been hypothesised that this highly localised exposure could lead to an enhanced cancer risk compared to the (more common) situation in which the dose is relatively uniformly distributed within a tissue. Recently, however, the human epidemiological data and that from animal experiments on localised radiation exposures have been reviewed (Charles et al., 2003). This review concluded that:

> ... in very broad terms (within a factor $\sim \pm 3$) the results of a large number of animal studies and a growing number of *in vitro* studies are in agreement regarding the lack of evidence to support a large hot particle enhancement factor. Human evidence is limited but does not support any significant hot particle enhancement. All of this is in stark contrast to the claims made more than 30 years ago, which fuelled so much concern, that [the hot particle enhancement factor] could be as much as five orders of magnitude.

Thus, the cancer risk from hot particles is best predicted by assuming that the highly localised exposure is averaged over the whole tissue. For example, the exposure from a hot particle containing 100 Bq of a radionuclide deposited in the lung would be equivalent to an exposure due to 100 Bq of that radionuclide evenly distributed in the lung tissue.

It should be noted, however, that fuel particles may remain in the body for different time periods than radionuclides inhaled or ingested in more chemically available forms. For example, Salbu et al. (1995) administered fuel particles released by the Chernobyl accident to six goats. In one of the goats, a particle was retained in the reticulum (second stomach compartment) for 3.5 months.

Detecting the Chernobyl 'signature' in low fallout areas

Outside the high fallout areas in the former Soviet Union (fSU), ^{137}Cs deposition from Chernobyl was often at similar levels to that which was already present in the environment as a result of nuclear weapons testing. The Chernobyl ^{137}Cs fallout could be distinguished from weapons test ^{137}Cs using the ^{134}Cs isotope. Since ^{134}Cs has a relatively short half-life (2.1 years), in 1986 none of the ^{134}Cs deposited during the nuclear weapons testing period (ca. 1957–1963) was still present in the environment. Thus, measurements of Chernobyl-derived ^{134}Cs were commonly used to estimate the fraction of ^{137}Cs originating from Chernobyl, using the ratio of ^{134}Cs : ^{137}Cs of approximately 0.55 (Cambray et al., 1987; Mück et al., 2002), though slightly different ratios have been used in other studies. For example, if the total ^{137}Cs activity concentration measured was 20,000 Bq m^{-2}, and the total ^{134}Cs was 7,000 Bq m^{-2}, the Chernobyl-derived ^{137}Cs is estimated to be 12,730 Bq m^{-2} (7,000/0.55), the remaining 7,270 Bq m^{-2} being attributed to weapons testing (all inventories decay corrected to the time of Chernobyl fallout). Around 10 years after the accident, however, this method became inaccurate because most of the Chernobyl-derived ^{134}Cs had decayed away.

2.1.2 Break up of hot particles

Rates of degradation of fuel particles can be estimated (Konoplev et al., 1992; Kashparov et al., 1999) using a simple decay equation:

$$A(t) = A_0 \exp(-k_p t) \tag{2.3}$$

where $A(t)$ and A_0 are the radionuclide activities in particles at time t and at the time of deposition respectively. The dissolution rate k_p (yr^{-1}) can be expressed as a half-time $T^p_{1/2} (= \ln 2/k_p)$, the time for the activity in the fuel particle to reduce by one-half.

Assuming that approximately 100% of ^{90}Sr in the 30-km zone was in fuel particles at the time of fallout, Kashparov et al. (1999) found that rates of dissolution were related to the type of fuel particle deposited and the pH of the soil. As shown in Figure 2.3, particles deposited to the west of the reactor during the initial release phase took much longer to dissolve ($T^p_{1/2} = 4.95 - 17.33$ years) than those deposited during the subsequent fire in a north–south plume ($T^p_{1/2} = 1.65 - 4.95$ years) (Kashparov et al., 1999). This difference was attributed by these workers to the lower degree of UO$_2$ oxidation of the particles released to the west of the reactor during the initial explosion.

These rates of fuel particle decay imply that at present the majority of fuel particles in terrestrial systems have dissolved. By 1997, less than 20% of ^{90}Sr remained in fuel particle form (Kashparov et al., 2004). Fuel particles in the sediments of lakes have, however, been observed to decay much more slowly than those in soils. In the Chernobyl Nuclear Power Plant (NPP) cooling pond, a significant proportion of the deposited radioactivity remains in fuel particle form (Voitsekhovitch et al., 2002).

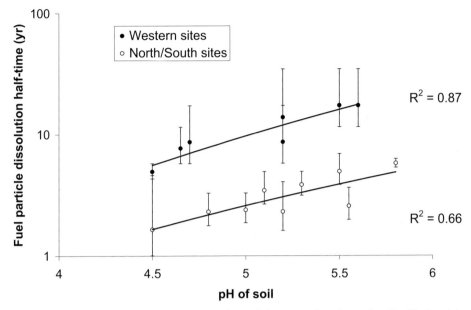

Figure 2.3. Half-time of dissolution of fuel particles as a function of soil pH. Particles deposited to the west of the reactor (black circles) during the initial release phase took much longer to dissolve than those deposited during the subsequent fire in a north–south plume (open circles).
From data in Kashparov et al. (1999).

2.2 ENVIRONMENTAL TRANSFERS OF RADIONUCLIDES

2.2.1 Migration of radionuclides in the soil

Initial movement of radioactivity deposited to the soil surface is believed to be relatively rapid as a result of infiltration of rainwater. Subsequently, vertical migration is much slower since, for the majority of radionuclides, a significant amount of the fallout is sorbed to the soil matrix. For some radionuclides, most notably radiocaesium, slow sorption reactions may result in decreasing mobility over a period of years after fallout. The high proportion of fallout in the form of fuel particles in the 30-km zone was an important feature of the Chernobyl accident. The physico-chemical form of the fallout therefore influenced its migration in the soil (Salbu et al., 1994b).

Vertical migration of radionuclides in soils has a direct effect on levels of external radiation exposure to people in radioactively contaminated areas. The amount and distribution of radioactivity in the plant rooting zone also affects transfers of radioactivity to the food chain and can influence the runoff of radio-activity from contaminated soils to surface waters. Ploughing of a soil obviously redistributes the radioactivity approximately evenly to a depth equal to the plough depth (typically 20–30 cm), so we will here consider vertical migration in undisturbed soils.

The factors that determine the variation in radionuclide activity with depth in a soil include the rainfall intensity and soil moisture content at the time of fallout, the soil's structure and the rate at which water can infiltrate as well as the sorption characteristics of the particular radionuclide. Studies in the Chernobyl zone in the late summer of 1986 showed that all radionuclides had penetrated the soil to a maximum depth of only 1.5 cm (Ivanov et al., 1997). This low migration was attributed to the very dry weather during the period since deposition (Ivanov et al., 1997) but may also have been influenced by the fraction of radioactivity deposited as fuel particles. An experimental study in the UK in which ^{134}Cs was artificially applied to a soil to simulate deposition during a storm event showed that, a short time after deposition, most of the ^{134}Cs was distributed within the surface 2-cm layer in four different soils (Owens et al., 1996). A simulation of deposition over a 10-month period by the same authors showed negligible ^{134}Cs below a depth of 3 cm.

Once the radioactivity has infiltrated into the soil and attached to soil particles, long-lived radionuclides such as ^{90}Sr, ^{137}Cs, ^{241}Am and isotopes of Pu are relatively immobile in soils over long periods of time. This relative immobility is illustrated in Figure 2.4(a) using measurements presented by Bunzl et al. (1994, 1995) for an undisturbed grassland soil in Germany in 1990. Almost 30 years after their deposition onto the soil, the majority of ^{137}Cs, ^{237}Np, $^{239+240}$Pu and ^{241}Am from nuclear weapons test (NWT) fallout has remained in the surface 15 cm of the soil profile. The similarity of the $^{239+240}$Pu and ^{241}Am distributions is perhaps not surprising since ^{241}Am originates from post-depositional ingrowth from ^{241}Pu. In addition, actinides have generally similar behaviours in soils. As noted by Bunzl et al. (1994), the ^{137}Cs (NWT) is slightly depleted in the surface 0–5-cm layer, but the profile is generally similar to the other NWT fallout radionuclides (Figure 2.4(a)).

There is evidence for slightly greater migration rates of ^{90}Sr than other radionuclides. In a sandy soil in the USA in 1970, 87% of NWT ^{90}Sr was retained in the 0–15-cm layer compared to 96% and 95% of ^{137}Cs and ^{239}Pu respectively (Hardy and Krey, 1971). In Nagasaki, a study was carried out (during 1984) of radioactivity in soils contaminated (in approximately equal amounts) by the 1945 atomic bomb and the global NWT fallout (Mahara, 1993). The study showed that 70% of ^{90}Sr and 95% of ^{137}Cs was still within the surface 0–10-cm soil layer between 20 and 40 years after their deposition.

In 1990, the Chernobyl-derived ^{137}Cs in a grassland soil (Bunzl et al., 1994; Figure 2.4(a)) had not migrated as far down the soil profile as the NWT ^{137}Cs, as was observed in a range of soils around Europe (Livens et al., 1991; Isaksson and Erlandson, 1995; Realo et al., 1995). Similarly, Chernobyl ^{137}Cs observed in the contaminated areas of the fSU was retained in the surface soil layer (Ivanov et al., 1997; Arapis et al., 1997). In these studies, carried out from 1–7 years after the accident, it was observed that most (>90%) of the ^{137}Cs and ^{90}Sr was found in the surface 0–10-cm layer. It should be noted that in these soils relatively close to the reactor a significant proportion of the radioactivity was deposited as fuel particles, thus reducing vertical migration.

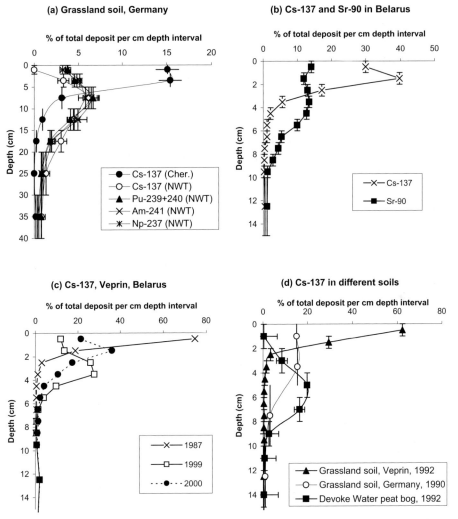

Figure 2.4. Examples of activity–depth profiles of various radionuclides in soils: (a) Chernobyl and NWT fallout in a grassland soil in Germany; (b) representative profiles of Chernobyl ^{90}Sr and ^{137}Cs in soils classed as soddy–podzolic, sandy and sandy loams in the Gomel region of Belarus; (c) migration of ^{137}Cs in a grassland soil, Veprin, Belarus; (d) ^{137}Cs activity–depth profiles in different soil types.
(a) Bunzl et al. (1994, 1995); (b) Arapis et al. (1997); (c) Kudelsky, pers. commun.; (d) Bunzl et al. (1995), Kudelsky, pers. commun., Smith et al. (1995).

There was evidence (Ivanov et al., 1997; Arapis et al., 1997) for slightly greater migration of ^{90}Sr compared to ^{137}Cs (see, e.g., Figure 2.4(b)) as had previously been observed for nuclear weapons fallout (Hardy and Krey, 1971). In some peat soils in Belarus in 1993, significant quantities of ^{90}Sr were observed at depths of 20–50 cm (Arapis et al., 1997).

2.2.2 Rates of vertical migration

It is important to consider how the migration of radionuclides changes over time after fallout. To quantify radionuclide mobility in soils, the common practice is to fit mathematical models to the radionuclide activity–depth profile. The models fall broadly into three classes (Smith and Elder, 1999):

(1) models based on solutions to the advection–dispersion equation (e.g., Kirk and Staunton, 1989; Kirchner and Baumgartner, 1992; Smith *et al.*, 1995; Kudelsky *et al.*, 1996; Ivanov *et al.*, 1997);
(2) models based on exponential fitting functions (Chamard *et al.*, 1993; Blagoeva and Zikovsky, 1995); and
(3) compartmental models based on residence times of different nuclides in different soil layers (Kirchner and Baumgartner, 1992; Bunzl *et al.*, 1994) often termed 'residence time' or 'box' models.

These models typically implicitly assume that migration rates are constant over time since they are calibrated using measurements of activity–depth profiles at a particular time after fallout. The models are therefore useful for characterising radioactivity profiles in the soil such as those presented in Figure 2.4. However, they cannot necessarily accurately predict the future migration of radionuclides in soils since in many soils migration rates change significantly over time. This is illustrated (Figures 2.3(c) and 2.5) using measurements from a grassland soil from Veprin, Belarus, approximately 250 km from Chernobyl (and therefore largely unaffected by fuel particle fallout).

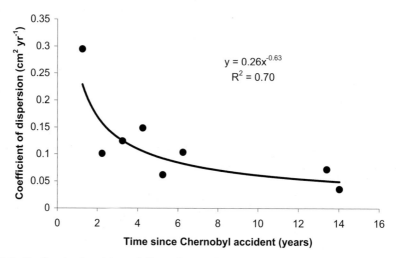

Figure 2.5. Decline in the observed dispersion coefficient as a function of time in a grassland soil at Veprin, Belarus.
From data supplied by A.V. Kudelsky (pers. commun.).

In the Veprin soil, the vast majority of radioactivity was in the 0–2-cm layer in 1987, whilst in 1999 and 2000 ^{137}Cs had migrated down to a depth of 5 cm or so. From Figure 2.4(c) and other similar observations for both ^{137}Cs and ^{90}Sr (Kagan and Kadatsky, 1996), it appears that there was a continuing downward migration in the long-term (1987 onwards) after fallout. This migration was quite variable between different soil cores measured at the same location: in Figure 2.4(c) the ^{137}Cs profile measured in 1999 is more dispersed than that measured in 2000.

The continuing long-term migration of radiocaesium was at a much slower rate than the migration rate during the early period (weeks–months) after fallout. Figure 2.5 shows the decrease in the dispersion of activity–depth profiles measured at the Veprin site from 1987–2000 implying that the dispersion prior to 1987 was at a much greater rate than that after 1987.

It is possible that the difference in activity–depth profiles of Chernobyl and NWT-derived radiocaesium (Figure 2.4(a)) is largely due to a much longer time period which has passed since the weapons test fallout. In other words, it is possible that the depth distribution of Chernobyl ^{137}Cs, through continuing slow dispersion, will in time resemble that of the NWT fallout. However, it is also possible, particularly in mineral soils, that the different activity–depth distributions of both Chernobyl and NWT fallout were largely determined during a short time period after fallout and were therefore dependent on soil and rainfall conditions during the time of deposition. In this case, the initial distributions in the soil would have been different. The Chernobyl radioactivity may remain near the soil surface and may not in time resemble the typical NWT profile.

Although it is typically seen that most of the ^{137}Cs and ^{90}Sr in soil remains in the top 10–15 cm for long periods of time, distributions vary at different sites. As indicated in Figure 2.4(d) the grassland soil at Veprin, Belarus (peaty and fine sand soil) shows less vertical migration of ^{137}Cs than at the site in Munich, Germany (mainly clay and silt). This difference (in this case) does not appear to be linked to the strength of sorption of ^{137}Cs to the soil since it would be expected that caesium would bind more strongly to the clay/silt soil in Germany than to the peat/fine sand soil in Belarus.

In some highly organic, saturated peat soils, however, the strength of ^{137}Cs binding to the soil has been shown to significantly influence its vertical migration. As illustrated in Figure 2.4(d), the ^{137}Cs in a saturated peat soil (>90% organic matter) at Devoke Water, UK is significantly more disperse than in the unsaturated mineral soils. Modelling of ^{137}Cs activity–depth profiles in saturated peats (Smith et al., 1995; Kudelsky et al., 1996) has shown that its vertical migration is consistent with rates of diffusion in the solution phase. In these saturated systems, the soil-water distribution coefficient (K_d) values are relatively low (of order 10^3 l kg^{-1}) (Kudelsky et al., 1996) compared with unsaturated soils, so a much greater proportion of radiocaesium diffuses in the aqueous phase. Such behaviour was observed after NWT fallout: at peat soil sites across the UK, Cawse and Baker (1990) reported that about 50% of ^{137}Cs deposited from NWT had been lost over a period of 20 years as a consequence of water flow and lack of clay minerals.

In unsaturated, mineral soils, long-term vertical migration is believed to be primarily due to the movement of radioactivity attached to soil particles rather than by diffusion in solution (Anspaugh et al., 2002). Müller-Lemans and van Dorp (1996) argue that the bioturbation of soil by earthworms can significantly redistribute radionuclides in the topsoil. These workers estimate that 'in grasslands it will take around 5–20 years for the earthworms to turn over the topsoil once, resulting in an intensive and more or less homogeneous mixing'. Most earthworm activity is in the soil surface to a depth of 10 or 20 cm, though some species can be found at depths of up to 3 m (Müller-Lemans and van Dorp, 1996). It was shown for unsaturated soils that, given even relatively low binding of a radionuclide to soil particles, advection and dispersion of radionuclides in soil water would play a minor role in their long-term redistribution in comparison to bioturbation (Müller-Lemans and van Dorp, 1996).

In summary, the available empirical evidence suggests a relatively rapid infiltration of radionuclides into the soil during and shortly after deposition. This is likely to have depended in a complex way on initial environmental and soil conditions at the time of fallout. During the first year after Chernobyl, the vast majority of the radioactivity was observed within the surface (top 5-cm layer or less) of the soil. Following this initial period, there was a long-term redistribution of radionuclides in the soil, though the majority of radioactivity is expected to remain in the top 0–15-cm layer (and therefore potentially available for root uptake by plants) for long periods (decades) after fallout. In unsaturated mineral soils, where long-lived radionuclides (e.g., ^{134,137}Cs, ^{237}Np, $^{239+240}$Pu and ^{241}Am) are strongly bound to soil particles, bioturbation is believed to be the key transport mechanism. In saturated, highly organic soils, redistribution by advection and dispersion in solution has been found to be important for radiocaesium. In some peat and sandy soils, radiostrontium was observed to be relatively mobile, with significant quantities observed at depths of around 50 cm.

2.2.3 Change in external dose rate over time

Radioactive decay and the migration of radioactivity in the soil significantly reduces the external dose rate over time after deposition. As was shown in Chapter 1 (Figure 1.15), the external dose to the population declined by more than an order of magnitude from late 1986 to 1993. This was mainly due to the physical decay of relatively short-lived radionuclides such as ^{106}Ru and ^{134}Cs. After this period, ^{137}Cs formed the major part of the external dose, so declines in external dose (due to physical decay and vertical migration) were relatively slow.

Measurements of external dose in a number of different locations in Belarus (Timms et al., 2004) during the year 2000 showed a very strong correlation between external dose rate and radiocaesium inventory in the soil (Figure 2.6). The observed relationship was:

$$Dose\ (\mu Sv\ hr^{-1}) = 1.14 D \qquad (2.4)$$

where D is the deposition (inventory) of ^{137}Cs in the soil in MBq per square metre.

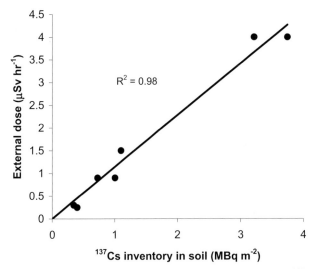

Figure 2.6. External dose rate 0.05 m above the ground as a function of ^{137}Cs inventory in the soil.
From Timms et al. (2004).

The dose was measured at 0.05 m above the ground: dose measured at a height of 1 m was approximately 80% of the value at 0.05 m. Note that this relationship and Figure 2.6 are for observed external dose rates above soils – they do not represent average external doses received by people, since they do not account for occupancy factors (i.e., proportion of time spent in a contaminated area, see Figures 1.15 and 2.7). The strong correlation between external dose rate and soil ^{137}Cs inventory indicates that, for the sites studied, different vertical migration rates played only a minor role in influencing external dose rate.

Using estimates of the long-term migration of radiocaesium, Balonov et al. (1995) and Jacob et al. (1996) calculated the average annual dose to various population groups in rural areas of Ukraine, Belarus and Russia. Doses per unit of deposition were approximately 25% higher relatively close to the Chernobyl NPP (<100 km) than further away (100–1000 km) because of the presence of fuel particles close to the reactor.

Figure 2.7 shows the predicted long-term change in external exposure of rural indoor workers living in wood-framed houses within 100 km of Chernobyl (Jacob et al., 1996). Calculations were made for outdoor workers, pensioners and children living in wooden, brick or multi-storey houses and accounted for occupancy times in various areas. The highest doses were received by adult outdoor (agricultural) workers living in wooden houses: these were approximately a factor 1.36 higher than those for indoor workers shown in Figure 2.7. Doses to forestry workers were not calculated in this study (these were higher than doses to typical agricultural workers, see Chapter 3).

Because most of the radiocaesium is retained in the surface 0–10-cm layer of soil, the external dose rate above undisturbed soils was much greater than above

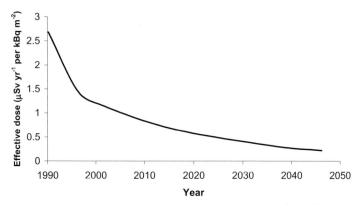

Figure 2.7. Change in annual effective external dose (per kBq m^{-2} of ^{137}Cs deposition) for rural indoor workers living in wood-framed houses. The calculation includes estimates of occupancy factors.
Graph constructed from models and data presented in Jacob *et al.* (1996).

ploughed soils where the radioactivity is relatively evenly distributed within the plough layer (typical plough depth is 20–30 cm). Calculations (Timms *et al.*, 2004) showed that the external dose rate from ^{137}Cs (per unit of deposition) in Belarus during the year 2000 was approximately 3–5 times higher in undisturbed soils than in soils ploughed to a depth of 30 cm.

2.2.4 Resuspension of radioactivity

Radioactivity deposited to the soil surface may become resuspended in the air and dispersed from the area of contamination. Radioactivity resuspended as dust or smoke from forest fires may be inhaled by people living in contaminated areas. Contaminated soil particles resuspended during the operation of agricultural machinery and vehicles can also present a potential risk when inhaled by the operator.

The resuspension of radioactivity attached to soil particles is dependent on local atmospheric conditions such as wind speed, rainfall and humidity. Also important are soil conditions such as vegetation cover, soil type and particle size distribution as well as soil moisture content (wet soils tend to be more cohesive and so less prone to resuspension than dry). Resuspension is expected to decrease over time after a fallout event as radioactivity is transferred from the soil surface to deeper layers.

The resuspension of radionuclides is commonly quantified using a resuspension factor K (m^{-1}) defined as the ratio of the concentration of radioactivity in air (Bq m^{-3}) to the mean surface contamination (Bq m^{-2}):

$$K = \frac{\text{Mean concentration in air (Bq m}^{-3}\text{)}}{\text{Surface contamination per unit area (Bq m}^{-2}\text{)}} \quad (2.5)$$

Figure 2.8 (from Garger *et al.*, 1997) summarises the measurements of the

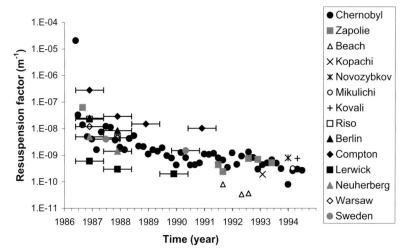

Figure 2.8. Change in resuspension factor as a function of time after fallout at a number of sites around Europe.
From data in Garger et al. (1997), with thanks to Jochen Tschiersch, GSF, Germany.

resuspension factor of ^{137}Cs in Europe over the years since the accident. Relatively good agreement is seen between different studies at different sites: Garger et al. (1997) concluded that 'all estimations of the resuspension factor indicate that, in the moderate climate of northern Europe, it has the same range for wind driven conditions' (as opposed to conditions of mechanical resuspension). Following the rapid decline in airborne radioactivity concentrations in the immediate aftermath of the accident, there was a steady decline in concentrations during the period 1987–1989 followed by a period of almost constant concentrations from 1990 onwards. It can be assumed (Garland and Pomeroy, 1994) that deposition of the initial release continued for a period of two months or so following the accident. Therefore during the first two months, airborne concentrations were dominated by initial dispersion of the release plume. After this period, it is assumed that atmospheric radioactivity was primarily resuspended material.

At a range of sites, mean values of the resuspension factor for ^{137}Cs and $^{239+240}$Pu were measured simultaneously between 1992–1994 (Garger et al., 1996). For three sites in the 30-km Chernobyl exclusion zone and one 45 km to the north (Kovali), the mean ^{137}Cs resuspension factor varied in the range 2.0–6.3×10^{-10} m^{-1} and was of the same order as that for $^{239+240}$Pu which varied in the range 1.1–11.0×10^{-10} m^{-1}. Resuspension rates of ^{144}Ce, ^{103}Ru, ^{95}Zr and ^{95}Nb were also similar to ^{137}Cs (Garger, 1994), however measurements of radionuclides other than caesium are limited, and these estimates close to Chernobyl may have been influenced by resuspension of hot particles.

Measurements of ^{137}Cs, ^{103}Ru and ^{144}Ce made in September 1986 at Zapolie, 14 km south of Chernobyl showed that the maximum concentrations in air were observed at ground level (1 m), with concentrations decreasing by a factor of 3–4

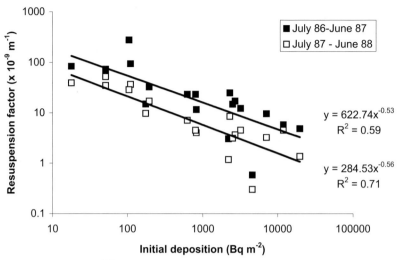

Figure 2.9. Annual mean ^{137}Cs resuspension factors measured at 20 different sites around Europe at large distances from Chernobyl for two different time periods. The strong inverse relationships between resuspension factors and initial deposition indicate non-local sources of ^{137}Cs in air in low Chernobyl fallout areas.
From data reviewed by Garland and Pomeroy (1994).

at a height of 12 m (Garger, 1994). Studies at this site also showed that radioactivity in air was much higher when the wind was blowing from the north (past the Chernobyl site) than from the south where fallout was much lower.

Measurements of the resuspension factor close to the ground in different European countries at large distances from Chernobyl showed a clear inverse relationship between the resuspension factor and the total Chernobyl fallout at a site (Garland and Pomeroy, 1994; see also Figure 2.9). If resuspended radioactivity was always primarily from local sources, one would expect that the resuspension factor would be independent of the total amount of Chernobyl fallout (since it is the ratio of air concentration to fallout in the local area, see Equation 2.5). The inverse relationship observed implies that air concentrations at the low-deposition sites are not determined solely by resuspension close to the site. Garland and Pomeroy (1994) hypothesised that radioactivity in air at sites of low fallout may largely be due to atmospheric transport from high-deposition areas. They also suggested that a portion of the initial radioactive release could have been 'stored' in the stratosphere, then subsequently leaked to the lower atmosphere. These processes would be expected to increase the radioactivity concentration in air at areas of low Chernobyl depositon, and therefore increase the observed resuspension rate (Garland and Pomeroy, 1994).

Resuspension of hot particles

Resuspension of hot particles has been observed in the Chernobyl zone, particularly resulting from agricultural activities. Maximum concentrations of hot particles at three sites in the 30-km zone were 2.6 particles per $1,000\,\mathrm{m}^3$ of air under wind resuspension conditions and 36 particles per $1,000\,\mathrm{m}^3$ during agricultural activity (Wagenpfeil and Tschiersch, 2001). The diameter of resuspended hot particles was in the range 6–12 μm and each particle contained between 1 and 12 Bq of ^{137}Cs (Wagenpfeil and Tschiersch, 2001).

Anthropogenic resuspension and forest fires

A number of studies have quantified the influence of human activity (road traffic and agricultural activity) on resuspension of dust and therefore associated radionuclides. Measurements (Höllander and Garger, 1996; see also Table 2.1) show that human activity can lead to resuspension factors up to several thousand times higher than wind-borne resuspension. Elevated concentrations of radionuclides in air are, however, relatively localised – Garger *et al.* (1998) report an increase in air concentrations by a factor of several thousand at a distance of 30–35 m from agricultural activity. This drops to an increase of 10–100 times normal concentrations at a distance of 100 m. These studies (Höllander and Garger, 1996) concluded that dust contamination from agricultural activity would decrease by two orders of magnitude at a distance of 200 m from the source of activity. Further studies by Höllander and Garger (1996) showed that resuspension following agricultural work decreased from $80 \times 10^{-9}\,\mathrm{m}^{-1}$ to $6 \times 10^{-9}\,\mathrm{m}^{-1}$ as the relative soil humidity increased from 1.7% to 5.3%.

Road traffic may locally increase atmospheric concentrations of radionuclides, though the observational data on this appears to be relatively sparse. Garland and Pomeroy (1994) report measurements by Aarkrog *et al.* (1988) which showed a weekly cycle in radioactivity concentrations in air in Denmark during July and August 1986. Lower values were observed during the weekend than on weekdays which was attributed to higher road traffic density during the weekdays. This pattern disappeared later in the year, in line with the observation (Nicholson *et al.*, 1989) that resuspension from roads declines rapidly over time after a fallout event.

Smoke from forest fires can remobilise radioactivity and transport it over relatively large distances. Measurements 17 km from a forest fire showed radioactivity concentrations in air of $1,000$–$2,000\,\mathrm{\mu Bq\,m}^{-3}$ compared to 10–$100\,\mathrm{\mu Bq\,m}^{-3}$ from wind-borne resuspension and 200–$400\,\mathrm{\mu Bq\,m}^{-3}$ from agricultural activity (Garger *et al.*, 1996). Modelling studies (Höllander and Garger, 1996) estimated resuspension during a forest fire of approximately 5% of the total ^{137}Cs in the area of the fire. During the summer of 1992, forest fires in the 30-km zone doubled activity concentrations in air at Chernobyl (Höllander and Garger, 1996). The concentration, however, was still several orders of magnitude lower than concentrations in air shortly after the accident. A study by Kashparov *et al.* (2000) found that the resuspension of ^{137}Cs from experimental and real forest fires led to a potential inhalation dose to firemen. This dose, however, was several orders of magnitude lower than local external exposures and therefore was not significant. These workers suggested, however, that resuspension of alpha-emitting radionuclides in the 30-km exclusion zone could potentially lead to significant inhalation doses.

Table 2.1. Radionuclide resuspension factors from agricultural activity, traffic and forest fires compared with natural wind resuspension.

Action	Resuspension factor (m^{-1})	Radionuclide studied	Location and date of study	Ref
Rotivation of soil to a depth of 5 cm	$(1,600–4,000) \times 10^{-9}$	^{137}Cs, ^{144}Ce	Site 7 km from Chernobyl 1990–1991, sampler 5 m downwind of rotivator, height 1 m	(1)
No activity – wind resuspension	1.2×10^{-9}*	^{137}Cs	Zapolie, 14 km south of Chernobyl during May 1993. Samplers were 45 m and 120 m from source of agricultural activity, 3–3.5 m height.	(2)
Grass cutting	3.9×10^{-9}*			
Harrowing, small tractor	$(10.5–94.8) \times 10^{-9}$*			
Harrowing, big tractor	155×10^{-9}*			
Large trucks simulating traffic on unpaved roads	$(89.3–384) \times 10^{-9}$*			
Traffic – sampler close to car park	$1,327 \times 10^{-9}$*† 119×10^{-9}**	^{137}Cs	Harwell, UK 1986–1987 and 1990–1991, sampler next to car park, 1 m height	(3)
Not near traffic – wind resuspension	73×10^{-9}† 35×10^{-9}**		Harwell, UK 1986–1987 and 1990–1991, sampler several hundred metres from car park, 1 m height‡	
Experimental fire Burning phase Smoldering phase Post-fire phase	$(10–100) \times 10^{-9}$ $(5–20) \times 10^{-9}$ $(2–5) \times 10^{-9}$	^{137}Cs	100 × 200 m pine forest. Measurements made 15–270 m downwind of the fire at 1 m height.	(4)
Real fire Burning phase Smoldering phase Post-fire phase	$(70–110) \times 10^{-9}$ $(30–60) \times 10^{-9}$ $(5-8) \times 10^{-9}$		Real fire 1997 10 ha of *Pinus silvestris* forest. Measurements made downwind of the fire.	

*Assumed mean areal deposition = 0.44 MBq m^{-2}, the average inventory in the top 5 cm of soil in two strips of land along which vehicles were driven. † Measured July 1986–June 1987. ** Measured July 1990–June 1991. ‡Likely to be over-estimate of real wind resuspension since this is a low-deposition site (see Figure 2.9).
(1) Ter-Saakov et al. (1994); (2) Garger et al. (1998); (3) Garland and Pomeroy (1994); (4) Kashparov et al. (2000).

Long term transport by airborne resuspension

The long-term transport of radionuclides by airborne resuspension may be estimated using measurements of the resuspension rate, R_a (yr^{-1}), defined as the fraction of the radioactivity in the soil which is resuspended per year:

$$R_a = \frac{\text{Resuspension flux (Bq m}^{-2}\text{ yr}^{-1})}{\text{Surface contamination of soil (Bq m}^{-2})} \qquad (2.6)$$

Measurements of resuspension rate of ^{137}Cs and $^{239+240}$Pu at 5 sites at distances of 4–150 km from Chernobyl gave values in the range $1.4\text{–}35 \times 10^{-5}\,\text{yr}^{-1}$ (Höllander and Garger, 1996). From these estimates, therefore, only 0.0014–0.035% of the total radionuclide deposit was resuspended per year. As discussed above, the majority of this resuspended material is likely to be re-deposited locally. This transport compares with estimates of long-term riverine transport ^{137}Cs of around 0.015–2% per year (Helton et al., 1985; Smith et al., 1997a; see also Chapter 4). It appears, therefore, that riverine transport was a more significant mechanism for long-term distribution of radionuclides than airborne transport. In a study of resuspension from forest fires, Kashparov et al. (2000) concluded 'even for the most unfavourable conditions, radionuclide resuspension during forest fires will not provide a significant contribution to terrestrial contamination. The additional terrestrial contamination due to the forest fire can be estimated [to be] in the range $10^{-4}\text{–}10^{-5}$ of its background value.'

2.2.5 Transport of radioactivity by rivers

Transport of radioactivity by rivers is discussed in detail in Chapter 4. As discussed above, erosional transport of radionuclides is relatively more significant (in terms of redistribution of deposited radioactivity) than atmospheric resuspension. Over long periods of time, soil erosion can redistribute significant amounts of radioactivity within a catchment. It is, however, important to note that thus far soil erosion processes played a relatively minor role in determining large-scale changes in radiocaesium and radiostrontium in the terrestrial ecosystem in comparison with physical decay (Kudelsky et al., 1998). Runoff of ^{90}Sr was slightly more significant than ^{137}Cs, but (from 1987 onwards) was still only a small fraction of the total losses from the terrestrial system by runoff and physical decay together (Kudelsky et al., 2002). It should be noted, however, that ^{90}Sr (Smith et al., 2004) and ^{137}Cs (Cawse and Baker, 1990) mobility in very organic saturated soils (peat soils), can be high, so losses in runoff water from these soils may be significant over a period of decades after fallout.

As discussed in Chapter 4, riverine input of radioactivity (primarily from the Danube and Dnieper Rivers) to the Black Sea was much less significant than direct atmospheric fallout to the sea surface. Over the period 1986–1995, riverine input for ^{137}Cs was only 4% of the initial atmospheric deposition, though ^{90}Sr inputs were more significant, being approximately 25% of the total inputs from atmospheric deposition and rivers combined (Kanivets et al., 1999). For the Baltic Sea, riverine inputs were similar to those in the Black Sea, being approximately 4% and 35% of atmospheric fallout for ^{137}Cs and ^{90}Sr, respectively (Nielsen et al., 1999).

In the long term, erosional redistribution processes of ^{137}Cs and ^{90}Sr played a relatively minor role in determining changes in their concentrations in biota and foodstuffs. Vertical migration, physical decay and changes in exchangeability/bioavailability of these radionuclides were much more important. For very long-lived radionuclides such as ^{241}Am and isotopes of Pu, however, slow erosional transfers over very long time periods will significantly alter their distribution in the landscape.

2.3 BIOAVAILABILITY, BIOACCUMULATION AND EFFECTIVE ECOLOGICAL HALF-LIVES

During the years after Chernobyl, radioactivity concentrations in foodstuffs and water declined significantly. This was primarily due to physical decay of short-lived radionuclides, but physical redistribution and changes in bioavailability were also important for long-lived radionuclides. In general, radiocaesium activity concentrations in foodstuffs declined during the years after the accident. Activity concentrations of radiostrontium also declined in general though there are reports of increases due to increased bioavailability of ^{90}Sr as hot particles broke up.

2.3.1 Aggregated Transfer Factor and Concentration Ratio

The transfer of radionuclides to biota is commonly estimated using an Aggregated Transfer Factor (T_{ag}, m^2 kg^{-1}):

$$T_{ag} = \frac{\text{Concentration of radionuclide in biota (Bq kg}^{-1})}{\text{Density of deposited radioactivity (Bq m}^{-2})} \text{ m}^2 \text{ kg}^{-1} \quad (2.7)$$

where the concentration of radionuclide in biota may be expressed per unit weight of fresh (wet) or dry matter. Since the transfer of radionuclides to foodstuffs is dependent on a number of factors (including transfer through the soil, soil composition and chemistry, time after fallout) the T_{ag} is clearly a simplified concept. It is, however, a valuable tool for estimating activity concentrations in foodstuffs from maps of contamination density. In addition, some models (e.g., Wright et al., 2003) can improve estimates of the T_{ag} by accounting for changes over time and using soil specific T_{ag} values. More complex models use soil data, such as percentage clay, potassium content and percentage organic matter, to estimate radiocaesium transfer to food crops and grass (e.g., Absalom et al., 2001).

An alternative way of parameterising the transfer of radionculides to biota is the Concentration Ratio (CR), defined as the activity concentration in biota (Bq kg^{-1}) divided by the activity concentration in surface soil (Bq kg^{-1}):

$$\text{CR} = \frac{\text{Concentration of radionuclide in biota (Bq kg}^{-1})}{\text{Concentration of radioactivity in dry soil (Bq kg}^{-1})} \quad (2.8)$$

where the concentration of radionuclide in biota is (usually, but not always) expressed per unit weight of dry matter. Observed CRs for a given radionuclide can vary over a range covering several orders of magnitude (Sheppard and Evenden, 1996). The CR is, like the T_{ag}, a simplified concept. For example, the mean concentration of radioactivity in soil is dependent on the depth to which the soil is sampled, since activity concentrations change with depth. Commonly, soils are sampled to a depth approximating to the depth of rooting of many plants (5–30 cm).

For aquatic systems, transfer of radionuclides to biota is commonly parameterised by a concentration factor (CF) where:

$$\text{CF} = \frac{\text{Activity concentration per kg of fish (wet wt)}}{\text{Activity concentration per litre of water}} \; 1 \, \text{kg}^{-1} \quad (2.9)$$

2.3.2 Physical, biological and ecological half-lives

Physical half-life

The physical decay of a radionuclide can be calculated relatively simply (Box 2.2). The activity concentration $C(t)$ of a radionuclide at time t is given by:

$$C(t) = C(0) \exp(-\lambda_p t) \quad (2.10)$$

where $C(0)$ is the activity concentration at time $t = 0$ and λ_p is the radioactive decay constant. The radioactive decay constant is related to the physical half-life T_p by:

$$T_p = \frac{\ln 2}{\lambda_p} \quad (2.11)$$

where $\ln 2 (= 0.693)$ is the natural logarithm of 2.

Box 2.2. Radioactive decay.

The physical or radioactive half-life is defined as the time taken for one-half of a given amount of a radioactive element to decay (i.e., transform into another isotope or element by emission of radiation). For example, 2.6×10^{17} Bq of ^{131}I were emitted from Chernobyl. ^{131}I has a physical half-life of 8.05 days so after 8.05 days (one half-life), 1.3×10^{17} Bq of ^{131}I remained in the environment. After 16.1 days (two half-lives), there was 0.65×10^{17} Bq ($\frac{1}{2} \times \frac{1}{2} = \frac{1}{4}$ of the original amount). After 10 half-lives (80.5 days) only about 0.1% of the ^{131}I remained, the rest having turned into stable ^{131}Xe by the radioactive emission of a beta-particle.

Biological half-life

The biological half-life is a measure of the rate of excretion of a radionuclide or stable element from an organism (or tissue/milk of an organism). It is the time taken for the amount of radioactivity in the organism to reduce by one-half and does not include reductions due to physical decay of the radionuclide. If, for example, there are 100 units of a stable element or relatively long-lived radionuclide in an organism, and 24 hours later this has reduced to 50 units, then the biological half-life is 24 hours. This calculation assumes that there has been no further uptake of the compound by the organism during the intervening 24-hr period. In addition, if there has been significant physical decay of the radionuclide, this must be accounted for when estimating the biological half-life.

If the initial concentration of radionuclide in an organism is $C(0)$ at a given time, then after time t the concentration $C(t)$ is given by an equation similar to Equation (2.10) above:

$$C(t) = C(0)\exp[-(\lambda_p + \lambda_{bio})t] \qquad (2.12)$$

where λ_{bio} is the rate of excretion of the radionuclide from the organism and the biological half-life T_{bio} is defined as:

$$T_{bio} = \frac{\ln 2}{\lambda_{bio}} \qquad (2.13)$$

The effective half-life $T_{eff}(bio)$ is the time taken for the amount of radioactivity in an organism to reduce by one-half by both physical decay and excretion and is therefore a combination of the physical and biological half-lives:

$$T_{eff}(bio) = \frac{\ln 2}{\lambda_p + \lambda_{bio}} \quad \text{or alternatively} \quad \frac{1}{T_{eff}(bio)} = \frac{1}{T_p} + \frac{1}{T_{bio}} \qquad (2.14)$$

Effective ecological half-life

The effective ecological half-life is the time taken for the amount of radioactivity in an environmental component (e.g., milk, meat or drinking water) to reduce by one-half following exposure in the natural environment. Assuming that the decline in radioactivity concentration C is exponential:

$$C(t) = C(0)\exp(-\lambda_{eff} t) \qquad (2.15)$$

The rate of decline λ_{eff} is often quoted as an effective ecological half-life T_{eff} (years) where:

$$T_{eff} = \frac{\ln 2}{\lambda_{eff}} \qquad (2.16)$$

The effective ecological half-life is a measure of the total change in the amount of radioactivity in an environmental component and therefore includes declines due to the physical decay of the radionuclide. The ecological half-life T_{eco} is the change in radioactivity due only to environmental processes and does not include physical decay:

$$\frac{1}{T_{eff}} = \frac{1}{T_p} + \frac{1}{T_{eco}} \qquad (2.17)$$

Since the effective ecological half-life is dependent on the source and bioavailability of the environmental contamination, it should be used and interpreted with care. In addition, as will be discussed below, the change in radioactivity in ecosystem components does not necessarily represent an exponential decline as given in Equation (2.15). For radiocaesium, for example, the rate of decline in activity concentrations reduces over time after fallout (see below). Therefore the observed effective ecological half-life of radiocaesium increased over time after the Chernobyl accident.

2.3.3 Changes in radiocaesium bioavailability over time

During the years after Chernobyl, the bioavailability and environmental mobility of radiocaesium declined markedly, resulting in large changes in contamination of foodstuffs, vegetation and surface waters. Laboratory studies on the sorption of radiocaesium to soils and sediments quantified the selective binding of Cs to specific sorption sites ('Frayed Edge Sites', FES) on illitic clay minerals (Cremers et al., 1988) (Figure 2.10). On these sorption sites, radiocaesium is available for ion-exchange with ions which have a similar hydrated radius, specifically potassium and ammonium. Over time, however, radiocaesium slowly diffuses into the illite lattice (Comans and Hockley, 1992) becoming unavailable for direct ion-exchange, a process commonly known as 'fixation' (Figure 2.11).

During the weeks to months after a fallout event, radiocaesium activity concentrations in both vegetation and surface waters are determined by short-term processes. Activity concentrations in plants are determined by interception and washoff of the initial fallout, as well as uptake by the roots (e.g., Fesenko et al., 1997). Similarly, in rivers and lakes, activity concentrations are initially high as a result of direct deposition to the water surface and rapid runoff of ^{137}Cs before it is sorbed to catchment soils. Activity concentrations then decline over a period of

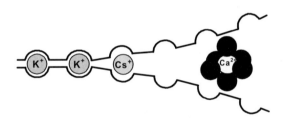

Figure 2.10. Illustration of Cs sorption to specific FES on illitic clay minerals and competition for sorption sites by ions of similar hydrated radius (K^+, NH_4^+) but not ions with much larger hydrated radius such as Ca^{2+}, Mg^{2+}.
Reproduced with the kind permission of Rob Comans, as modified from Jackson (1968).

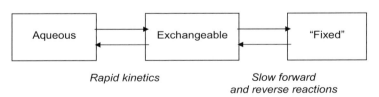

Figure 2.11. Illustration of the dynamic model for radiocaesium sorption to illitic clay minerals showing rapid uptake to 'exchangeable' sites and slower 'fixation' in the mineral lattice.

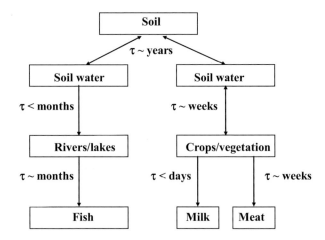

Figure 2.12. Schematic diagram indicating timescales τ of release of radiocaesium from soils to terrestrial and aquatic ecosystems during the years after a fallout event. The timescale of 'fixation' in soils is significantly longer than rates of retention and release of radioactivity in many other parts of the ecosystem. Thus, in the long term, changes in the soil–soil water partitioning controls changes in radioactivity concentration in surface waters, vegetation, etc. Adapted from Smith et al. (1999).

weeks to months as a result of reduced runoff from catchments and, for lakes, loss of ^{137}Cs through the outflow and deposition to bottom sediments (Chapter 4).

On long timescales (years), however, radiocaesium transfers to vegetation are controlled by the uptake from soil solution by plant roots. Similarly, concentrations of dissolved radiocaesium in rivers and many lakes are controlled by runoff of dissolved radioactivity from catchment soils. Thus changes in radiocaesium in the ecosystem are controlled to a large extent by slow changes in its bioavailabilty in the soil ('fixation').

As shown in Figure 2.12, radiocaesium transfers through a number of different ecosystem components are controlled by processes which operate on widely different timescales. Following a fallout event, some of these processes will achieve equilibrium more rapidly than others. For example, the rate at which radiocaesium in vegetation is transferred to milk has a timescale of days. If the activity concentration in vegetation is constant, or changing on a timescale very much greater than days, then the activity concentration in milk will rapidly reach equilibrium with respect to the activity concentration in vegetation. Thus, the activity concentration in milk will be a constant multiple of the (slowly changing) activity concentration in vegetation. This constant ratio is often described by the CR: the equilibrium ratio of activity concentration in milk to that in vegetation.

These concepts of differing kinetic rates may be understood in terms of the first-order differential equations which describe transfers of activity between the different ecosystem components. For example, the activity concentration in vegetation C_v (Bq kg^{-1}) declines exponentially with rate constant r_v (d^{-1}), so that

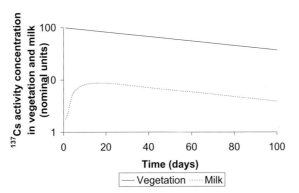

Figure 2.13. Illustration of changes in radiocaesium in milk in a system with declining activity concentrations in vegetation ($r_v = 0.01\,\text{d}^{-1}$) and relatively rapid rates of uptake and removal ($r_{mu} = 0.02\,\text{d}^{-1}$; $r_{mr} = 0.2\,\text{d}^{-1}$) from milk. Parameter values are for illustrative purposes only. A real example of time changes in vegetation and milk is given in Chapter 3.

$C_v = C_v(0)e^{-r_v t}$. The uptake and removal of activity from milk may be described by uptake and removal rate constants: r_{mu} and r_{mr} (d^{-1}), the activity concentration in milk, C_m being given by:

$$\frac{dC_m}{dt} = r_{mu}C_v - r_{mr}C_m \tag{2.18}$$

Equation (2.18), with $C_v = C_v(0)e^{-r_v t}$, has solution:

$$C_m = \frac{r_{mu}C_v(0)}{(r_{mr} - r_v)}\left(e^{-r_v t} - e^{-r_{mr} t}\right) \tag{2.19}$$

If the timescale ($t = 1/r$) of changes in activity concentration of vegetation is significantly greater than the timescale of uptake and removal of activity from milk, then, as time increases, the activity concentration in milk will change at the same rate as the activity concentration in vegetation (i.e., $\exp(-r_{mr}t)$ tends to zero, therefore C_m decreases in direct proportion to $\exp(-r_v t)$). This is illustrated graphically in Figure 2.13. On this timescale, therefore, the effective ecological half-life measured in milk will be identical to that measured in vegetation, since the rate of decline of ^{137}Cs in the two components is the same.

On a sufficiently long timescale (weeks–months) after a radioactive fallout event, radiocaesium activity concentrations in a wide range of foodstuffs should change at the same rate as the radiocaesium bioavailability in the soil. Note, however, that some ecosystem components such as trees have very slow biological transfers (of order years) and so may not reach equilibrium concentrations relative to the slow sorption processes in the soil.

Changing radiocaesium bioavailability in the environment

Smith et al. (1999) analysed many long-term field studies of temporal changes in radiocaesium in three different ecosystem components: vegetation, surface waters

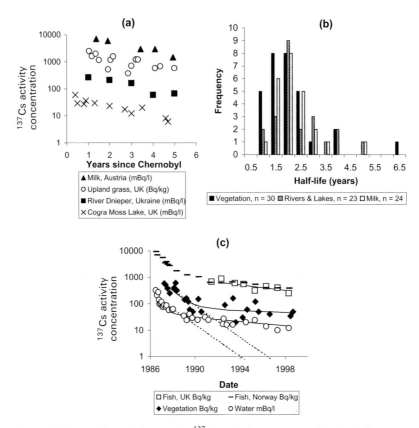

Figure 2.14. (a) Examples of changes in ^{137}Cs activity concentration in different ecosystem components after Chernobyl (Smith et al., 1999). (b) Frequency distribution of effective ecological half lives in different ecological components during the first five years after Chernobyl (Smith et al., 1999). (c) Long-term changes in ^{137}Cs in brown trout, Norway (Jonsson et al., 1999), perch, terrestrial vegetation and water, UK (Smith et al., 2000a). The decline in ^{137}Cs in immature fish, water and vegetation during the first five years has a T_{eff} of 1–4 yr as a result of 'fixation'. The dotted lines demonstrate a hypothetical continuation of irreversible fixation.

(dissolved phase) and milk following the Chernobyl accident (examples shown in Figure 2.14(a)). Figure 2.14(b) shows a histogram of measured T_{eff} values of radiocaesium in three different ecosystem components, vegetation, milk and water. Combining results for all three ecosystem components gives a mean $T_{eff} = 1.7$ years, with 91% of all measurements falling within the range 1–4 years. In Table 2.2, these rates of decline of radiocaesium in water and foodstuffs are compared with studies of the change in bioavailable ('exchangeable') radiocaesium in the soil and with the rate of diffusion ('fixation') of potassium-40 (a close analogue of caesium) into the lattice of illitic clay minerals. The observation that rates of change in ^{137}Cs in the ecosystem are similar to changes in its chemical availability

Table 2.2. Summary of mean values of rate of decline in ^{137}Cs activity concentrations in different environmental compartments, and comparison with rate of diffusion of ^{40}K into the illite lattice.

Adapted from Smith *et al.* (1999) who analysed data from a variety of sources.

Description	Mean rate of decline λ_{eff} (y^{-1})	Mean effective half-life T_{eff} (y)
Rivers and lakes (n = 23)	0.38	1.8
Vegetation (n = 30)	0.46	1.5
Milk (n = 24)	0.39	1.8
Exchangeable ^{137}Cs (n = 3)	0.35	2.0
^{40}K sorption to illite (n = 2)*	0.55	1.3

* ^{40}K diffusion into the illite lattice is considered as an analogue to Cs diffusion.

and 'fixation' implies that the slow sorption ('fixation') of radiocaesium to soil minerals controls its change in availability for transfer to many ecosystem components.

Variations in effective ecological half-lives

Variation in rates of decline in ^{137}Cs availability to plants and animals may be due to a number of factors including soil type and mineralogy, seasonal effects, agricultural practices and application of countermeasures (Sanzharova *et al.*, 1994; Fesenko *et al.*, 1997). A number of authors (Squire and Middleton, 1966; Sandalls and Bennett, 1992; Fesenko *et al.*, 1997) have suggested that the rate of decline in ^{137}Cs in vegetation is related to soil type. In particular, it is hypothesised that in the first years after fallout, rates of decline are greater in vegetation grown on soils containing significant clay fractions and lower in highly organic soils.

Using measurements from 11 upland sites in Cumbria, UK (Horrill and Howard, 1991; Sandalls and Bennett, 1992), rates of decline in radiocaesium activity concentrations in vegetation were found to be related to organic matter content of the soil (Smith *et al.*, 1999). For these sites, with a range of soil organic matter content of 19 to 88%, these workers observed a significant ($r = 0.8$, $p < 0.001$) negative correlation between the rate of decline in ^{137}Cs concentration and the organic matter content. Rates of decline in ^{137}Cs activity concentrations were twice as high in vegetation grown on soils with organic matter content <50% than on soils with organic matter content >80%. It should be noted, however, that the highly organic soils showed very large variability in rates of decline.

A study by Valcke and Cremers (1994) showed that in soils of low to medium organic matter content (<40%) the majority of the ^{137}Cs is sorbed on illitic FES: it is only in soils with high organic matter content (>80%) that the majority is associated with regular ion-exchange sites. These results indicate that low illite contents in highly organic soils will lead to a slower rate of reduction in ^{137}Cs availability, in agreement with observations. They are also consistent with observations that in the majority of soils, rates of fixation are similar. As long as there is a significant illite mineral component, the majority of the ^{137}Cs will be sorbed to FES (Valcke and Cremers, 1994). Thus, rates of fixation will be controlled by (approximately) the same diffusion rate into the illite lattice, regardless of soil type.

There is evidence for slower rates of decline in radiocaesium concentrations in forest and semi-natural ecosystems as compared with agricultural systems (see Chapter 3). For example, Zibold et al. (2001) observed T_{eff} in roe deer in a spruce forest in Germany of 3.5 years, similar to the T_{eff} of green plants on which they grazed (2–6 years, Klemt et al., 1999). In a pine/spruce forest on peat bog soil, the T_{eff} of roe deer was even higher, approximately 17 years. In Scandinavia, effective ecological half-lives in reindeer were in the range 2–5 years (Åhman and Åhman, 1994).

Clearly, there is natural variation in T_{eff} values which may be linked to soil characteristics. However, to our knowledge, there is currently no systematic way of predicting such variation, except that it is known that semi-natural and forest ecosystems may show high T_{eff} values. Nevertheless, for the time period 0.5–5 years after fallout, the range in rates of decline observed for a number of ecosystem components ($T_{eff} = 1$–4 y) is relatively small. A value of $T_{eff} = 2$ y will describe declines in components (grass, milk, meat, water, fish, etc.) in most agricultural and surface water systems during this time period after fallout.

Potential changes of ^{137}Cs in the absence of slow fixation

It is valuable to consider the relative importance of soil 'fixation' of ^{137}Cs compared to other processes acting to reduce its mobility and bioavailability. Two other processes have been hypothesised which may reduce radiocaesium activity concentrations following a fallout event: (1) the reduction in inventory of radiocaesium as a result of losses of activity from soils in runoff water; (2) transport of activity down the soil profile.

If there is no slow fixation of radiocaesium, the activity concentration in soil water would be a constant fraction of that sorbed to the soil. Thus, the activity concentration in runoff water would decline only as a function of declining catchment ^{137}Cs inventory. In this case, the rate of change λ in ^{137}Cs activity concentrations in runoff water would be given by: $\lambda = f_c$, where f_c is the fractional loss of ^{137}Cs from the catchment per year (y^{-1}). Using measured values of radiocaesium loss from typical catchments of ca. 0.5–2% per year (Helton et al., 1985; Smith et al., 1997a), gives values of $\lambda = 0.005$–0.02 y^{-1} ($T_{eff} = 139$–35 y) which is more than one order of magnitude slower than the rate of decline observed after Chernobyl (Table 2.2). Over a period of several years, the decline in inventory of catchment soils as a result of removal in runoff water is negligible. Thus the *rates of change* in activity concentrations in surface waters and vegetation ($T_{eff} \sim 2$ years) cannot be due to loss of inventory, but must be due to a reduction in radiocaesium mobility and bioavailability.

In the absence of slow fixation (and negligible reduction due to runoff), the change in radiocaesium activity concentration in vegetation would decline only as a result of declining activity in the rooting zone as it migrates down the soil profile. However, rates of vertical migration of ^{137}Cs are slow (see above) and cannot alone explain a T_{eff} value of around 2 years. In a study of radiocaesium migration in meadow ecosystems, Fesenko et al. (1996) showed that the residence time of radiocaesium in the soil surface was very high. They concluded that 'this indicates that the contribution of vertical migration to the decrease of ^{137}Cs in the root zone on mineral soils is negligible. On the contrary, on wet meadow and on peatland it can be an important factor which influences the decrease of the transfer of ^{137}Cs into foodchains.'

Longer term changes in radiocaesium bioavailability

In the longer term, (more than, say, 5 years after fallout), observations showed (Jonsson et al., 1999; Smith et al., 2000a) that the rate of decline in ^{137}Cs activity concentrations in surface water, terrestrial vegetation and fish reduced significantly (Figure 2.14(c)). The effective ecological half-life in young fish, water and terrestrial vegetation increased from 1–4 years during the first five years after Chernobyl to 6–30 years in the late 1990s. From the observed persisting mobility of radiocaesium in the environment, and particularly the increase of T_{eff} towards the physical decay rate of ^{137}Cs ($T_{1/2} = 30.2$ years), it was concluded (Smith et al., 2000a) that the sorption–desorption process of radiocaesium in soils and sediments was tending towards a reversible steady state. In other words, ^{137}Cs is not irreversibly 'fixed' in the soil matrix and remains available in the environment for long periods of time after fallout (Figure 2.14(c)). Thus, as illustrated in Figure 2.14(c), activity concentrations of radiocaesium in vegetation and foodstuffs has remained at much higher levels than would be expected if the 'fixation' continued irreversibly.

Temporal behaviour of Chernobyl ^{137}Cs compared to weapons test fallout

The time changes in ^{137}Cs in foodstuffs after Chernobyl was compared to NWT fallout by Mück (1997) who found that T_{eff} values were higher after NWT than Chernobyl, implying a slower rate of decline. This, however, was attributed to differences in the pattern of fallout between the two events (the NWT fallout mainly occurred over several years), rather than to any differences in the environmental behaviour of ^{137}Cs (Mück, 1997). In a study of long-term ^{137}Cs in Finnish rivers after both NWT and the Chernobyl accident, Smith et al. (2000b) accounted for differences in the two source terms. Once the different levels of fallout and the different temporal patterns had been accounted for, there were no significant differences in the radiocaesium activity concentration in rivers after the two deposition events. It is therefore likely that, in the event of any future atmospheric fallout of radiocaesium, the changes in its mobility and bioavailability would follow similar patterns to those observed after Chernobyl and NWT.

2.3.4 Temporal changes in radiostrontium bioavailability

There is much less information on time changes of radiostrontium in the environment than for radiocaesium. In the initial period after fallout (weeks–months), radiostrontium activity concentrations are expected to decline relatively rapidly as a result of physical redistribution processes (washoff from plant surfaces, infiltration and sorption to the soil). In the longer term, in contrast to radiocaesium, radiostrontium is expected to remain relatively bioavailable in the environment (e.g., Tikhomirov and Sanzharova, 1978, quoted in Coughtrey and Thorne, 1983). Changes in activity concentrations of radiostrontium in the ecosystem may, however, occur by very slow declines in its bioavailability, or removal of ^{90}Sr from the surface layers of soil (by vertical migration or transport to rivers).

Results of lysimeter experiments (Nisbet and Shaw, 1994) showed declines in concentrations of ^{90}Sr in crops grown on loam and sand soils of up to a factor of 3 in cabbage, but relatively little decline in concentrations in barley and carrots. No significant declines were observed in crops grown on peat soils. In a variety of crops grown on loam, clay and sand soils, ^{90}Sr uptake showed significant seasonal variation in concentrations, but no significant changes in mean activity concentrations in crops over a 7-year period (Noordijk et al., 1992). Earlier lysimeter studies by Squire (1966) showed only a slight reduction in uptake over an 8-year period, and this was largely due to vertical migration rather than to any reduction in bioavailability.

Studies of NWT fallout in Greenland (Aarkrog, 2000) showed declines in ^{90}Sr concentrations in grass, lamb, reindeer and muskox with an effective ecological half-life of 5–18 years. A study of measurements of NWT fallout in various countries showed the mean effective ecological half-life in milk was 9.5 years (±SD of 4.4 years) (Cross, 2001). This compares with estimates of declines in Chernobyl ^{90}Sr in two Finnish rivers with T_{eff} of 4.9 and 9.9 years (Cross et al., 2002).

It appears that there may be a slow long-term decline in ^{90}Sr activity concentrations over time after fallout. This may be due to vertical migration, transport processes or very slow 'fixation' in the soil: probably it is some combination of all three. In areas close to Chernobyl, however, any small declines in ^{90}Sr availability due to these processes were masked by the *increasing* bioavailability of ^{90}Sr in the soil due to dissolution of hot particles (Krouglov et al., 1997). This is illustrated by the contrasting change in the CR of ^{137}Cs (largely deposited in chemically available forms) and ^{90}Sr (largely deposited as hot particles) in soils in the Ukraine (Figure 2.15; Krouglov et al., 1997). The ^{137}Cs CR declines significantly over time as a result of 'fixation', whilst the ^{90}Sr CR increases as its bioavailability increases due to break-up of the hot particles (Krouglov et al., 1997). These workers concluded that the '^{90}Sr content in crops increased with a mean half-life of 17 months [1.42 years]'. This is lower than, but of the same order as, the leaching rates determined for fuel particle dissolution by Kashparov et al. (1999) (Figure 2.3).

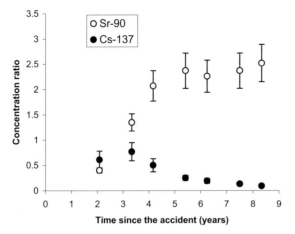

Figure 2.15. Change in CR of wheat on soddy–podzolic soil (Krouglov *et al.*, 1997). The CR of ^{137}Cs tends to decrease as a result of increasing 'fixation' to the soil, whilst the ^{90}Sr CR increases as bioavailability increases due to dissolution of fuel particles.

2.4 CHARACTERISTICS OF KEY CHERNOBYL RADIONUCLIDES

2.4.1 Radioiodine

Isotope data

Isotope		Half-life	Emission energy* MeV (intensity)
I-131		**8.04 d**	
Principal radioactive emissions		β-	0.069 (2%)
		β-	0.097 (7%)
		β-	0.192 (90%)
		γ	0.284 (6%)
		γ	0.365 (82%)
		γ	0.637 (7%)
		e-(I)	0.046 (4%)
		e-(I)	0.330 (2%)
Decays to		*Xe-131*	
I-129		**1.6 × 10^7 y**	
Principal radioactive emissions		β-	0.041 (100%)
		γ	0.040 (7.5%)
Decays to		*Xe-129*	

(I) internal conversion electron.
* For beta-particles, mean emission energy of beta-groups is given.

Chemical properties and general environmental behaviour

Iodine (stable isotope: ^{127}I) is found in trace quantities in the environment, most importantly in seawater. Atmospheric transport of iodine from seawater is an important source of iodine to terrestrial and freshwater environments, though weathering from rocks also provides a source. Short-lived ^{131}I is the most important radioactive form of iodine, having been the major contributor to radiation exposure of the population in the early period after Chernobyl. However, in the first 1–2 days after deposition ^{133}I ($T_{1/2}$ = 21 hours) also contributed to dose. Very long-lived ^{129}I was of negligible radiological significance, but it has been studied as an environmental tracer and as an indicator of ^{131}I deposition patterns.

Iodine is soluble in water, occurring as iodide (I^-) and iodate (IO_3^-). It binds strongly to soils: a study of ^{129}I from a reprocessing facility in the USA showed that 85–90% was bound in the top 15 cm of soil (Rao and Fehn, 1999). These authors quote residence times of approximately 30 years in the top 30 cm of soil, though studies near the Savannah River site estimate a mean residence time of 80 years (Kocher, 1991). Iodine in soils is believed to be strongly adsorbed to organic matter (Koch-Steindl and Pröhl, 2001), being found in plant litter, humic matter and colloidal organic molecules (Oktay et al., 2001). Another study (Kocher, 1991) cites experimental evidence that 'iodine is rapidly fixed in soil by association with organic matter, clays, and aluminium and iron oxides'. For the short-lived ^{131}I, physical distribution processes following deposition (deposition to and washoff from plant surfaces, infiltration to the soil) are of major importance in determining transfers to the food chain.

Iodine is not believed to be an essential plant nutrient (Whitehead, 2000), though it is absorbed by plants via the roots and atmospheric deposition on leaves. Iodine is an important nutrient for humans and many animals, being essential for operation of the thyroid gland. A large proportion of the iodine in humans and grazing animals is found in the thyroid gland and ingested iodine is rapidly transferred to the thyroid. Vinogradov (1953) reported high concentrations of iodine in the thyroid of fish. Both a lack and an excess of iodine in the diet can lead to a number of health problems in humans, the best known of which is swelling of the thyroid (goitre).

Inhalation and ingestion (of milk and, to a lesser extent, leafy vegetables) were believed to have formed the major pathways for radioiodine dose to the thyroid in the populations affected by Chernobyl (Pröhl et al., 2002). Incorporation of ^{131}I in the thyroid (in humans and animals) is increased when dietary stable iodine is low, and excess stable iodine can significantly reduce ^{131}I uptake. Thus, potassium iodide pills may be administered to people during or shortly after radioactive releases, though this measure was not systematically taken immediately after Chernobyl.

Examples of stable iodine concentrations in the environment

Stable iodine (^{127}I) concentration		
Range in freshwaters	0.5–211 µg l^{-1}	Moran et al. (2002)
Seawater	60 µg l^{-1}	CRC (1988)
Marine algae (f.w.)	10–2000 mg kg^{-1}	Vinogradov (1953)
Freshwater fish, muscle (f.w.)	0.01–0.24 mg kg^{-1}	Vinogradov (1953)
Marine fish, muscle (f.w.)	0.1–5.7 mg kg^{-1}	Vinogradov (1953)
Soil	0.1–50 mg kg^{-1}	Vinogradov (1959)
Milk (f.w.)	0.3 mg kg^{-1}	MAFF (2000)
Meat products (f.w.)	0.05 mg kg^{-1}	EC (2002)
Vegetables (f.w.)	0.03 mg kg^{-1}	EC (2002)

f.w. = fresh weight.

2.4.2 Radiostrontium

Isotope data

Isotope	Half-life	Emission energy* MeV (intensity)
Sr-89	50.5 d	
Principal radioactive emissions	β-	0.585 (100%)
Decays to	*Y-89*	
Sr-90	28.8 y	
Principal radioactive emissions	β-	0.196 (100%) (Sr-90 → Y-90)
	β-	0.934 (100%) (Y-90 → Zr-90)
Decays to	*Y-90 then Zr-90*	

* For beta-particles, mean emission energy of beta-groups is given.

Chemical properties and general environmental behaviour

Stable strontium is found in the minerals celestite and strontianite. It occurs in the environment in trace quantities compared to other group II (alkaline earth) elements, calcium and magnesium. It exists as four stable isotopes: ^{88}Sr (abundance 82.58%), ^{86}Sr (abundance 9.86%), ^{87}Sr (abundance 7.00%) and ^{84}Sr (abundance 0.56%). Radioactive isotopes of strontium occur in the environment as a result of above-ground testing of nuclear weapons, routine discharges from nuclear installations and from nuclear accidents. Environmental concentrations of radioactive ^{89}Sr and ^{90}Sr are several orders of magnitude lower than those of stable strontium. Because of their low concentrations relative to the stable isotope, the chemical and biological properties of radioactive strontium isotopes are not expected to be influenced by the amount of radioactivity deposited.

Strontium occurs in aqueous solution as a free Sr^{2+} ion and generally binds to solids by electrostatic 'ion exchange' processes. Because of the high charge and hence large hydrated radius of the strontium ion it (in contrast to caesium) is not believed to be 'fixed' in mineral lattices: it is absorbed to relatively easily available sorption sites on mineral surfaces and organic particles. The strength of its sorption to solids is determined primarily by the cation exchange capacity (CEC) of the soil or sediment (which is highest in soils and sediments with high organic matter content) and on the concentration of other ions in solution (in particular Ca^{2+} and Mg^{2+}). In a study of radiostrontium sorption to Japanese soils, Yasuda and Uchida (1993) found that the solids–aqueous partitioning of Sr increased in proportion to the soil CEC and was inversely proportional to the electrical conductivity (a measure of ionic strength) of the solution. In contrast to radiocaesium, radiostrontium is believed to remain predominantly reversibly sorbed to ion exchange sites in soils (e.g., Tikhomirov and Sanzharova, 1978, quoted in Coughtrey and Thorne, 1983), hence bioavailability is not expected to change significantly over time.

Strontium is bioaccumulated by the same mechanisms as (chemically similar) calcium, an important nutrient. It is believed to have a biological role in bone and teeth formation in animals. Uptake to plants is related to the calcium concentration, with the Sr:Ca ratio in plants being approximately equal to that in soil (Tikhomirov and Sanzharova, 1978, quoted in Coughtrey and Thorne, 1983). In aquatic environments, ^{90}Sr accumulation in fish is in inverse proportion to the calcium concentration of the water (Blaylock, 1982). Similarly, ^{90}Sr absorption in animals and the subsequent transfer to animal derived food products was shown to be inversely proportional to the dietary calcium intake with Sr:Ca ratios in tissues and milk being lower than those in the diet (e.g., the ratio Sr:Ca in milk/ Sr:Ca in the diet is approximately 0.1 for mammals) (Comar et al., 1966; Beresford et al., 1998). An important feature of Sr transfers to animals and humans is its incorporation into bone, and hence its long retention time in the body. Similarly, strontium is incorporated into the shells of invertebrate species such as mussels.

Examples of stable strontium concentrations in the environment

Stable strontium concentration		
Range in freshwaters	0.006–0.5* mg l^{-1}	Hilton et al. (1997)
Seawater	8.1 mg l^{-1}	CRC (1988)
Freshwater fish, muscle (f.w.)	1.2–3.7 mg kg^{-1}	Coughtrey and Thorne (1983)
Freshwater fish, bone (f.w.)	170–320 mg kg^{-1}	Coughtrey and Thorne (1983)
Marine fish, muscle (f.w.)	1–5 mg kg^{-1}	Coughtrey and Thorne (1983)
Marine fish, bone (f.w.)	67–390 mg kg^{-1}	Coughtrey and Thorne (1983)
Soil	10–280 mg kg^{-1}	Vinogradov (1959)
Fruits and vegetables (d.w.)	0.42–64 mg kg^{-1}	ATSDR (2004)
Cow milk (f.w.)	0.42 mg kg^{-1}	Anderson (1992)
Human, muscle (f.w.)	0.05 mg kg^{-1}	ATSDR (2004)
Human, bone (f.w.)	63–281 mg kg^{-1}	

*A few significantly higher values were reported in this review. f.w. = fresh weight; d.w. = dry weight.

2.4.3 Radiocaesium

Isotope data

Isotope	Half-life		Emission energy* MeV (intensity)
Cs-134	**2.065 y**		
Principal radioactive emissions		β-	0.210 (70%)
		β-	0.023 (27%)
		β-	0.123 (2.5%)
		γ	0.605 (97.6%)
		γ	0.796 (85.5%)
		γ	1.365 (3.0%)
Decays to	*Ba-134*		
Cs-137	**30.17 y**		
Principal radioactive emissions		β-	0.174 (94.4%)
		β-	0.416 (5.6%)
		γ	0.662 (85.1%)
		e-(I)	0.624 (8%) 0.656 (1%)
Decays to	*Ba-137m then Ba-137*		

(I) internal conversion electron.
* For beta-particles, mean emission energy of beta-groups is given.

Chemical properties and general environmental behaviour

Stable caesium-133 is an alkali metal which can be found at relatively high concentrations in the minerals lepidolite and pollucite. Rich sources of pollucite can contain 20% caesium, but typically it is found in trace concentrations in the environment. Radioactive isotopes of caesium occur in the environment as a result of aboveground testing of nuclear weapons, routine discharges from nuclear installations and from nuclear accidents. Environmental concentrations of ^{134}Cs and ^{137}Cs are several orders of magnitude lower than those of stable ^{133}Cs. Because of their low concentrations relative to the stable isotope, the chemical and biological properties of radioactive caesium isotopes are not expected to be influenced by the amount of radioactivity deposited.

The chemical properties of caesium are similar to those of the much more abundant potassium. Caesium is very soluble, existing in dissolved form as the monovalent cation Cs^+. It readily absorbs to negatively charged mineral surfaces, its small hydrated radius making it more strongly attracted to minerals than the other group I alkali metals Na and K. Caesium is relatively immobile in soils, absorbing strongly to clay minerals (particularly illites) (Jacobs and Tamura, 1960). It competes for binding sites on the clay mineral lattice with K^+ and NH^{4+} ions which have a similar hydrated radius. Thus, a common measure of the 'bioavailability' or 'exchangeability' of radiocaesium sorption to soils and sediments is to extract the radiocaesium with (typically 0.1–1 M) ammonium acetate or ammonium chloride solution. In soils with significant clay content, typically less than 10% of radiocaesium is found in exchangeable form, the rest being 'fixed' to the mineral lattice. Environmental studies of global NWT fallout, for example, showed that transfers of radiocaesium to milk were much higher in the Faroe Islands than in Denmark. This was attributed to greater availability of ^{137}Cs in the organic soils of the Faroe Islands than in the more mineral soils of Denmark (Aarkrog, 1979).

Although it has no known biological role in organisms, caesium is bioaccumulated by the same mechanisms as (chemically similar) potassium, an important nutrient. If potassium is in short supply, plants need to actively accumulate it and as a result more strongly accumulate caesium. In terrestrial systems, for example, addition of fertiliser containing K^+ decreases ^{137}Cs uptake by plants in some soils, though the effectiveness of application is reduced by the potential for K^+ ions to desorb ^{137}Cs from soil binding sites (Coughtrey and Thorne, 1983).

Uptake of radiocaesium to biota is therefore expected to be highest in soils of low nutrient status (particularly low K^+) and low clay mineral content. These effects are illustrated by data collected by the United Nations Scientific Committee on the Effects of Atomic Radiation (UNSCEAR) on the transfers of radiocaesium from nuclear weapons test fallout to humans. Whole-body radiocaesium activity concentrations were approximately two orders of magnitude greater in people living in 'marginal' arctic and sub-arctic environments with nutrient poor soils (e.g., Finnish reindeer herders, populations in northern Russia and Alaska) than in populations whose diet originated predominantly from intensive agriculture (UNSCEAR, 1977).

Examples of stable caesium concentrations in the environment

Stable caesium (^{133}Cs) concentration		
Range in freshwaters	$(1.3–57) \times 10^{-3} \, \mu g \, kg^{-1}$	Hilton et al. (1997)
Seawater	$0.5 \, \mu g \, kg^{-1}$	CRC (1988)
Freshwater fish, *Salvelinus malma* from L. Dal'nii (f.w.)	$35–140 \, \mu g \, kg^{-1}$	Fleishman (1973)
Marine fish, cod *Gadus morhua*, Newfoundland (f.w.)	$25–65 \, \mu g \, kg^{-1}$	Hellou et al. (1992)
Soil	$(0.3–25) \times 10^3 \, \mu g \, kg^{-1}$	Vinogradov (1959)
Mushrooms (d.w.)	$(0.1–1.0) \times 10^3 \, \mu g \, kg^{-1}$	Yoshida et al. (2000)
Trees (various parts, d.w.)	$(0.01–1.0) \times 10^3 \, \mu g \, kg^{-1}$	Yoshida et al. (2000)
Grass (d.w.)	$100–400 \, \mu g \, kg^{-1}$	Oughton (1989)
Sheep (f.w.)	$60–540 \, \mu g \, kg^{-1}$	Oughton (1989)

f.w. = fresh weight; d.w. = dry weight.

2.4.4 Plutonium and americium

Isotope data

Isotope	Half-life	Emission energy* MeV (intensity)
Pu-238	**87.7 y**	
Principal radioactive emissions	α	5.456 (28.98%)
	α	5.499 (70.91%)
Decays to	U-234	
Pu-239**	**2.41 × 10⁴ y**	
Principal radioactive emissions	α	5.157 (73.3%)
	α	5.144 (15.1%)
	α	5.106 (11.5%)
Decays to	U-235	
Pu-240**	**6.54 × 10³ y**	
Principal radioactive emissions	α	5.168 (72.8%)
	α	5.124 (27.1%)
Decays to	U-236	
Pu-241	**14.4 y**	
Principal radioactive emissions	β-	0.0052 (100%)
Decays to	Am-241	
Am-241	**432.2 y**	
Principal radioactive emissions	γ	0.0595 (35.9%)
	α	5.486 (84.5%)
	α	5.442 (13.0%)
	α	5.388 (1.6%)
Decays to	Np-237	

* For beta-particles, mean emission energy of beta-groups is given.
** Activity concentrations of these isotopes are often measured and quoted as '$^{239+240}$Pu' since their alpha emission energies are almost identical.

Chemical properties and general environmental behaviour

Plutonium is produced and used in nuclear reactors and the vast majority of environmental plutonium comes from nuclear wastes, explosions and accidents. Trace amounts (mainly ^{239}Pu) are, however, naturally found in uranium-bearing rocks, formed as a result of neutron capture in uranium. Pu has no stable isotopes. Americium does not occur naturally, and has no stable isotopes. The most important isotope, ^{241}Am is formed from the decay of ^{241}Pu. Most americium in the environment is formed from releases of ^{241}Pu (following Chernobyl and NWT), though some direct releases of ^{241}Am occur in nuclear wastes. Americium-241 is most commonly found in household smoke detectors. Following a release of ^{241}Pu, activity concentrations of ^{241}Am gradually increase as the ^{241}Pu decays with a half-life of

14.4 years. Maximum americium concentrations in the environment occur approximately 73 years after the ^{241}Pu release, reaching a maximum activity of approximately 3% of the initial ^{241}Pu release.

The aqueous chemistry of plutonium is complex, with a number of different species being potentially present in solution. Plutonium has a high affinity for soils, being associated to a large degree with organic matter (Livens et al., 1987), though other studies have shown associations with other soil fractions (Alberts and Orlandini, 1981; Bunzl et al., 1994). Affinity for sediments is also very high: there is evidence for remobilisation of small amounts of plutonium from anoxic lake sediments (Sholkovitz et al., 1982), though other studies have shown the effect to be minimal (Alberts and Orlandini, 1981). There is evidence of mobility of small amounts of plutonium attached to colloids (e.g., Orlandini et al., 1990).

Plutonium and americium present a low external dose risk since they emit primarily alpha activity (excepting ^{241}Pu which emits only a weak beta). The primary health risk is therefore from ingestion and inhalation. Am and Pu have relatively low mobility due to their strong tendency to sorb onto soil particles. Americium generally exists in the +III valence state, whereas Pu can be found in any of four oxidation states (+III, +IV, +V, +VI) depending on the redox conditions of the soil system. Changes in soil pH and Eh and the presence of complexing agents such as dissolved organic compounds and extracellular metabolites of microflora are known to increase the mobility of Pu in soils (e.g., Negri and Hinchman, 2000). Plant uptake of Am has been shown to increase with soil pH (Frissel et al., 1990). Like plutonium, americium is relatively immobile in the environment, being strongly absorbed by soils (Bunzl et al., 1994; Smith et al., 1997b) and sediments (Alberts and Orlandini, 1981). The dry weight ratio of plant to soil (CR) activity concentrations of Pu are typically in the order 10^{-6} to 10^{-4} (IAEA, 1994). For americium, the values range from 10^{-5} to 10^{-3} (IAEA, 1994).

Bioaccumulation of Pu in aquatic systems is relatively low, with CFs in freshwater fish (wet weight basis) being in the range 4–300 l kg^{-1} (IAEA, 1994). Accumulation factors of freshwater plants and invertebrates may be much higher and there is evidence of decreasing accumulation at higher trophic levels (Blaylock, 1982). A study of accumulation by various marine organisms found that 'plutonium concentrations in benthic (bottom-dwelling) biota are generally 1–2 orders of magnitude lower than in surface sediments. Furthermore, a significant part of this plutonium is probably not metabolised but rather associated with particles in the guts and adhering to the surface structure of the animals' (Dahlgaard et al., 2001). There are relatively few data of ^{241}Am accumulation in aquatic biota, however studies suggest that accumulation in fish is similar to plutonium (IAEA, 1994).

Plutonium is not readily absorbed by animals, although relatively few studies have considered environmentally relevant sources of Pu to grazing animals, work with sheep suggest that less than 0.01% of ingested Pu is absorbed from the gastrointestinal tract into the blood circulation system (Beresford et al., 2000). This compares to values of 0.1–0.001% which have been determined in monogastric (non-ruminant) species administered 'soluble' forms of Pu (for comparison the absorption of Cs is in the range 80–100% and that of iodine is complete). Absorption of Am from the gastrointestinal tract is similar to that of Pu. Once absorbed, plutonium and americium are largely deposited in bone and liver, each of these tissues contributing approximately 50% of the body burden of Pu and Am (Coughtrey et al., 1984).

2.5 REFERENCES

Aarkrog, A. (1979) *Environmental studies on radioecological sensitivity and variability with special emphasis on the fallout nuclides Sr-90 and Cs-137*. Risø National Laboratory Report, Risø, Denmark.

Aarkrog, A. (1988) The radiological impact of the Chernobyl debris compared with that from Nuclear Weapons Fallout. *Journal of Environmental Radioactivity*, **6**, 151–162.

Aarkrog, A., Dahlgaard, H. and Nielsen, S.P. (2000) Environmental radioactive contamination in Greenland: A 35 years retrospect. *Science of the Total Environment*, **245**, 233–248.

Absalom, J.P., Young, S.D., Crout, N.M.J., Sanchez, A., Wright, S.M., Smolders, E., Nisbet, A.F. and Gillett, A.G. (2001) Predicting the transfer of radiocaesium from organic soils to plants using soil characteristics. *Journal of Environmental Radioactivity*, **52**, 31–43.

Åhman, B. and Åhman, G. (1994) Radiocesium in Swedish reindeer after the Chernobyl fallout: Seasonal variations and long-term decline. *Health Physics*, **66**, 503–512.

Alberts, J.J. and Orlandini, K. (1981) Laboratory and field studies of the relative mobility of ^{239,240}Pu and ^{241}Am from lake sediments under oxic and anoxic conditions. *Geochimica et Cosmochimica Acta*, **45**, 1931–1939.

Anderson, R.R. (1992) Comparison of trace elements of milk of four species. *Journal of Dairy Science*, **75**, 3050–3055.

Anspaugh, L.R., Simon, S.L., Gordeev, K.I., Likhtarev, I.A., Maxwell, R.M. and Shinkarev, S.M. (2002) Movement of radionuclides in terrestrial ecosystems by physical processes. *Health Physics*, **82**, 669–679.

Arapis, G., Petrayev, E., Shagalova, E., Zhukova, O., Sokolik, G. and Ivanova, T. (1997) Effective migration velocity of Cs-37 and Sr-90 as a function of the type of soils in Belarus. *Journal of Environmental Radioactivity*, **34**, 171–185.

ATSDR (2004) *Toxicological Profile for Strontium*. U.S. Department of Health and Human Services: Agency for Toxic Substances and Disease Registry, Atlanta (available online at: http://www.atsdr.cdc.gov/toxpro2.html).

Balonov, M.I., Bruk, G.Ya., Golikov, V.Yu., Erkin, V.G., Zvonova, I.A., Parchomenko, V.I. and Shutov, V.N. (1995) Long-term exposure of the population of the Russian Federation as a consequence of the accident at the Chernobyl nuclear power plant. In: *Environmental Impact of Radioactive Releases*, pp. 397–411 (IAEA-SM-339/115). IAEA: Vienna.

Beresford, N.A., Mayes, R.W., Hansen, H.S., Crout, N.M.J., Hove, K. and Howard, B.J. (1998) Generic relationship between calcium intake and radiostrontium transfer to milk of dairy ruminants. *Radiation Environmental Biophysics*, **37**, 129–131.

Beresford, N.A., Mayes, R.W., Cooke, A.I., Barnett, C.L., Howard, B.J., Lamb, C.S. and Naylor, G.P.L. (2000) The importance of source dependent bioavailability in determining the transfer of ingested radionuclides to ruminant derived food products. *Environmental Science and Technology*, **34**, 4455–4462.

Blagoeva, R. and Zikovsky, L. (1995) Geographic and vertical distribution of Cs-137 in soils in Canada. *Journal of Environmental Radioactivity*, **27**, 269–274.

Blaylock, B.G. (1982) Radionuclide data bases available for bioaccumulation factors for freshwater biota. *Nuclear Safety*, **23**, 427–438.

Bobovnikova, Ts.I., Virchenko, Ye.P., Konoplev, A.V., Siverina, A.A. and Shkuratova, I.G. (1991) Chemical forms of occurrence of long-lived radionuclides and their alteration in soils near the Chernobyl Nuclear Power Station. *Soviet Soil Science*, **23**, 52–57.

Bunzl, K., Förster, H., Kracke, W. and Schimmack, W. (1994) Residence times of fallout $^{239+240}$Pu, ^{238}Pu, ^{241}Am and ^{137}Cs in the upper horizons of an undisturbed grassland soil. *Journal of Environmental Radioactivity*, **22**, 11–27.

Bunzl, K., Kofuji, H., Schimmack, W., Tsumura, A., Ueno, K. and Yamamoto, M. (1995) Residence times of global weapons testing fallout Np-237 in a grassland soil compared to Pu-239,240, Am-241 and Cs-137. *Health Physics*, **68**, 89–93.

Cambray, R.S., Cawse, P.A., Garland, J.A., Gibson, J.A.B., Johnson, P., Lewis, G.N.J., Newton, D., Salmon, L. and Wade, B.O. (1987) Observations on radioactivity from the Chernobyl accident. *Nuclear Energy*, **26**, 77–101.

Cawse, P.A. and Baker, S.J. (1990) *The migration of Cs-137 in peat*. AEA Technology Report AEA-EE-0014. AEA Technology, Harwell.

Chamard, P., Velasco, R.H., Belli, M., Di Silvestro, D., Ingrao, G. and Sansone, U. (1993) Caesium-137 and Strontium-90 distribution in a soil profile. *The Science of the Total Environment*, **136**, 251–258.

Chamberlain, A.C. (1970) Aspects of the deposition of radioactive and other gases and particles. *Atmospheric Environment*, **4**, 63–88.

Charles, M.W., Mill, A.J. and Darley, P.J. (2003) Carcinogenic risk of hot particle exposures. *Journal of Radiological Protection*, **23**, 5–28.

Comans, R.N.J. and Hockley, D.E. (1992) Kinetics of cesium sorption on illite. *Geochimica et Cosmochimica Acta*, **56**, 1157–1164.

Comar, C.L., Wasserman, R.H. and Lengemann, F.W. (1966) Effect of dietary calcium on secretion of strontium into milk. *Health Physics*, **12**, 1–6.

Coughtrey, P.J. and Thorne, M.C. (1983) *Radionuclide distribution and transport in terrestrial and aquatic ecosystems. A critical review of data* (Volume 1). A.A. Balkema, Rotterdam.

Coughtrey, P.J., Jackson, D., Jones, C.H. and Thorne, M.C. (1984) *Radionuclide distribution and transport in terrestrial and aquatic ecosystems. A critical review of data* (Volume 5). A.A. Balkema, Rotterdam.

CRC (1988) *Handbook of Chemistry and Physics* (68th Edition). CRC Press Inc., Boca Raton, Florida.

Cremers, A., Elsen, A., De Preter, P. and Maes, A. (1988) Quantitative analysis of radiocaesium retention in soils. *Nature*, **335**, 247–249.

Cross, M.A. (2001) Mathematical modelling of ^{90}Sr and ^{137}Cs in terrestrial and freshwater environments. PhD thesis, University of Portsmouth, UK.

Cross, M.A., Smith, J.T., Saxén, R. and Timms, D.N. (2002) An analysis of the time dependent environmental mobility of radiostrontium in Finnish river catchments. *Journal of Environmental Radioactivity*, **60**, 149–163.

Dahlgaard, H., Eriksson, M., Ilus, E., Ryan, T., McMahon, C.A. and Nielsen, S.P. (2001) Plutonium in the marine environment at Thule, NW-Greenland after a nuclear weapons accident. In: Kudo, A. (ed.), *Plutonium in the Environment*, pp. 15–30. Elsevier, Amsterdam.

Devell, L., Tovedal, H., Bergström, U., Appelgren, A., Chyssler, J. and Andersson, L. (1986) Initial observations of fallout from the reactor accident at Chernobyl. *Nature*, **321**, 192–193.

EC (2002) *Opinion of the Scientific Committee on Food on the Tolerable Upper Intake Level of Iodine*. European Commission, Brussels.

Fesenko, S.V., Bunzl, K., Belli, M., Ivanov, Yu., Spiridonov, S., Velasco, H. and Levchuk, S. (1996) Mathematical modelling of radionuclide migration in components of meadow ecosystems. In: Karaoglou, A., Desmet, G., Kelly, G.N. and Menzel, H.G. (eds), *The Radiological Consequences of the Chernobyl Accident*, pp. 197–200. European Commission, Brussels.

Fesenko, S.V., Spiridonov, S.I., Sanzharova, N.I. and Alexakhin, R.M. (1997) Dynamics of ^{137}Cs bioavailability in a soil–plant system in areas of the Chernobyl nuclear power plant accident zone with a different physico-chemical composition of radioactive fallout. *Journal of Environmental Radioactivity*, **34**, 287–313.

Fleishman, D.G. (1973) Radioecology of marine plants and animals. In: Klechkovskii, V.M., Polikarpov, G.G. and Aleksakhin, R.M. (eds), *Radioecology*, pp. 347–370. Wiley, New York.

Frissel, M.J., Noordijk, H., and Van Bergeijk, K.E. (1990) The impact of extreme environmental conditions, as ocurring in natural ecosystems, on the soil-to-plant transfer of radionuclides. In: Desmet, G., Nassimbeni, P. and Belli, M. (eds), *Transfer of Radionuclides in Natural and Semi-natural Environments*, pp. 40–47. Elsevier Applied Science, London and New York.

Garger, E.K. (1994) Air concentrations of radionuclides in the vicinity of Chernobyl and the effects of resuspension. *Journal of Aerosol Science*, **25**, 745–753.

Garger, E.K., Gordeev, S., Holländer, W., Kashparov, V., Kashpur, V., Martinez-Serrano, J., Mironov, V., Peres, J., Tschiersch, J., Vintersved, I. and Watterson, J. (1996) Resuspension and deposition of radionuclides under various conditions. *Proceedings of the conference 'The radiological consequences of the Chernobyl accident'*, Minsk, CEC publication EUR 16544EN, pp. 109–120. CEC Luxembourg.

Garger, E.K., Kashpur, V., Belov, G., Demchuk, V., Tschiersch, J., Wagenpfeil, F., Paretzke, H.G., Besnus, F., Holländer, W., Martinez-Serrano, J. and Vintersved, I. (1997) Measurement of resuspended aerosol in the Chernobyl area. 1. Discussion of instrumentation and estimation of measurement uncertaintly. *Radiation and Environmental Biophysics*, **36**, 139–148.

Garger, E.K., Paretzke, H.G. and Tschiersch, J. (1998) Measurement of resuspended aerosol in the Chernobyl area. Part III: Size distribution and dry deposition velocity of radioactive particles during anthropogenic enhanced resuspension. *Radiation and Environmental Biophysics*, **37**, 201–208.

Garland, J.A. and Pomeroy, I.R. (1994) Resuspension of fall-out material following the Chernobyl accident. *Journal of Aerosol Science*, **25**, 793–806.

Hardy, E. and Krey, P.W. (1971) Determination of the atmospheric deposition of radionuclides by soil sampling and analysis. *Proceedings of the Environmental Plutonium Symposium, LA-4756*, pp. 37–42.

Hellou, J., Warren, W.G., Payne, J.F., Belkhode, S. and Lobel, P. (1992) Heavy metals and other elements in three tissues of cod, *Gadus morhua* from the Northwest Atlantic. *Marine Pollution Bulletin*, **24**, 452–458.

Helton, J.C., Muller, A.B. and Bayer, A. (1985) Contamination of surface-water bodies after reactor accidents by the erosion of atmospherically deposited radionuclides. *Health Physics*, **48**, 757–771.

Hilton, J., Cambray, R.S. and Green N. (1992) Chemical fractionation of radioactive caesium in airborne particles containing bomb fallout, Chernobyl fallout and atmospheric material from the Sellafield site. *Journal of Environmental Radioactivity*, **15**, 103–111.

Hilton, J., Nolan, L. and Jarvis, K.E. (1997) Concentrations of stable isotopes of cesium and strontium in freshwaters in northern England and their effect on estimates of sorption coefficients (K_d). *Geochimica et Cosmochimica Acta*, **61**, 1115–1124.

Holländer, W. and Garger, E. (eds) (1996) *Contamination of surfaces by resuspended material*, Report EUR 16527 EN. European Commission, Luxembourg.

Horrill A.D. and Howard D.M. (1991) Chernobyl fallout in three areas of upland pasture in West Cumbria. *Journal of Radiological Protection*, **11**, 249–257.

IAEA (1986) *Summary report on the post-accident review meeting on the Chernobyl accident*, Safety series No. 75-INSAG-1. IAEA, Vienna.

IAEA (1994) *Handbook of transfer parameter values for the prediction of radionuclide transfer in temperate environments*, Technical Report Series No. 364. IAEA, Vienna.

Ivanov, Y.A., Lewyckyj, N., Levchuk, S.E., Prister, B.S., Firsakova, S.K., Arkhipov, N.P., Arkhipov, A.N., Sandalls, J. and Askbrant, S. (1997) Migration of ^{137}Cs and ^{90}Sr from Chernobyl fallout in Ukraine, Belarussian and Russian soils. *Journal of Environmental Radioactivity*, **35**, 1–21.

Isaksson, M. and Erlandson, B. (1995) Experimental Determination of the vertical and horizontal distribution of ^{137}Cs in the ground. *Journal of Environmental Radioactivity*, **271**, 141–160.

Jackson, M.L. (1968) Weathering of primary and secondary minerals in soils. *Transactions of the International Society of Soil Science* **4**, 281–292.

Jacob, P., Roth, P, Golikov V., Balonov, M., Erkin, V., Likhtariov I., Garger, E. and Kashparov, V. (1996) Exposures from external radiation and from inhalation of resuspended material. In: Karaoglou, A., Desmet, G., Kelly, G.N. and Menzel, H.G. (eds), *The Radiological Consequences of the Chernobyl Accident*, pp. 251–260. European Commission, Brussels.

Jacobs, D.G. and Tamura, T. (1960) The mechanism of ion fixation using radio-isotope techniques. *Proc. 7th International Congress of Soil Science, Madison USA*, pp. 206–214.

Jonsson, B., Forseth, T. and Ugedal, O. (1999) Chernobyl radioactivity persists in fish. *Nature*, **400**, 417.

Kagan, L.M. and Kadatsky, V.B. (1996) Depth migration of Chernobyl orginated ^{137}Cs and ^{90}Sr in soils of Belarus. *Journal of Environmental Radioactivity*, **33**, 27–39.

Kanivets, V.V., Voitsekhovitch, O.V., Simov, V.G. and Golubeva, Z.A. (1999) The post-Chernobyl budget of ^{137}Cs and ^{90}Sr in the Black Sea. *Journal of Environmental Radioactivity*, **43**, 121–135.

Kashkarov, L.L., Kalinina, G.V. and Perelygin, V.P. (2003) α-track investigation of the Chernobyl Nuclear Power Plant accident region soil samples. *Radiation Measurements*, **36**, 529–532.

Kashparov, V.A., Oughton, D.H., Zvarich, S.I., Protsak, V.P. and Levchuk, S.E. (1999) Kinetics of fuel particle weathering and ^{90}Sr mobility in the Chernobyl 30 km exclusion zone. *Health Physics*, **76**, 251–299.

Kashparov, V.A., Lundin, S.M., Kadygrib, A.M., Protsak, V.P., Levtchuk, S.E. and Yoschenko, V.I. (2000) Forest fires at the territory contaminated as a result of Chernobyl accident: Radioactive aerosols resuspension and firemen exposure. *Journal of Environmental Radioactivity*, **51**, 281–298.

Kashparov, V.A., Ahamdach, N., Zvarich, S.I., Yoschenko, V.I., Maloshtan, I.M. and Dewiere, L. (2004) Kinetics of dissolution of Chernobyl fuel particles in soil in natural conditions. *Journal of Environmental Radioactivity*, **72**, 335–353.

Kirchner, G. and Baumgartner, D. (1992) Migration rates of radionuclides deposited after the Chernobyl accident in various North German soils. *The Analyst*, **117**, 475–479.

Kirk, G.J.D. and Staunton, S. (1989) On predicting the fate of radioactive caesium in soil beneath grassland. *Journal of Soil Science*, **40**, 71–84.

Klemt, E., Drissner, J., Kaminski, S., Miller, R. and Zibold, G. (1999) Time-dependency of the bioavailability of radiocaesium in lakes and forests. In: Linkov, I. and Schell, W.R. (eds), *Contaminated Forests*, pp. 95–101. Kluwer, Dordrecht.

Kocher, D.C. (1991) A validation test of a model for long-term retention of ^{129}I in surface soils. *Health Physics*, **60**, 523–531.

Koch-Steindl, H. and Pröhl, G. (2001) Considerations on the behaviour of long-lived radionuclides in the soil. *Radiation and Environmental Biophysics*, **40**, 93–104.

Konoplev, A.V., Bulgakov, A.A., Popov, V.E. and Bobovnikova, Ts.I. (1992) Behaviour of long-lived radionuclides in a soil–water system. *Analyst*, **117**, 1041–1047.

Krouglov, S.V., Filipas, A.S., Alexakhin, R.M. and Arkhipov, N.P. (1997) Long-term study on the transfer of ^{137}Cs and ^{90}Sr from Chernobyl-contaminated soils to grain crops. *Journal of Environmental Radioactivity*, **34**, 267–286.

Krouglov, S.V., Kurinov, A.D. and Alexakhin, R.M. (1998) Chemical fractionation of ^{90}Sr, ^{106}Ru, ^{137}Cs and ^{144}Ce in Chernobyl-contaminated soils: An evolution in the course of time. *Journal of Environmental Radioactivity*, **38**, 59–76.

Kudelsky, A.V., Smith, J.T., Ovsiannikova, S.V. and Hilton, J. (1996) Mobility of Chernobyl-derived ^{137}Cs in a peatbog system within the catchment of the Pripyat River, Belarus. *The Science of the Total Environment*, **188**, 101–113.

Kudelsky, A.V., Smith, J.T., Zhukova, O.M., Matveyenko, I.I, and Pinchuk, T.M. (1998) Contribution of river runoff to the natural remediation of contaminated territories (Belarus). *Proceedings of the Academy of Sciences of Belarus*, **42**, 90–94 (in Russian).

Kudelsky, A.V., Smith, J.T., Zhukova, O.M., Rudaya, S.M. and Sasina, N.V. (2002) The role of river discharge in ^{90}Sr decontamination within the polluted areas in Belarus. *Proceedings of the Academy of Sciences of Belarus*, **46**, 95–99 (in Russian).

Livens, F.R., Baxter, M.S. and Allen, S.E. (1987) Association of plutonium with soil organic matter. *Soil Science*, **144**, 24–28.

Livens F.R., Horrill A.D., Singleton D.L. (1991) Distribution of radiocaesium in the soil–plant systems of upland areas of Europe. *Health Physics*, **60**, 4, 539–545.

Livens, F.R., Fowler, D. and Horrill A.D. (1992) Wet and dry deposition of ^{131}I, ^{134}Cs and ^{137}Cs at an upland site in Northern England. *Journal of Environmental Radioactivity*, **16**, 243–254.

MAFF (2000) *Iodine in Milk*. MAFF Surveillance Information Sheet No. 198. Ministry of Agriculture, Fisheries and Food, London.

Mahara, Y. (1993) Storage and migration of fallout strontium-90 and cesium-137 for over 40 years in the surface soil of Nagasaki. *Journal of Environmental Quality*, **22**, 722–730.

Moran, J.E., Oktay, S.D., Santschi, P.H. (2002) Sources of iodine and iodine 129 in rivers. *Water resources research*, **38**, 24-1 to 24-10.

Mück, K. (1997) Long-term effective decrease of cesium concentration in foodstuffs after nuclear fallout. *Health Physics*, **72**, 659–673.

Mück, K., Pröhl, G., Likhtarev, I., Kovgan, L., Meckbach, R. and Golikov, V. (2002) A consistent radionuclide vector after the Chernobyl accident. *Health Physics*, **82**, 141–156.

Müller-Lemans, H. and van Dorp, F. (1996) Bioturbation as a mechanism for radionuclide transport in soil: relevance of earthworms. *Journal of Environmental Radioactivity*, **31**, 7–20.

Negri, M.C. and Hinchman, R.R. (2000) The use of plants for the treatment of radionuclides. In: Raskin, I. and Ensley, B.D. (eds), *Phytoremediation of Toxic Metals: Using Plants to Clean Up the Environment*, pp. 107–132. Wiley, New York.

Nicholson, K.W., Branson, J.R., Geiss, P. and Cannell, R.J. (1989) The effects of vehicle activity on particle resuspension. *Journal of Aerosol Science*, **20**, 1425–1428.

Nielsen, S.P., Bengtson, P., Bojanowsky, R., Hagel, P., Herrmann, J., Ilus, E., Jakobson, E., Motiejunas, S., Panteleev, Y., Skujina, A. and Suplinska, M. (1999) The radiological exposure of man from radioactivity in the Baltic Sea. *Science of the Total Environment*, **237/238**, 133–141.

Nisbet, A.F. and Shaw, S. (1994) Summary of a 5-year lysimeter study on the time-dependent transfer of ^{137}Cs, ^{90}Sr, 239,240Pu and ^{241}Am to crops from three contrasting soil types. 1: Transfer to the edible portion. *Journal of Environmental Radioactivity*, **23**, 1–17.

Noordijk, H., van Bergeijk, K.E., Lembrechts, J. and Frissel, M.J. (1992) Impact of ageing and weather conditions on soil-to-plant transfer of radiocesium and radiostrontium. *Journal of Environmental Radioactivity*, **15**, 277–286.

Oktay, S.D., Santschi, P.H., Moran, J.E. and Sharma P. (2001) ^{129}I and ^{127}I transport in the Mississippi River. *Environmental Science and Technology*, **35**, 4470–4476.

Orlandini, K.A., Penrose, W.R., Harvey, B.R., Lovett, M.B. and Findlay, M.W. (1990) Colloidal behaviour of actinides in an oligotrophic lake. *Environmental Science and Technology*, **24**, 706–712.

Oughton, D.H. (1989) The environmental chemistry of radiocaesium and other nuclides. Ph.D Thesis, University of Manchester.

Owens, P.N., Walling, D.E. and He, Q. (1996) The behaviour of bomb-derived Cs-137 fallout in catchment soils. *Journal of Environmental Radioactivity*, **32**, 169–191.

Pröhl, G., Mück, K., Likhtarev, I., Kovgan, L. and Golikov, V. (2002) Reconstruction of the ingestion doses received by the population evacuated from the settlements in the 30-km Zone around the Chernobyl reactor. *Health Physics*, **82**, 173–181.

Rao, U. and Fehn, U. (1999) Sources and reservoirs of anthropogenic Iodine-129 in Western New York. *Geochimica et Cosmochimica Acta*, **63**, 1927–1938.

Realo, E., Jogi, J., Koch, R. and Realo, K. (1995) Studies on radiocaesium in Estonian soils. *Journal of Environmental Radioactivity*, **29**, 111–119.

Salbu, B., Krekling, T., Oughton, D.H., Østby, G., Kashparov, V.A., Brand, T.L. and Day, J.P. (1994a) Hot particles in accidental releases from Chernobyl and Windscale nuclear installations. *Analyst*, **119**, 125–130.

Salbu, B., Oughton, D.H., Ratnikov, A.V., Zhigareva, T.L., Kruglov, S.V., Petrov, K.V., Grebenshakikova, N.V., Firsakova, S.K., Astasheva, N.P., Loshchilov, N.A., Hove, K. and Strand, P. (1994b) The mobility of ^{137}Cs and ^{90}Sr in agricultural soils in the Ukraine, Belarus and Russia, 1991. *Health Physics*, **67**, 518–528.

Salbu, B., Krekling, T., Hove, K., Oughton, D., Kashparov, V.A. and Astasheva, N.A. (1995) Biological relevance of hot particles ingested by animals. In: *Proceedings of the Conference: Environmental Impact of Radioactive Releases*, pp. 695–697, IAEA-SM-339/198P. IAEA: Vienna.

Sandalls, J. and Bennett, L. (1992) Radiocaesium in upland herbage in Cumbria, UK: A three-year field study. *Journal of Environmental Radioactivity*, **16**, 147–165.

Sanzharova, N.I., Fesenko, S.V., Alexakhin, R.M., Anisimov, V.S., Kuznetsov, V.K. and Chernyayeva, L.G. (1994) Changes in the forms of ^{137}Cs and its availability for plants as dependent on properties of fallout after the Chernobyl nuclear power plant accident. *The Science of the Total Environment*, **154**, 9–22.

Schubert, P. and Behrend, U. (1987) Investigations of radioactive particles from the Chernobyl fallout. *Radiochimica Acta*, **41**, 149–155.

Sheppard, S.C. and Evenden, W.G. (1996) Variation in transfer factors for stochastic models: Soil-to-plant transfer. *Health Physics*, **72**, 727–733.

Sholkovitz, E.R., Carey, A.E. and Cochran, J.K. (1982) Aquatic chemistry of plutonium in seasonally anoxic lake waters. *Nature*, **300**, 159–161.

Smith, F.B. and Clark, M.J. (1989) *The transport and deposition of airborne debris from the Chernobyl nuclear power plant accident*. Meteorological Office Scientific Paper No. 42, HMSO, London.

Smith, J.T., Hilton, J. and Comans, R.N.J. (1995) Application of two simple models to the transport of Cs-137 in an upland organic catchment. *Science of the Total Environment*, **168**, 57–61.

Smith, J.T., Leonard, D.R.P, Hilton, J. and Appleby, P.G. (1997a) Towards a generalised model for the primary and secondary contamination of lakes by Chernobyl derived radiocaesium. *Health Physics*, **72**, 880–892.

Smith, J.T., Appleby, P.G., Hilton, J. and Richardson, N. (1997b) Inventories and fluxes of Pb-210, Cs-137 and Am-241 determined from the soils of three small catchments in Cumbria, UK. *Journal of Environmental Radioactivity*, **37**, 127–142.

Smith, J.T. and Elder, D.G. (1999) A comparison of different methods of characterising radionuclide activity-depth profiles in soils. *European Journal of Soil Science*, **50**, 295–307.

Smith, J.T., Fesenko, S.V., Howard, B.J., Horrill, A.D., Sanzharova, N.I., Alexakhin, R.M., Elder, D.G. and Naylor, C. (1999). Temporal change in fallout ^{137}Cs in terrestrial and aquatic systems: A whole-ecosystem approach. *Environmental Science and Technology*, **33**, 49–54.

Smith, J.T., Comans, R.N.J., Beresford, N.A., Wright, S.M., Howard, B.J. and Camplin, W.C. (2000a) Chernobyl's legacy in food and water. *Nature*, **405**, 141.

Smith, J.T., Clarke, R.T. and Saxén, R. (2000b) Time dependent behaviour of radiocaesium: a new method to compare the mobility of weapons test and Chernobyl derived fallout. *Journal of Environmental Radioactivity*, **49**, 65–83.

Smith, J.T., Wright, S.M., Cross, M.A., Monte, L., Kudelsky, A.V., Saxén, R., Vakulovsky, S.M. and Timms, D.N. (2004) Global analysis of the riverine transport of ^{90}Sr and ^{137}Cs. *Environmental Science and Technology*, **38**, 850–857.

Spezzano, P. and Giacomelli, R. (1991) Transport of ^{131}I and ^{137}Cs from air to cows' milk produced in North-Western Italian farms following the Chernobyl accident. *Journal of Environmental Radioactivity*, **13**, 235–250.

Squire, H.M. (1966) Long-term studies of ^{90}Sr in soils and pastures. *Radiation Botany*, **6**, 49–63.

Squire, H.M. and Middleton, L.J. (1966) Behaviour of ^{137}Cs in soils and pastures: A long-term experiment. *Radiation Botany*, **6**, 413–423.

Ter-Saakov, A.A., Glebov, M.V., Gordeev, S.K., Ermakov, A.I., Luchkin, Y.L. and Khilov, A.A. (1994) An experimental study of the radioactivity associated with soil and dust particles in the vicinity of the Chernobyl nuclear-power-plant. *Journal of Aerosol Science*, **25**, 779–787.

Tikhomirov, F.A. and Sanzharova, N.I. (1978) Update of ^{90}Sr by herbaceous plants as an index of root activity distribution over the soil profile. *Moscow University Soil Science Bulletin*, **33**, 54–58.

Timms, D.N., Smith, J.T., Cross, M.A., Kudelsky, A.V., Horton, G. and Mortlock, R. (2004) A new method to account for the depth distribution of radiocaesium in soils in the calculation of external radiation dose rate. *Journal of Environmental Radioactivity*, **72**, 323–334.

Valcke, E. and Cremers, A. (1994) Sorption–desorption dynamics of radiocaesium in organic matter soils. *Science of the Total Environment*, **157**, 275–283.

UNSCEAR (1977). *Sources, Effects and Risks of Ionizing Radiation*, Report to the General Assembly, with Annexes. United Nations Publication, New York.

Vinogradov, A.P. (1953) *The Elementary Composition of Marine Organisms*. Sears Foundation, New Haven, 647 pp.

Vinogradov, A.P. (1959) *The Geochemistry of Rare and Dispersed Chemical Elements in Soils* (2nd Edition). Consultants Bureau Inc., New York, 209 pp.

Voitsekhovitch, O.V., Dutton, M. and Gerchikov, M. (2002) *Experimental studies and assessment of the present state of Chernobyl cooling pond bottom topography and bottom sediment radioactive contamination*. Final report Contract C647/D0426 Center for monitoring Studies and Environmental Technologies/Ukrainian Hydrometeorological Institute.

Wagenpfeil, F. and Tschiersch, J. (2001) Resuspension of coarse fuel hot particles in the Chernobyl area. *Journal of Environmental Radioactivity*, **52**, 5–16.

Whitehead, D.C. (2000) *Nutrient Elements in Grassland: Soil–Plant–Animal Relationships*. CABI Publishing, Wallingford, UK, 369 pp.

Wright, S.M., Smith, J.T., Beresford, N.A. and Scott, W.A. (2003) Monte-Carlo prediction of changes in areas of west Cumbria requiring restrictions on sheep following the Chernobyl accident. *Radiation and Environmental Biophysics*, **42**, 41–47.

Yasuda, H. and Uchida, S. (1993) Statistical approach for the estimation of Strontium distribution coefficient. *Environmental Science and Technology*, **27**, 2462–2465.

Yoshida, S., Muramatsu, Y., Steiner, M., Belli, M., Pasquale, A., Rafferty, B., Rühm, W., Rantavaara, A., Linkov, I., Dvornik, A. and Zhuchenko, T. (2000) Relationship between radiocesium and stable cesium in plants and mushrooms collected from forest ecosystems with different contamination levels. *Proceedings of the International Radiation Protection Association 10th Congress, Hiroshima, Japan*. Paper ref. P-11-244, 6 pp (available online at: http://www.irpa.net/pub/pr/index.html).

Zheltonovsky, V., Mück, K. and Bondarkov, M. (2001) Classification of hot particles from the Chernobyl accident and nuclear weapons detonations by non-destructive methods. *Journal of Environmental Radioactivity*, **57**, 151–166.

Zibold, G., Drissner, J., Kaminski, S., Klemt, E. and Miller, R. (2001) Time-dependence of the radiocaesium contamination of roe deer: measurement and modelling. *Journal of Environmental Radioactivity*, **55**, 5–27.

3

Radioactivity in terrestrial ecosystems

Jim T. Smith, Nick A. Beresford, G. George Shaw and Leif Moberg

3.1 INTRODUCTION

Radioactivity from the Chernobyl accident affected food production systems throughout Europe. Large areas of agricultural land in the Ukraine, Belarus and Russia were abandoned and much of this land remains uninhabited and unused to this day. Parts of western Europe were also affected, with advice not to consume fresh vegetables being given in, for instance, parts of Italy and Germany. Most affected were the Scandinavian countries, where activity concentrations in reindeer, goat's milk, sheep, game animals and freshwater fish were above intervention levels, and (as with sheep in some upland areas of the UK) are still subject to restrictions (Howard *et al.*, 2001).

In 1986, intensive farming in the Ukraine, Belarus and Russia was primarily in the form of large state-owned collective farms. People also commonly grew vegetables on small private plots of land. Such 'private farms' may have one or two cows producing milk: these may often graze marginal land such as the edges of forests and roadside or river verges. In rural areas, self-production would have been the main source of food for the household. Although in western Europe growing vegetables for personal consumption is relatively common, the vast majority of food production is intensive agriculture. Extensive agriculture in western Europe tends to take place on nutrient poor, often (especially in north-western Europe) highly organic soils which are used for free-grazing of animals (e.g., upland or mountain pastures).

During the first few weeks after the Chernobyl accident, radioactivity deposited directly on plant surfaces formed the major transfer pathway to plants and animals in terrestrial ecosystems. Subsequent washoff of radionuclides from plant surfaces and adsorption to the soil matrix meant that (for long-lived radionuclides such as radiocaesium and radiostrontium) uptake via the roots gradually became the dominant contamination pathway.

Initially, radiation exposures via the terrestrial food chain were mainly due to short-lived ^{131}I in milk and to a lesser extent leafy vegetables (Drozdovitch et al., 1997; Pröhl et al., 2002). Radioiodine is accumulated in the thyroid, a problem which may have been exacerbated by low natural levels of (stable) iodine in people's diet of the affected regions of the former Soviet Union (fSU). The primary, well documented, health effect arising from the Chernobyl accident has been the development of thyroid cancers, mainly as a consequence of ^{131}I ingestion via contaminated milk (see Chapter 6).

Several months after fallout, short-lived radionuclides such as ^{131}I had decayed away, and radiocaesium formed the major part of the radiation dose to the population. As we will discuss below, radionuclide uptake to plants is influenced by soil characteristics and competition from other elements in the soil and soil solution. In general, intensive agricultural systems have nutrient-rich soils with a high binding capacity for radiocaesium. In contrast, extensive (or 'semi-natural') agricultural systems generally have poorer soils and may have low pH and a lower ability to bind radiocaesium. Thus, for example, radiocaesium accumulation in sheep grazing (extensive) upland organic soils in the UK and parts of Scandinavia was significantly greater than in sheep reared on lowland farms in intensive agricultural systems.

The level of contamination in foodstuffs was therefore not directly related to the amount of radiocaesium deposited, because foodstuffs grown on certain soil types had particularly high accumulation. For instance, in 1997 Prister et al. (1998) studied radiocaesium accumulation on different soil types in the fSU. They reported that in areas of soddy–podzolic soils a ^{137}Cs deposition in the range of 140–500 kBq m^{-2} would result in an annual ingested dose of 1 mSv. In areas of peaty soil, this level of dose was reached at ^{137}Cs depositions as low as 7–50 kBq m^{-2} (see Chapter 1 for deposition maps). Currently, it is the policy within the countries of the fSU to implement protective measures where the total (external + internal) dose exceeds 1 mSv y^{-1} (Balonov et al., 1999; Kenik et al., 1999).

As discussed in Chapter 2, the transfer of radionuclides to biota is commonly estimated using an aggregated transfer factor (T_{ag}, m^2 kg^{-1}):

$$T_{ag} = \frac{\text{Concentration of radionuclide in biota (Bq kg}^{-1})}{\text{Density of deposited radioactivity (Bq m}^{-2})} \text{ m}^2 \text{ kg}^{-1} \quad (3.1)$$

where the concentration of radionuclide in biota is commonly expressed per unit weight of dry matter for plants and plant products, and per unit fresh weight for milk and meat. Note, however, that in Ukrainian, Belarussian and Russian studies, aggregated transfer coefficients of vegetables are typically expressed per unit fresh weight.

An alternative parameter for the transfer of radionuclides to biota is the *concentration ratio* (CR), defined as the activity concentration in biota (Bq kg^{-1}) divided by the activity concentration in surface soil (typically defined as 0–10 cm for grasslands and 0–20 cm for agricultural soils; IAEA, 1994):

$$\text{CR} = \frac{\text{Concentration of radionuclide in biota (Bq kg}^{-1})}{\text{Concentration of radioactivity in dry soil (Bq kg}^{-1})} \quad (3.2)$$

Table 3.1. Average ratio of fresh weight : dry weight of various products.

Product	Ratio f.w. : d.w.	Reference
Barley, oats, rye, wheat	1.16	(1)
Grass (hay)	10	(1)
Carrot	6.25	(1)
Potato	4.8	(1)
Cabbage	8.3	(1)
Onion	9.1	(1)
Meat (lamb)†	3.6	(2)
Meat (beef)	3.6	(3)
Strawberries	10	(4)
Fungi (mushrooms)	10	(5)
Fish*	4.2	(6)

† Muscle; * soft parts and muscles; f.w. = fresh weight; d.w. = dry weight.
(1) IAEA (1994); (2) Beresford, unpublished; (3) Jacobs (1958); (4) Carini (2001); (5) Gillet and Crout (2000); (6) Vinogradov (1953).

where the concentration of the radionuclide in biota is commonly expressed per unit weight of dry matter for grains (cereals) and grasses/fodder crops but per unit fresh weight for milk and meat. Both of the above parameters are discussed in Chapter 2, however we have redefined them here for convenience. Ratios of fresh : dry weight of various products are given in Table 3.1.

Both transfer parameters are simplifications of the complex processes of radionuclide accumulation by biota. For example, bioaccumulation is expected to change significantly over time due to the dynamics of uptake/loss and changes in bioavailability. These equilibrium ratios are, however, useful parameters for predicting radioactivity concentrations in foodstuffs. Most of the data from the Ukraine, Belarus and Russia is expressed using the T_{ag} which is most convenient for estimating concentrations in foodstuffs on the basis of fallout per unit area. We have therefore presented, where possible, data as T_{ag} values.

Illustrative radiocaesium transfer factors from different terrestrial systems are shown in Table 3.2. The relatively high radiocaesium activity concentration of forest products compared to agricultural foodstuffs is illustrated in Table 3.2 and Figure 3.1.

Radiation exposures via forest ecosystems occur as a result of external exposures from contaminated soil and trees as well as exposures from forest foods (fungi, berries, game animals) and timber products including their use for heat (energy) production. A large proportion of the areas most affected by Chernobyl are covered by forest and before the accident these provided an important timber resource. Gathering of fungi and berries (so-called 'forest gifts') is of great cultural and economic importance in the Ukraine, Belarus and Russia. Hunting for deer and wild boar is also common, both in the fSU and in some countries of western Europe.

Table 3.2. Illustrative productivities and radiocaesium transfer factors of food products derived from different ecosystem types.†
Adapted from data of Zhuchenko (1998) presented in Kryshev and Ryazantsev (2000).

Ecosystem	Component	Productivity (tonne ha^{-2})	Aggregated transfer factor ($\times 10^{-3}$ m^2 kg^{-1})
Arable land	Cereals	3.5	0.1
	Potato	20	0.05
Unimproved pasture	Hay	1.0	6.4
	Milk	0.83	0.64
	Beef	0.06	2.6
Improved pasture	Hay	3.0	2.2
	Milk	2.5	0.22
	Beef	0.18	0.88
Drained peat	Hay	4.0	23
	Milk	3.3	2.3
	Beef	0.24	9.2
Forest	Fungi	0.006 4	30
	Berries	0.006	10
	Game	0.005	3

† The values of the aggregated transfer factors are based on measurements in the Gomel region (Belarus) during 1994–1995. Aggregated transfer factors and productivities of grains and hay are calculated on a dry weight basis; values for beef, potato, fungi and berries are calculated on a fresh weight basis (Kryshev, pers. commun.). Note that values presented are multiplied by 10^{-3} to give the T_{ag}.

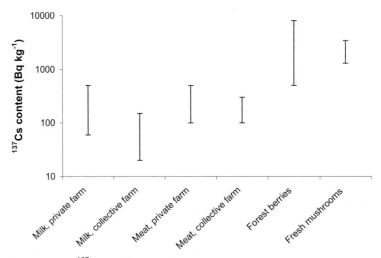

Figure 3.1. Ranges in ^{137}Cs activity concentration (fresh weight basis) in various products from the Luginsk district, Zhitomir region, Ukraine in 1995. Products from private farms tended to be more contaminated than from collective farms, but forest berries and fungi were about ten times more contaminated than milk and meat.
From data presented in Prister et al. (1996).

As with agricultural systems, radiocaesium (particularly long-lived ^{137}Cs) was of most significance in determining radiation exposures from forest systems in the long term. Because of its importance in determining long-term radiation exposures, the majority of studies of radionuclide transfer in forest ecosystems have been focused on ^{137}Cs. As had been observed after the atmospheric weapons tests era, forest ecosystems efficiently accumulate and recycle radiocaesium.

3.2 AGRICULTURAL ECOSYSTEMS

3.2.1 Interception of radioactive fallout by plants

The interception of deposited radionuclides by vegetation determines initial contamination and uptake by absorption through plant surfaces (foliar absorption). As noted by Anspaugh et al. (2002) this is a '...poorly understood process. Empirical observations of this factor have been noted to vary substantially, and uncertainty in this factor is a primary contributor to uncertainty in calculations of dose from the ingestion of terrestrial foodstuffs.'

Deposition to plant surfaces occurs during dry weather ('dry deposition') or during rainfall events ('wet deposition'). It is believed that dry deposition is more efficiently intercepted by vegetation than wet deposition (Pröhl and Hoffman, 1996). The concentration of a radionuclide in vegetation (Bq kg^{-1}) due to surface deposition is given by (e.g., Pröhl et al., 2002):

$$C_v = \frac{f D_i}{m} \quad (3.3)$$

where f (the interception factor) is the fraction of radioactivity intercepted by vegetation, D is the deposition of the radionuclide (Bq m^{-2}) and m is the mass of vegetation per unit area (kg m^{-2}). The interception factor has been observed to increase as the mass of vegetation per unit surface area increases (Chamberlain and Chadwick, 1966).

For dry deposition, the ratio f/m (the mass interception factor) is approximately 3 m^2 kg^{-1} (Pröhl and Hoffman, 1996). Measurements of dry deposition of ^{137}Cs and ^{131}I to grass in the UK after Chernobyl gave values of f in the range 0.75–0.91 (Smith and Clarke, 1989). This study, and studies prior to Chernobyl (Chamberlain and Chadwick, 1966) imply that, for relatively dense herbage, the majority of dry deposited radioactivity is initially intercepted by vegetation.

Interception of radionuclides deposited in rainfall events can be lower than in dry conditions since rainfall can run off the plant surfaces before radionuclides are adsorbed. The fraction of radioactivity intercepted therefore decreases as rainfall intensity increases. As with dry deposition, interception is also greater for high-density vegetation. Anspaugh et al. (2002) highlights the wide variation in the interception factor during wet deposition. In this study a mean value of $f = 0.31$ was indicated, but 'it is reasonable to expect variation from 0.1 to 0.9' (Anspaugh et al., 2002).

Washoff of radionuclides deposited on plant surfaces by dry or wet deposition is dependent on weather conditions, but usually occurs during the weeks after fallout. The time taken for the radioactivity deposited on plant surfaces to decline by one-half (the 'weathering half-life') is typically of the order of 12–17 days (SCOPE, 1993). The activity concentrations in grass will reduce as a consequence of dilution due to growth in addition to weathering, and it is not always clear whether this is taken into account in the derivation of 'weathering half-lives'. In grass in the UK, Cambray *et al.* (1987) observed a slightly faster half-time (excluding physical decay) of reduction in the activity concentrations in grass of approximately 11 days for both ^{131}I and ^{137}Cs. In a number of different countries, half-times of activity concentrations in grass had a mean value of 9 days for ^{131}I and 11 days for ^{137}Cs (Kirchner, 1994). Washoff rates will clearly vary according to weather conditions, though studies in different countries showed relatively good agreement between measured values, and were in agreement with experiments conducted prior to Chernobyl (Kirchner, 1994).

3.2.2 Transfer of radionuclides to crops and grazed vegetation

Transfer mechanisms

Radionuclides deposited to plant surfaces may also be absorbed by the plant ('foliar absorption') or particles may adhere to plant surfaces. Surface absorption and adsorption is the dominant contamination pathway in the short term (Chamberlain, 1970). Radionuclide uptake by transfer from the soil via the root system increases in importance after the initial deposit of radioactivity is washed from plant surfaces. Radionuclides absorbed by the plant (via surfaces or root uptake) may subsequently be translocated to other parts of the plant.

For radiocaesium and radiostrontium (following the initial surface deposition period) root uptake is the dominant uptake mechanism. Several years after the Chernobyl accident, Hinton *et al.* (1996) studied the relative importance of root uptake of radiocaesium in comparison with surface adhered soil particles (and foliar absorption from surface particles). At this time after the accident, contamination of plant surfaces occurred by atmospheric resuspension of radioactivity rather than direct fallout. At both sites in the fSU studied, root uptake was the dominant contamination pathway (Table 3.3). The study authors (Hinton *et al.*, 1996) concluded that, for radiocaesium contamination after the initial period of direct fallout, 'foliar absorption of resuspended ^{137}Cs does not need to be considered as a critical pathway in routine radionuclide transport models'. Subsequent work by Malek *et al.* (2002) on ^{90}Sr showed an even greater dominance of root uptake over the other pathways than was observed for ^{137}Cs (Table 3.3).

For radionuclides with a low soil-to-plant transfer, resuspension of contaminated soil and deposition to vegetation surfaces is likely to be the predominant mechanism of plant contamination. For example, assuming a low value of soil adhesion to vegetation surfaces of 1% by dry matter then adhered soil contributes >50% of the plant's radionuclide content for soil-to-plant (root uptake) concentration ratios of <0.01. At soil-to-plant CRs (by root uptake) of <0.001 the adhered soil

Table 3.3. Percentage of the total plant contamination from different contamination pathways.
From Hinton et al. (1996) and Malek et al. (2002).

Site/radionuclide	Root uptake	Surface adhesion	Foliar absorption
Polesskoye/^{137}Cs	94%	3%	3%
Polesskoye/^{90}Sr	89%	1%	10%
Chistogalovka/^{137}Cs	70%	22%	8%
Chistogalovka/^{90}Sr	99%	<0.1%	1%

would contribute >90% of the plants radionuclide content. In some circumstances, soil adhesion can form the predominant contribution of plant radiocaesium depending upon factors such as soil type, stocking rate and weather conditions (Beresford and Howard, 1991).

Influence of soil properties

Plants absorb caesium and strontium by the same uptake mechanisms as their competitor ions (potassium and calcium respectively). Both potassium and calcium are important plant nutrients and are therefore actively taken up by the plant. Where their competitor ions are abundant and bioavailable, radiocaesium and radiostrontium accumulation by plants is expected to be relatively low. Accumulation of radiocaesium is negatively related to the concentration of potassium in the soil interstitial water, and the same applies to radiostrontium in relation to calcium. Different soil properties, however, can strongly influence the concentrations of radiocaesium and radiostrontium (and their competitor ions) in the soil water and observed uptake to plants can vary within a wide range of values (Sheppard and Evenden, 1997).

Post-Chernobyl studies confirmed previous work showing the influence of soil properties (clay mineral content and exchangeable potassium concentration) on radiocaesium uptake by the food chain (e.g., Squire and Middleton, 1966). For example, ^{137}Cs CRs in various crops in Finland were influenced by the clay mineral content and exchangeable soil potassium (Paasikallio et al., 1994). In a study of uptake of radiocaesium to vegetation in Cumbria, UK, Sandalls and Bennett (1992) observed a significant negative correlation between the ^{134}Cs soil–plant CR and the potassium concentration of soil interstitial water. In the fSU, one of the most commonly used countermeasures to reduce ^{137}Cs in foodstuffs was to increase the potassium content of soils using potassium based fertilisers (Chapter 5).

In mineral soils, ^{137}Cs is adsorbed to so-called 'Frayed Edge Sites' (FES) on illitic clay minerals (most mineral soils in Europe contain illitic clays) (Cremers et al., 1988). Studies have shown that in soils of low to medium organic matter content (<40%) the majority of the ^{137}Cs is sorbed on illitic FES: it is only in soils with very high organic matter content (>80%) that the majority is associated with regular ion-exchange sites (Valcke and Cremers, 1994).

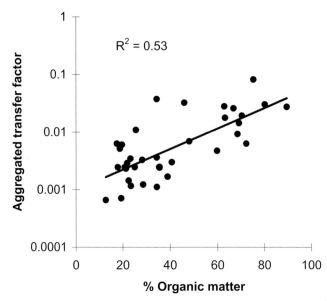

Figure 3.2. Plot of T_{ag} (dry weight basis) vs. organic matter content in soils in 5 catchments in Cumbria, UK.

Measurements of radiocaesium uptake to plants in organic (>15% organic matter) upland soils of Cumbria showed a highly significant ($r^2 = 0.53$, $p < 0.01$) positive correlation between T_{ag} and organic matter content of the soil (Figure 3.2; see also Sandalls and Bennett, 1992). It was also found (Smith et al., 1999) that rates of decline in ^{137}Cs activity concentration in vegetation were slower in highly organic soils (>80% organic matter) than in more mineral soils. It can be seen from Figure 3.2 that (in upland Cumbria at least) soil organic matter is a reasonable predictor of T_{ag} in soils of medium to high organic matter content. In soils of low organic matter content (<10–15% organic matter), FES (and therefore T_{ag}) will be related to the clay mineral content of the particular soil, and thus the direct relationship with organic matter is not expected to hold in mineral soils. Relatively high T_{ag} values have been observed, for example, in sandy soils of low organic matter content (see Table 3.4).

In contrast to radiocaesium, radiostrontium is not 'fixed' to the soil: the vast majority remains in available form (though note the influence of fuel particles close to Chernobyl, Chapter 2). As was also observed in pre-Chernobyl studies (e.g., Squire, 1966), ^{90}Sr uptake to plants after Chernobyl was influenced by calcium in the soil. In Finland, a strong negative relationship was observed between the concentration factor (CF) of ^{90}Sr in various crops and the calcium content of the soil (Paasikallio et al., 1994). The sorption of ^{90}Sr to soil is not specific, so it competes with other cations for binding sites. In a study of radiostrontium sorption in 36 Japanese soils (Yasuda and Uchida, 1993), it was found that the ^{90}Sr solid–aqueous distribution coefficient (K_d) could be described by a

Table 3.4. Soil–grass[††] aggregated transfer factor for radiocaesium (T_{ag}, m² kg⁻¹, dry weight basis). Values are averages over the time period given.

Soil type	Time period[†]	Organic matter (%)	T_{ag}* (×10⁻³ m² kg⁻¹)	Notes	Reference
Loam	1988–1991	8.0[a]	1.95	Poland, uncultivated grassland.	(1)
Peat		52.0[a]	49		
Sandy		2.1[a]	18		
Sandy		1.7[a]	8.1		
Podzolic gravelly & sandy moraine	1990–1993	6.0–10.8[a]	41–150 ($n = 7$ sites)	Sweden, cultivated grass and mountain pasture.	(2)
Deep peat	1988–1992	88[b]	85 ($n = 1$)	UK upland pasture, Cumbria.	(3, 4)
Peat ranker		88[b]	45 ($n = 3$)		
Peaty gley		38[b]	15 ($n = 1$)		
Brown earth		19–25[b]	7.4 ($n = 3$)		
Automorphic**	1988–1991	1.5[c]	1.4 ($n = 11$)	15–50 km from Chernobyl NPP. Former cultivated pastures and natural meadows.	(5)
Hydromorphic**		2.1[c]	3.0 ($n = 6$)		
Hydromorphic (peaty)		12.2[c]	5.2 ($n = 5$)		
Peaty	1994–1995[‡]	Not given	33.1 (9.4–121)	Dubrovitsa, Ukraine. Grass on collective and private farms, no countermeasures.	(6)
Soddy–podzols			3.2 (0.34–8.7)		
Gleys			1.2 (0.33–3.6)		
Alluvium			0.24 (0.1–0.7)		

[†] As far as possible, data was selected from similar time periods after the accident to facilitate comparison.
* Where values are from several different sites, the number of sites n or the range in measured values is given in brackets. ** Automorphic soils are those formed from normal parent materials and under normal drainage conditions, hydromorphic soils are formed under conditions of periodic or permanent saturation.
[‡] Note that T_{ag} declines with time, so these values would have been higher in 1988–1992, the period covered by most of the other studies. [a] 0–10-cm layer; [b] 0–6-cm layer; [c] depth not given. NPP = nuclear power plant. [††] In some instances this will be a mixture of graminaceous species.
(1) Calculated for Chernobyl ¹³⁴Cs from data in Pietrzak-Flis et al. (1994); (2) Rosén et al. (1995); (3) Sandalls and Bennett (1992); (4) Pomeroy et al. (1996); (5) Sanzharova et al. (1994); (6) Howard et al. (1996).

simple relationship:

$$K_d = 2.1 \frac{[CEC]}{[EC]} \quad (3.4)$$

where [CEC] is the cation exchange capacity of the soil (equivalents kg⁻¹) and [EC] is the electrical conductivity of the solution (equivalents l⁻¹). The constant in Equation (3.4) can be interpreted as a factor ('selectivity coefficient') which accounts for the preference of the sorption sites for Sr over the sum of competing cations. Highly organic soils tend to have greater [CEC] than more mineral soils and hence ^{90}Sr uptake to plants (in contrast to radiocaesium) may be lower in organic soils than in mineral ones.

Cultivation of the soil also influences radioactivity in crops grown on agricultural land. Shortly after fallout, radionuclides were held in the top few centimetres of soil: ploughing redistributed this radioactivity approximately evenly throughout the ploughed layer (in approximately the surface 25–30 cm of soil) consequently reducing the activity concentration in plants.

Transfer of radionuclides to grass

The transfer of radionuclides to grass controls the intake of radioactivity by grazing animals and hence the activity concentrations in milk and meat. The soil–grass aggregated transfer factor of radiocaesium varied widely around Europe (Table 3.4). As was expected from pre-Chernobyl studies, there was a general tendency to higher values in soils of high organic matter content (peaty soils). In Dubrovitsa, Ukraine, it was found that ^{137}Cs in milk was highest in those areas where the cows grazed peaty soils (Howard et al., 1996). Over Europe in general, areas most vulnerable to radiocaesium contamination of milk and meat were those with nutrient poor soils of low clay mineral content (e.g., Desmet et al., 1990). In the Ukraine, Belarus and Russia, private farms were particularly affected since animals in such farms often graze soils of lower nutrient status, including land on roadside verges and the edges of forests. In western Europe, the areas most affected by Chernobyl were (and still are) 'marginal' lands with extensive agriculture characterised by poor unfertilised soils.

In the Bragin district of Belarus, ^{90}Sr transfer coefficients to vegetation were of the same order or in some cases higher than those for ^{137}Cs (Table 3.5). Due to the

Table 3.5. ^{137}Cs and ^{90}Sr soil-to-grass aggregated transfer coefficient for different soil groups, Bragin, Belarus, 1994–1995.
From Aslanoglou et al. (1996).

	Aggregated Transfer Factor† ($\times 10^{-3}$ m^2 kg^{-1})	
Soil type	^{137}Cs mean ± std dev.	^{90}Sr mean ± std dev.
Peaty	15.4 ($n=2$)	7.3 ($n=2$)
Soddy–podzol	0.26 ± 0.22	3.7 ± 0.7
Soddy–podzol swampy	0.23 ± 0.15 (ploughed) 43.7 ($n=3$) (undisturbed)	4.3 ± 2.0*
Peaty–swampy	29.6 ($n=3$)	
Soddy–swampy	0.67 ± 0.43	3.5 ± 1.5

†At this time after the accident, fuel particles are not expected to significantly influence the T_{ag} of either radionuclide. Where sample numbers n were insufficient to calculate standard deviation, values of n are given. T_{ag} values are assumed to be on a dry weight basis as this is the standard method for grass, though this is not stated in the reference. * Not specified if ploughed or undisturbed.

much lower fallout of ^{90}Sr (outside the 30-km zone), however, activity concentrations in vegetation were lower than for ^{137}Cs.

Transfers to crops

Initial activity concentrations of radionuclides in crops are determined by deposition to the plant surfaces and subsequent translocation to edible parts. For short-lived radionuclides such as ^{131}I, this is the dominant mechanism of entry into the human food chain. Short-lived radionuclides are likely to have decayed away before significant amounts of radioactivity are taken up by roots. Following the surface deposition period (typically some months after fallout), root uptake from contaminated soils determines activity concentrations of longer lived radiocaesium and radiostrontium in crops as well as grasses for grazing, hay or silage. Note that here we will not cover transfers to agricultural fruit crops. A comprehensive review of radionuclide transfer to fruit can be found in Carini (2001). Transfers of radionuclides to forest fruits (berries) are discussed in Section 3.3.2.

Table 3.6 gives examples of ^{90}Sr and ^{137}Cs transfers to various crops. Whilst the data in this table infers a lower uptake by potatoes compared with cereals this is largely a consequence of cereals being presented as dry matter and potatoes as fresh (fresh weight basis), potatoes being around 80% water. Studies of the CR (all on a dry weight basis) of ^{137}Cs in different crops in Finland (Figure 3.3), showed concentration ratios decreasing in the following order: lettuce, cabbage > carrot, potato > cereals, onion (Paasikallio et al., 1994).

There is wide variation between radionuclide accumulation in crops according to soil characteristics and agricultural practices. Analysis of a large database on the accumulation of radiocaesium in various crops (Frissel et al., 2002), however, showed that 'it was clear that if the uptake for a specific soil system was relatively high for one crop, it was high for all crops, or if it was low for one crop, it was low for all crops'. On the basis of this observation, the authors developed a system of conversion factors which allow transfer factors or CRs for different crop groups to be calculated if the transfer factor or CR for one crop group is known. The developed system is based on cereals: if the T_{ag} or CR is known (or can be estimated) for cereals, the T_{ag} or CR for other crops can be estimated by multiplying by the conversion factors given in Table 3.7 (reproduced from Frissel et al., 2002). For systems in which the transfer factor to cereals is not known, Frissel et al. (2002) give recommended values for generic soil types and conditions.

Note that Table 3.7 implies greater accumulation of ^{137}Cs in potatoes than in cereals, in contrast to findings in Ukrainian, Belarussian and Russian systems, but in agreement with studies in Finland (Paasikallio et al., 1994). This again may be due to the fact that transfer factors for potatoes in the Ukraine, Belarus and Russia are commonly presented per unit fresh weight, whereas the ratios in Table 3.7 and in the Finnish study are based on dry weights of crops. The CF for potatoes (tubers) on a fresh weight basis would be approximately 5 times lower than the value for dry weight presented in Table 3.7.

Table 3.6. Aggregated transfer factors of ^{90}Sr and ^{137}Cs to various crops. T_{ag} for cereals and rape are on a dry weight basis, for potato and other vegetables fresh weight is used.

Crop	Time period	T_{ag} ($\times 10^{-3}$ m^2 kg^{-1})	Notes	Reference
		^{90}Sr		
Cereal grain	1994	0.7†	Bragin region, Belarus	(1)
Potato	1994	0.06†	Bragin region, Belarus	(1)
	1993	0.056–0.095*	Collective farm, Russia	(2)
		0.068–0.176*	Private farm, Russia	(2)
		^{137}Cs		
Cereal grain	1994	0.1	Bragin region, Belarus	(1)
	1992–1994	0.2–1.3	Bryansk region, Russia	(3)
	1993	0.005	Central Europe	(4)
	1992–1994	0.11**	Sweden, peat soil	(5)
		0.03	Sweden, sandy loam	(5)
Potato	1994	0.05	Bragin region, Belarus	(1)
	1992–1994	0.03–0.05	Bryansk region, Russia	(3)
	1993	0.0075	Central Europe	(4)
	1993	0.028–0.034	Collective farm, Russia	(2)
		0.019–0.05	Private farm, Russia	(2)
Vegetables	1993	0.0035	Central Europe	(4)
Oilseed rape	1992–1994	0.55**	Sweden, peat soil	(5)

† The influence of fuel particles on ^{90}Sr CFs is expected to be minor at this time after the accident (see Chapter 2). * Negligible influence of fuel particles. ** Mean value estimated from data given for a number of farms in Rosén et al. (1996). (1) Zhuchenko et al. (2002); (2) Korobova et al. (1998); (3) Alexakhin et al. (1996); (4) Mück (2003); (5) Rosén et al. (1996).

In the study in Finland discussed above (Paasikallio et al., 1994), CR values for ^{90}Sr were generally significantly higher than for ^{137}Cs (Figure 3.3) and varied significantly between different soil groups. ^{90}Sr CRs in organic soils were approximately 5 times lower than in mineral soils (Paasikallio et al., 1994).

Different crop types exhibit differing accumulation of ^{90}Sr, even given the same soil conditions. Conversion factors relating radiostrontium T_{ag} and CR for cereals to values for other crops were estimated by Frissel (2001), as shown in Table 3.8. In contrast to these estimates, the ^{90}Sr values estimated for the Ukraine, Belarus and Russia (Table 3.6), as with ^{137}Cs, showed a much lower T_{ag} for potatoes (tubers) than cereals. This is again likely to be due to the fact that transfer factors for potatoes in the Ukraine, Belarus and Russia are presented per unit fresh weight, whereas the ratios here are based on dry weights of crops. The ratios in Table 3.8 broadly agree with differences in CR shown in the Finnish study (Figure 3.3;

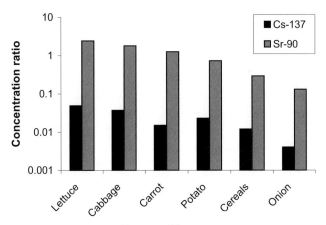

Figure 3.3. CR (dry mass basis) of ^{90}Sr and ^{137}Cs in various vegetables in Finland, clay and silt soils, 1987.
From data in Paasikallio et al. (1994).

Table 3.7. Recommended factors for radiocaesium to convert CR or T_{ag} values (dry weight basis) for cereals to values for other crops.
From Frissel et al. (2002).

Radiocaesium	Cereals	Cabbage	Green veg.	Legumes (pods)	Tubers	Root crops	Grass	Fodder crops	Fruit[a]	Onions
Conversion factor	1	7	9	5	4	3	4.5[b]	4	5	1
Range		3–11	4–14	1–11	1–7	1–5	1–10			

Frissel et al. (2002) note that: [a] 'the use of one factor for fruit is a simplification; it may have to be split into fruit categories'; [b] for application to T_{ag} rather than CR, the conversion for grass is 9 instead of 4.5 because of an assumed rooting depth for grass of 10 cm (cf. 20 cm for crops). Frissel et al. (2002) further note that 'the conversion factor of Cs for tea, herbs of woody species, leaves of trees and new wood seems to be about 20'.

Paasikallio et al., 1994) though in the latter, onions show much lower uptake than other crops. It is clear from the large ranges in values given in Table 3.8, that there is significant variation around the recommended values.

3.2.3 Transfers to animal-derived food products

In the most affected areas of the Ukraine, Belarus and Russia, milk and (to a lesser extent) meat are the main contributors to an internal dose from agricultural systems (Alexakhin et al., 1996). Ingestion of contaminated feed by farm animals results in transfers of radionuclides to meat and dairy products. Though inhalation of airborne radionuclides is also a potential uptake pathway, it is of minor significance in

Table 3.8. Recommended factors for radiostrontium to convert CR or T_{ag} values (dry weight basis) for cereals to values for other crops. From Frissel (2001).

Radiostrontium	Cereals	Cabbage	Green veg.[a]	Legumes (pods)	Tubers	Root crops	Grass	Fodder	Fruit[b]	Onion
Conversion factor	1	12	10	5	1	7	5	4	2	7
Range			2.5–79	2.6–71						

Frissel (2001) note that: [a] 'values for spinach are usually higher than for other vegetables'; [b] 'the use of one factor for fruit is a simplification; it may have to be split into fruit categories'.

comparison with ingestion. Following ingestion, radionuclides may be absorbed via the gastrointestinal tract and transferred to various organs via the blood. The transfer of radionuclides from an animal's diet to milk or meat is most often expressed as the equilibrium transfer coefficient (F_f or F_m for meat or milk respectively, units: $d\,kg^{-1}$), defined as the ratio of the activity concentration in a tissue to the rate of radionuclide ingestion:

$$F_f(\text{or } F_m) = \frac{\text{Activity concentration in tissue (or milk) (Bq kg}^{-1})}{\text{Radionuclide ingestion rate (Bq d}^{-1})} = \frac{C_{f(m)}}{C_v I_f} \quad (3.5)$$

or

$$C_{f(m)} = F_{f(m)} C_v I_f \quad (3.6)$$

where C_f is the activity concentration in the tissue (Bq kg^{-1}, fresh weight) (C_m for milk), C_v is the activity concentration in vegetation (i.e., feed, Bq kg^{-1}, dry matter) and I_f (kg d^{-1}, dry matter) is the feed intake rate. Thus, the activity concentration of the radionuclide in meat or milk may be predicted using the transfer coefficient and the daily radionuclide ingestion rate.

Table 3.9 presents transfer coefficients for radiocaesium as advised for a range of animal-derived food products (IAEA, 1994). These imply, for animals ingesting the same amount of radiocaesium daily, that the radiocaesium activity concentration in meat will be higher than that in milk or eggs. The lower CR of radiocaesium in milk and eggs mirrors the lower potassium concentration in these foodstuffs in comparison to meat (Table 3.9).

It can also be seen that radiocaesium transfer coefficients generally decrease with increasing animal mass (and consequently dry matter intake). However, this does not necessarily imply that radionuclide transfer to larger animals is less efficient. If we rearrange Equation 3.6 in terms of the equilibrium CR (kg kg^{-1}), the ratio of activity concentration in tissue (fresh weight) to that in feed, then:

$$\text{CR} = \frac{\text{Activity concentration in tissue (or milk) (Bq kg}^{-1})}{\text{Activity concentration in feed (Bq kg}^{-1})} = \frac{C_{f(m)}}{C_v} = F_{f(m)} I_f \quad (3.7)$$

Equation 3.7 implies that, unless CR changes in inverse proportion to dry matter intake, then the transfer coefficient would be expected to decrease with increasing dry

Table 3.9. Recommended transfer coefficients for radiocaesium and dry matter feed intake rates. These are used to calculate typical radiocaesium CRs (see Equation 3.7) for various animal products. Typical potassium concentrations of these products are also shown. From IAEA (1994).

Foodstuff	Feed ingestion rate by animal I_f (kg d^{-1})	Transfer coefficient F_f or F_m (d kg^{-1})	CR (dimensionless)	Typical potassium concentration (mg kg^{-1})
Milk				
Cow	16.1	7.9×10^{-3}	0.13	1,430
Goat	1.3	0.1	0.13	1,930
Sheep	1.3	5.8×10^{-2}	0.075	1,370
Meat				
Beef	7.2	5×10^{-2}	0.36	3,040
Lamb	1.1	0.49	0.54	3,060
Pork	2.4	0.24	0.58	3,765
Chicken	0.07	10	0.7	2,570
Eggs				
Hen	0.1	0.4	0.04	1,260

matter ingestion rate. The transfer coefficient is not a measure of any one process but an amalgam of processes including absorption and tissue turnover rates: radiocaesium absorption from feed does not appear to be influenced by animal size.

Using advised feeding rate values (IAEA, 1994), Table 3.9 also presents estimated CR values: these can be seen to vary considerably less between species than transfer coefficients. Thus, for example, the observation that chickens tend to have much higher F_f values than cows (IAEA, 1994) may primarily be because their dry matter intake rate is much lower. Therefore, the apparent differences in transfer between animals of different size (which may be implied by different F_f) are largely a consequence of the method of calculation of the transfer coefficient. A comparison of potassium concentrations in Table 3.9 further demonstrates that the large differences in radiocaesium uptake between animals which are sometimes implied by the use of transfer coefficients are a consequence of the method of their calculation. Another study on individual species has demonstrated that the transfer coefficient of radiocaesium to sheep tissues decreases with increasing dry matter intake (Beresford et al., 2002a)

Beresford (2003) has hypothesised that CR should be a constant for a range of radionuclides across animal species. It should also be acknowledged, however, that other authors have reported that estimation of transfer as F_f rather than CR reduced variability within study groups of animals of the same species (Ward et al., 1965); this being the initial justification for using transfer coefficients rather than the CR.

The transfer of radiostrontium to milk is strongly influenced by the amount of calcium in the diet. Studies prior to the Chernobyl accident (Comar, 1966) showed

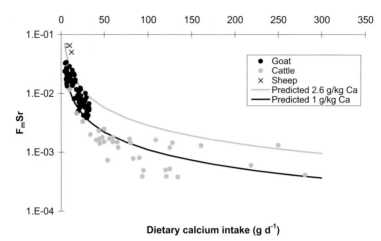

Figure 3.4. Comparison between calcium intake and F_m for strontium* with additional recent data for cattle†. The lines represent predicted values from Equation 3.9 based upon calcium contents in milk of $1\,g\,kg^{-1}$ (typical for cattle) and $2.6\,g\,kg^{-1}$ (typical for sheep).
*Adapted from Beresford et al. (1998). † From Beresford et al. (2000a).

that the radiostrontium feed–milk transfer coefficient F_m was proportional to the feed–milk transfer coefficient of calcium:

$$F_m = \frac{[Sr]_{milk}}{I_{Sr}} = \text{OR}\,\frac{[Ca]_{milk}}{I_{Ca}} \qquad (3.8)$$

where I_{Sr} is the ingestion rate $(Bq\,d^{-1})$ of ^{90}Sr, I_{Ca} is the ingestion rate $(g\,d^{-1})$ of calcium and $[Sr]_{milk}$ and $[Ca]_{milk}$ are the concentrations of Sr and Ca in milk in units of $Bq\,l^{-1}$ and $g\,l^{-1}$ respectively. OR is the 'observed ratio', an empirically determined coefficent with mean value 0.11 and range 0.03–0.4 (Howard et al., 1997) for cow milk. The fact that OR is less than 1.0 implies that calcium is preferentially transferred to milk compared to strontium. Since the calcium concentration in cow milk varies within a relatively narrow range, and assuming an average OR value of 0.11, Howard et al. (1997) obtained the following relationship between F_m of strontium in cows and dietary calcium intake:

$$F_m = \frac{0.127}{I_{Ca}} \qquad (3.9)$$

Predictions using this model were in good agreement with empirical observations, as shown in Figure 3.4 which also expands the comparison to sheep and goat milk.

In contrast to radiostrontium, there is little evidence that the uptake of radiocaesium to milk is significantly influenced by the amount of its competitor ion (potassium) in the diet.

As with humans, uptake and retention of ^{131}I in animals is dependent on dietary stable iodine intake. Iodine is an important trace element required by the thyroid for hormone synthesis, and the metabolism and excretion of radioiodine is controlled by the individual's stable iodine status. Absorption of radioiodine in the gut is complete, regardless of dietary iodine intake rates, and there is a subsequent rapid transfer to the thyroid and milk (Beresford et al., 2000b). As the stable dietary iodine increases, the proportion of the daily radioiodine intake which is transferred to the thyroid declines because of the constant rate of uptake of iodine by the thyroid. Thus, a smaller fraction of the daily iodine intake (stable iodine and radioiodine isotopes) is transferred to the thyroid and the proportion available for transfer to the mammary gland increases. At high stable iodine daily intakes, this effect is offset by the saturation of the transfer from plasma to milk (Crout et al., 2000). This was demonstrated in a study in which goats fed 0.5 mg of stable iodine six hours before being fed ^{131}I had approximately 50% less radioiodine in their milk than under normal conditions (Crout et al., 2000). In another study, cows given different levels of stable iodine in their diet (from 4 to 75 mg d^{-1}) showed only a marginal difference in the ^{131}I transfer coefficient to milk (Vandecasteele et al., 2000).

Distribution of radioactivity within the animal

Studies of radiocaesium in various organs of cows, goats and sheep (Table 3.10) showed similar transfer coefficients in muscle and most internal organs, with slightly elevated levels in the kidney and significantly lower levels in the blood, brain and fat. Whilst radiocaesium is relatively evenly distributed throughout most organs of the animal, other radionuclides may be concentrated in particular tissues.

As is the case with humans, radioiodine is strongly concentrated in the thyroid of animals: Kirchner (1994) quotes studies showing that 'between 10% and 50% of the ^{131}I ingested by a cow accumulates in the thyroid'. Again, as with humans, radiostrontium is accumulated in the bones of animals and is only released from bone at a slow rate (although this is dependent on the calcium status of the animal).

Table 3.10. Radiocaesium transfer coefficients[†] (calculated using fresh weight of tissue) to various organs of cows, goats and sheep. From Assimakopoulos et al. (1995).

Compartment	Cows (d kg^{-1})	Goats (d kg^{-1})	Sheep (d kg^{-1})
Whole blood	1.9×10^{-3}	5.2×10^{-2}	3.1×10^{-2}
Blood cells	2.0×10^{-3}	6.8×10^{-2}	3.5×10^{-2}
Muscle	22×10^{-3}	84×10^{-2}	71×10^{-2}
Lung	24×10^{-3}	51×10^{-2}	41×10^{-2}
Liver	28×10^{-3}	53×10^{-2}	56×10^{-2}
Kidney	47×10^{-3}	89×10^{-2}	96×10^{-2}
Heart	31×10^{-3}	64×10^{-2}	53×10^{-2}
Brain	7.7×10^{-3}	33×10^{-2}	23×10^{-2}
Gut	13×10^{-3}	50×10^{-2}	34×10^{-2}
Fat	2.3×10^{-3}	10×10^{-2}	7.4×10^{-2}

[†] Data are averages of 6 lactating and non-lactating animals fed with contaminated hay at Pripyat, close to Chernobyl. Standard errors were less than 20% of the mean.

Table 3.11. Examples of radioactivity concentrations in milk and meat of domestic animals in various parts of Europe contaminated by the Chernobyl accident.†

Radionuclide	Date	Activity concentration (Bq kg^{-1})	Notes
		Cow milk (f.w.)	
^{90}Sr	1993	210	Cows fed on pasture grass 3.5 km from Chernobyl (1)
^{131}I*	May 1986	250	Germany (2)
	May 1986	870	Northern Italy (2)
	May 1986	300 (max.)	Greece (2)
^{137}Cs	May 1986	250	Germany (2)
	May 1986	180	Northern Italy (2)
	July 1986	Most <250, max. 375	Sweden (2)
	1987	Most <70, max. 120	Sweden (2)
	1993	1,520	Cows fed on pasture grass 3.5 km from Chernobyl (1)
		Meat of domestic animals (Bq kg^{-1}) (f.w.)	
^{137}Cs	September 1986	1,500 (sheep)	Northern England (2)
	July 1987	1,170 (sheep)	Northern England (2)
	1990	1,090 (sheep)	Northern Sweden (3)
	June 1986	70 (cow)	North-east Scotland (4)
	<2 yrs post Chernobyl	3,000 (cow)	Mountain pasture, Norway (5)
		8,000 (sheep)	Mountain pasture, Norway (5)
	1987	6,300 (cow/pig)	Average: Novozybkov, Russia (6)
	1992	590 (cow/pig)	Average: Novozybkov, Russia (6)

†Measurements in western Europe tend to emphasise high values. *We have not found measurements of ^{131}I activity concentrations in milk in the 30-km zone in the literature, but expect that activity concentrations would have been of order MBq kg^{-1} in some settlements in May 1986 (estimated from data in Pröhl et al., 2002; Spezzano and Giacomelli, 1991). f.w. = fresh weight.
(1) Beresford et al. (2000c); (2) Eisler (2003); (3) Rosén et al. (1995); (4) Martin et al. (1988); (5) Howard et al. (1991); (6) Alexakhin et al. (1996).

Approximately 90% of the body Sr would be expected to be found in bone (Coughtrey and Thorne, 1983).

Observations after the Chernobyl accident

Radioactivity concentrations in animals varied widely due to differing fallout levels and different environmental and animal management factors. Table 3.11 illustrates measurements of various radionuclides in milk and meat in different European countries after the Chernobyl accident. Note that measurements available, particularly in western Europe, tend to emphasise high values since monitoring tended to be focused on areas of particularly high fallout and/or high transfer of radioactivity to animals.

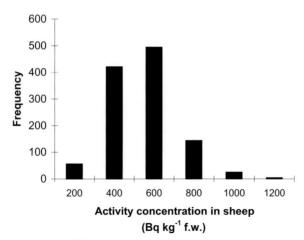

Figure 3.5. Variation in ^{137}Cs activity concentration in 1,144 sheep, Cumbria, UK.
From Beresford et al. (1996).

Within a given herd or flock of animals considerable variability between individuals has been observed as a consequence of variation in deposition over grazing areas, dietary and grazing habits (particularly for free-ranging animals in semi-natural ecosystems) and different metabolic characteristics. For instance, in studies of sheep flocks in upland west Cumbria following the Chernobyl accident, ^{137}Cs activity concentrations varied by around one-order of magnitude between individual animals (Beresford et al., 1996) (see Figure 3.5).

The transfer of ^{90}Sr to meat after the Chernobyl accident was not significant in comparison to ^{137}Cs due to its generally lower fallout and to the much lower accumulation of ^{90}Sr in meat. Most ^{90}Sr absorbed by animals is found in the bone (because it behaves like calcium) so transfer coefficients (F_f) to meat are generally a factor of 5–10 lower than for ^{137}Cs (IAEA, 1994; data in Mück et al., 2001).

A radioisotope of silver, ^{110m}Ag ($T_{1/2} = 250$ d), was found in fallout from the Chernobyl accident. In the UK, ^{110m}Ag activity concentrations in vegetation samples collected soon after deposition were typically 1–4% of ^{137}Cs activity concentrations (Beresford 1989a). However, unexpectedly high levels of ^{110m}Ag were measured in the liver of sheep (Beresford, 1989a; Beresford, 1989b; Martin et al., 1989). In the liver of some sheep, ^{110m}Ag activity concentrations were comparable to those of ^{134}Cs (Beresford, 1989b); Beresford (1989b) reported one sample from an upland farm in 1987 with a higher ^{110m}Ag activity concentration than that of ^{137}Cs. Similar observations were made for farm animals in Germany (Pfau et al., 1989). The observation of ^{110m}Ag in farm animals and other samples (e.g., fungi (Bryne, 1988), marine molluscs (Boccolini et al., 1988) and birds (Ruiz et al., 1987)) resulted in the inevitable newspaper headline *'every cloud has a silver lining'*.

Radionuclides deposited on grass may be rapidly transferred to the milk of grazing animals. Measurements of the value of F_m of ^{131}I, ^{90}Sr and ^{134,137}Cs in cows' milk after Chernobyl are summarised in Table 3.12. Observations shortly

Table 3.12. Feed–milk transfer coefficient following intake of contaminated herbage by cows. Values obtained from pre-Chernobyl studies are shown in italics for comparison.

Isotope	Period after accident	F_m (d l^{-1})	Notes	Reference
^{131}I	0–121 days	$0.0034 \pm 0.0004^{\dagger}$	$n = 15$ studies around Europe	(1)
	0–30 days	0.0046	1 study in Italy	(2)
	Pre-Chernobyl	*0.006–0.010*		*(3)*
^{134,137}Cs	0–196 days	$0.0054 \pm 0.0005^{\dagger}$	$n = 65$ studies around Europe	(1)
	0–30 days	0.0021	1 study in Italy	(2)
	12 days Jul–Aug 1993	$0.0057 \pm 0.0007^{\dagger}$	1 study 3.5 km from CNPP	(4)
	1986–1988	0.0035–0.0021	Farm in Bavaria, Germany	(5)*
	1989–1993	0.0072–0.010	Farm in Bavaria, Germany	(6)
	1989–1992	0.002–0.016	15 farms in Sweden	(7)
	'First hay cut after the accident'	0.003	1 study, Italy	(3)
	Pre-Chernobyl	*0.007–0.012*		
^{90}Sr	12 days July–Aug 1993	$0.43 \pm 0.03 \times 10^{-3\dagger}$	1 study 3.5 km from CNPP	(4)
	'First hay cut after the accident'	0.8×10^{-3}	1 study, Italy	(7)
	Pre-Chernobyl	*$(0.74–1.8) \times 10^{-3}$*		*(3)*

† Standard error. * This study was included in the Kirchner (1994) review, but is included here to illustrate the change in F_m over time. CNPP = Chernobyl Nuclear Power Plant.
(1) Kirchner (1994); (2) Spezzano and Giacomelli (1991); (3) Tracy et al. (1989); (4) Beresford et al. (2000c); (5) Voigt et al. (1996); (6) Karlén et al. (1995); (7) Fabbri et al. (1994).

after the accident showed that the feed–cow milk transfer coefficients (F_m) of radioiodine and radiocaesium were lower than those which had been observed after the nuclear weapons test (NWT) fallout (Kirchner, 1994). For comparison, values of F_m estimated from pre-Chernobyl studies are also shown in Table 3.12.

A number of authors reported that transfer coefficients for recently deposited radiocaesium from Chernobyl were lower than those for plant incorporated radiocaesium via root uptake, potentially explaining why transfer coefficients were often greater after NWT fallout than those measured shortly after Chernobyl. Lower transfer coefficients to animal tissues and milk were observed in 1986 compared to subsequent years (Ward et al., 1989; Beresford et al., 1989; Hansen and Hove, 1991; Selnaes and Strand, 1992; Voigt et al., 1996). For instance, Hansen and Hove (1991) found increasing transfer coefficients to goats' milk from 0.042 d l^{-1} in 1986 to 0.124 d l^{-1} in 1988, the latter value being similar to that observed for a (bioavailable) ^{134}Cs tracer. Similarly, a study of radiocaesium at a farm in Bavaria (Voigt et al., 1996) showed an increase in F_m over the years after Chernobyl (Figure 3.6) attributed to increasing bioavailability of the radiocaesium ingested. By 1989, the observed values of F_m were approximately equal to those observed after nuclear weapons

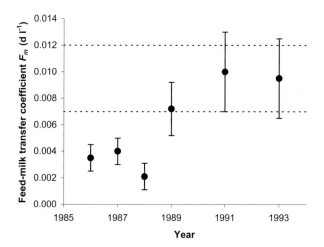

Figure 3.6. Increase in feed–milk transfer coefficient over time at a farm in Bavaria.* The range of typical estimates of the value of this parameter from pre-Chernobyl studies is shown by the dotted lines.†
*From data presented in Voigt et al. (1996). †From Tracy et al. (1989).

testing (Voigt et al., 1996). There is a mechanism for increased availability of radiocaesium between 1986 and 1987: in 1986 external contamination by intercepted deposition dominated animal intake whereas in 1987 more bioavailable radiocaesium incorporated into plants via root uptake would be expected to dominate intake. It is, however, somewhat more difficult to suggest a hypothesis as to why bioavailability increased over a period of 3 years as observed by Voigt et al. (1996), and shown in Figure 3.6.

Variation in F_m values may also have been caused by differences in physiological factors between animals, differences in farming practices and the influence of stable elements. A study of two farms in Bavaria (Voigt et al., 1996) showed that there was a substantial difference in ^{137}Cs activity in milk even though the farms were only 4 km apart and had similar levels of contamination and similar soils. The difference was partly attributed to differing grazing intensity. This study (Voigt et al., 1996) concluded that 'under normal agricultural conditions ... higher continuous grazing pressure resulted in lower activity concentrations in milk (in this case by a factor of 2).' In a study of ^{90}Sr transfer to cows' milk within the Chernobyl 30-km zone, Beresford et al. (2000c) observed a lower F_m value than was expected from pre-Chernobyl studies, but attributed this to differences in dietary Ca and ^{90}Sr between the experimental diet and the diet of the animals prior to the study.

The transfer of ^{131}I to milk was observed to be negatively correlated to milk yield, but no significant correlation was observed for ^{137}Cs (Kirchner, 1994). For ^{131}I, F_m values declined from approximately $6.2 \times 10^{-3}\, d\,l^{-1}$ at a milk yield of $10\, l\, d^{-1}$ to approximately $2.5 \times 10^{-3}\, d\,l^{-1}$ at a yield of $30\, l\, d^{-1}$.

Transfer coefficients of radionuclides from feed to the milk of goats and sheep are expected to be significantly higher than those observed in cows. Transfer coefficients (F_m) of radiocaesium in milk (Hansen and Hove, 1991) were in the range 0.042–0.124 d l^{-1} for goats (in Norway) and from 0.03–0.12 d l^{-1} for sheep (from a review of data from various countries). These values are approximately one order of magnitude higher than those typically observed for cows' milk, primarily due to the much lower (approximately a factor of 10) dry matter intake rates of sheep and goats in comparison to cows (see Equation 3.7). A study of the transfer of ^{131}I to milk of sheep fed ryegrass harvested from a pasture in Scotland in May 1986 (Howard et al., 1993) found F_m to be 0.29 ± 0.017 d l^{-1}. This is almost two orders of magnitude greater than the estimates for cows' milk (Table 3.12), and cannot be explained by different dry matter intake rates alone. However, as noted by Howard et al. (1993) the transfer of stable iodine from plasma to milk is 50-fold higher in sheep (and goats) compared to cattle.

Influence of feed characteristics on radionuclide transfer

The type of feed can make a difference to transfer coefficients to meat and dairy products, though this is usually minor. The transfer coefficient in goats fed with willow bark was 62% of that for transfer from hay, though only one animal was studied (Hansen and Hove, 1991). In the same study, the transfer coefficient of radiocaesium in goats fed with fungi was close to the coefficient for hay. The absorption of radiocaesium from heather (*Calluna vulagaris*) by sheep was observed to be lower (by about 20%) than that from grass (Beresford et al., 1992). Radiocaesium transfer to chicken muscle was significantly influenced by feed type: the transfer coefficient in chickens fed with wheat was approximately twice that in chickens fed with grass pellets (Voigt et al., 1993). The difference in this case was attributed to the way in which the feed became contaminated. Radiocaesium in the grass pellets was due to direct fallout to the plant surface, whereas that in the grain was internally translocated from foliage and hence was assumed to be more bioavailable.

The results of a study of radiocaesium uptake in ruminants (Beresford et al., 2000b) suggested a potentially lower transfer to animals grazing herbage species of relatively low digestibility. Kirchner (1994) cites pre-Chernobyl studies which indicate an influence of diet on transfer of radiocaesium to cows' milk, but in review of many post-Chernobyl studies Kirchner (1994) concluded that any influence of feed type on caesium transfer to milk was obscured by other physiological and environmental factors.

Ingestion of soil particles by grazing animals can also potentially lead to radionuclide uptake to milk and meat. This, however, only accounts for a small proportion of the total radiocaesium uptake. Beresford et al. (2002b) quote a review which found that 'the rate of soil ingestion has been estimated to be up to 30% and 18% of the daily dry matter intake of sheep and cattle respectively.' Absorption of radiocaesium from soil by ruminants is, however, believed to be only 4–15% of the absorption from vegetation (Beresford et al., 2000b). Goats fed with organic soil showed relatively low uptake of radiocaesium from the soil. Transfer coefficients

Table 3.13. Ratios of activity concentrations of ^{90}Sr and ^{137}Cs in milk products to those in milk.* Approximate ratios of concentrations of their stable analogues (Ca, K, respectively) in milk products to those in milk are shown in brackets.**

Product	Ratio of concentration in milk product (f.w.) to concentration in milk (f.w.)†	
	^{90}Sr (Ca)	^{137}Cs (K)
Milk	1.0 (1.0)	1.0 (1.0)
Skimmed milk	1.0 (1.1)	1.0 (1.1)
Cream	0.88 (0.61)	0.63 (0.68)
Butter	0.25 (0.21)	0.25 (0.17)
Buttermilk	1.5 (1.0)	1.3 (1.1)
Fresh cheese	3.8 (6.4)	0.58 (0.69)

*From Long et al. (1995). **From data in USDA (2004). † Midpoint of range quoted by Long et al. (1995) is presented where no mean was given. f.w. = fresh weight.

were approximately 7% of those observed when the goats were fed hay in which radiocaesium was in an available form (Hansen and Hove, 1991).

Milk products

Radionuclide activity concentrations in milk products can be significantly different to those in milk. Table 3.13 shows mean ratios of activity concentrations of ^{90}Sr and ^{137}Cs in milk products compared to those in milk. Note that there is significant variation about these mean values. The ratios of these radionuclides are similar to those of their stable analogues (e.g., Long et al., 1995), as illustrated in Table 3.13. The activity concentration of both radionuclides is significantly lower in butter than in the milk from which it was made. The relatively high concentration of calcium in cheese is mirrored by ^{90}Sr accumulation in cheese.

3.2.4 Time changes in contamination of agricultural systems

The transfer of radionuclides to agricultural products was rapid. The maximum contamination of grass was observed at the end of the period of significant atmospheric fallout from Chernobyl in early May (at this stage dominated by intercepted fallout). Radionuclides were rapidly transferred to the milk of grazing animals, so that maximum contamination of milk was observed a few days later (Figure 3.7). As discussed above, activity concentrations of ^{131}I and ^{134,137}Cs in milk and vegetation rapidly declined over the following weeks ($T_{eco} \sim 11$ days) as a result of washoff

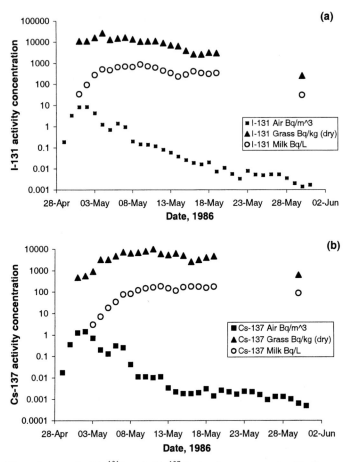

Figure 3.7. Time changes in (a) ^{131}I and (b) ^{137}Cs in air, grass and milk (from cows fed only with fresh grass) in north-western Italy during the first month after the accident.
Reproduced from Spezzano and Giacomelli (1991) using data kindly supplied by Pasquale Spezzano, ENEA, Italy.

('weathering') of radioactivity from vegetation to the soil. Measurements of ^{103}Ru and ^{137}Cs in a leafy vegetable (broccoli) showed maximum activity concentrations in early May which declined with a half-life (T_{eco}) of 6 and 11 days respectively (Martin et al., 1988).

The short-term dynamics of radionuclide uptake to crops and vegetation are strongly influenced by the amount of biomass (for interception) and growth rate (for root uptake and growth dilution). Thus initial contamination is to a large extent determined by the timing of fallout. The Chernobyl accident occurred in spring, at the start of the growing season in most affected areas. The fallout occurred after the soils were cultivated, so radionuclides tended to remain in the surface layers of soils. Ploughing of soils after the 1986 growing season played an important part in the reduction of activity concentrations in agricultural foodstuffs between 1986 and

1987. Typically, ploughing to a (normal) depth of 20–30 cm reduces the uptake of radiocaesium and radiostrontium by approximately 50% (Nisbet et al., 2004). Ploughing, combined with other land improvements, was used as a method of reducing the contamination of grass and crops in the fSU (see Chapter 5).

It is known that the transfer of radiocaesium to the meat of grazing animals is rapid (e.g., Coughtrey and Thorne, 1983). In Scotland, for example, the maximum concentration of radiocaesium in meat (lamb and beef) was observed shortly after fallout (Martin et al., 1988). In this study, activity concentrations in lamb grazing a contaminated pasture declined with a half-life (T_{eco}) of 25 days at a similar rate to the weathering half-life of ^{137}Cs in the grass (22 days). The longer half-time of ^{137}Cs in grass, observed by Martin et al. (1988), than those observed in a number of other studies was attributed to the fact that measurements did not begin until 2–3 weeks after Chernobyl (the early period showed the most rapid declines).

The ^{137}Cs activity concentration in lambs removed from a contaminated upland pasture in the UK to a lowland pasture (which received comparatively low levels of Chernobyl fallout) declined with a biological half-life (T_{bio}) of 11 days (Howard et al., 1987). In another study (Martin et al., 1988) the ^{137}Cs activity concentration in lambs removed from a contaminated pasture and fed uncontaminated feedstuffs declined with a biological half-life (T_{bio}) of 17 days.

The rapid uptake, and relatively short biological half-life of radiocaesium in grazing animals meant that within a week or so after the accident, the activity concentration in milk and meat was in approximate equilibrium with that in the animals' feed (see also Chapter 2). Radiocaesium activity concentrations in milk (Figure 3.7) and meat therefore initially declined rapidly, at a similar rate to declines in activity concentrations of pasture vegetation. In the long term, activity concentrations in crops, pasture vegetation, milk and meat continued to decline (at a slower rate) as radiocaesium became less available to plants. As discussed in Chapter 2, there was significant variation in rates of decline, but half-times of 1–4 years were typical of agricultural systems during the first 5 or so years after the accident. In the longer term, activity concentrations declined at a much slower rate. This is illustrated by the long-term changes of the ^{137}Cs aggregated transfer factor in the Bryansk region of Russia (Travnikova et al., 1999; see Figure 3.8). The significant declines in radiocaesium activity concentrations in agricultural systems were in contrast to very slow declines observed in forest ecosystems, as illustrated for mushrooms in Figure 3.8 (see also Section 3.3). Thus, over time, forest products accounted for an increasing fraction of the total ingestion dose to the population.

There is much less available evidence for the long-term decrease of ^{90}Sr after Chernobyl because changes in activity concentrations in foodstuffs were strongly influenced by the presence of hot particles: these degraded with time, acting as a source of ^{90}Sr (see Chapter 2). But studies of fallout ^{90}Sr following the period of atmospheric nuclear weapons testing (mainly 1958–1963) can be used to estimate long-term ecological half-lives. Work by Mück et al. (2001) estimated the long-term decline in ^{90}Sr activity concentration in milk, cabbage, cereals and apples in Austria to be in the range $T_{eff} = 8$–10 years. Studies were carried out of ^{90}Sr in milk and total diet (i.e., the average for all foodstuffs) in 12 different countries by Cross (2001) from

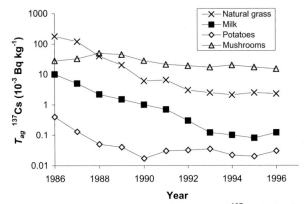

Figure 3.8. Time changes in the aggregated transfer factor of ^{137}Cs in the first decade after the accident.
From Travnikova *et al.* (1989).

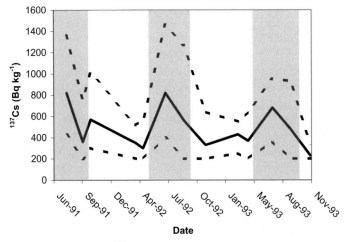

Figure 3.9. Seasonal trends in the ^{137}Cs activity concentrations of study ewes at one of the farms of Beresford *et al.* (1996); solid line denotes mean, dashed lines maximum and minimum values, 'summer' is indicated by grey shading.

data in UNSCEAR publications. Although there was significant variation in the data, this study showed average T_{eff} values of 9.5 years (milk) and 14.3 years (total diet) during the period 1970–1980.

Seasonal variation in the radiocaesium activity concentration of grass (and consequently animals) has been observed. Data for the British Isles showed ^{137}Cs activity concentrations in the muscle of free-ranging sheep to be generally higher in the summer months (i.e., July–September) than at other times of the year (e.g., Colgan, 1992; Howard and Beresford, 1994; Beresford *et al.*, 1996) (e.g., Figure 3.9). Many workers observed increased radiocaesium levels in upland vegetation during the same summer period (e.g., Horrill and Howard, 1991; Sandalls and Bennett, 1992). A study carried out in northern Greece (Papastefanou et al., 1991) showed a seasonal trend in ^{137}Cs in cows' milk: in this case, high values were observed in late winter and spring (approximately January–May). This was attributed to the fact that in the first 6 months of the year cattle were fed with stored feed from the previous year which was at higher contamination levels than grass from the current year.

3.2.5 Very long-lived radionuclides in agricultural systems

Because of the dominance of radiocaesium isotopes (and to a much lesser extent, ^{90}Sr) in determining long-term ingestion doses, other long-lived radionuclides have been studied relatively little. Isotopes of Pu will remain in the environment when ^{90}Sr and ^{137}Cs have decayed away and ^{241}Am will increase over the coming decades due to ingrowth from ^{241}Pu (see Chapter 1). Doses from these isotopes could potentially be significant in some small areas within the exclusion zone.

Plutonium and americium accumulation in grassy vegetation in Belarus was found to be inversely proportional to the measured radionuclide soil–water distribution coefficient (K_d) (Sokolik et al., 2004) of the soil. This result implies that root uptake will be dependent upon soil type; organic soil having the highest Pu and Am availabilities. The high uptake in organic soil may be due to its acidity: plant uptake of Am has been shown to increase with soil pH (Frissel et al., 1990). It should be noted, however, that many previous authors have suggested that resuspension of contaminated soil would be the dominant route of contamination of plants by Pu and Am (Hinton et al., 1995). In the study of Sokolik et al. (2004), the aggregated transfer factor (T_{ag}, dry weight basis) in 24 study sites varied in the range 3.5×10^{-5} to 2.5×10^{-3} m^2 kg^{-1} for $^{239+240}$Pu and 1.4×10^{-4} to 4.0×10^{-3} m^2 kg^{-1} for ^{241}Am (Sokolik et al., 2004). Maximum activity concentrations at the sites studied were 2.5 Bq kg^{-1} for $^{239+240}$Pu and 5.7 Bq kg^{-1} for ^{241}Am.

There are few measurements of Chernobyl derived actinide activity concentrations of animal derived food products within the fSU. If we use the values determined in grass by Sokolik et al. (2004) and assume diet to milk and meat transfer coefficients as summarised by IAEA (1994), we can predict activity concentrations in cows' milk and beef from hypothetical pastures at a very high level of contamination by $^{239+240}$Pu – 400 kBq m^{-2} (Table 3.14). Only approximately 2 km^2 of land contaminated by Chernobyl had contamination greater than this level of $^{239+240}$Pu (Kashparov et al., 2003). Note that this land is presently scrub vegetation and forest: this hypothetical scenario is an illustration only. Activity concentrations in vegetation in Table 3.14 are based on the range of soil-to-plant concentration values observed by Sokolik et al. (2004). Milk and beef concentrations are estimated from the vegetation concentrations using Equation 3.6 with recommended dry matter intake rates and transfer coefficients from IAEA (1994). For comparison, maximum permitted levels (Council Food Intervention Limits, CFILs) of d-emitting isotopes of Pu and Am in dairy products and other major foodstuffs are 20 Bq l^{-1} and 80 Bq kg^{-1} respectively.

Table 3.14. Estimated activity concentrations in milk and beef from a hypothetical pasture located in an area very highly contaminated by transuranium elements.

Radionuclide	Surface contamination† (kBq m^{-2})	Pasture vegetation (Bq kg^{-1}) (d.w.)	Milk (Bq l^{-1})	Beef (Bq kg^{-1}) (f.w.)
$^{239+240}$Pu	400	14–1,000	$(0.25–18) \times 10^{-3}$	$(0.01–72) \times 10^{-3}$
^{238}Pu	188*	6.6–470	$(0.12–8.3) \times 10^{-3}$	$(0.47–34) \times 10^{-3}$
^{241}Am	460*	64–1,800	$(1.6–44) \times 10^{-3}$	$(1.9–53) \times 10^{-3}$

† Surface contamination is for year 2000, but note that ^{241}Am will increase in the environment (see Chapter 1) until its maximum in 2058 when the amount will be approximately 1.7 times higher than that in year 2000. *Estimated from the ratio to $^{239+240}$Pu using data from Kashparov et al. (2003). f.w. = fresh weight; d.w. = dry weight.

Table 3.15. Estimated activity concentrations of cereals and potatoes grown on hypothetical agricultural land in an area very highly contaminated by transuranium elements.

Radionuclide	Surface contamination (kBq m^{-2})	Soil activity concentration (Bq kg^{-1})	Cereals (Bq kg^{-1}) (d.w.)	Potato (Bq kg^{-1}) (d.w.)
$^{239+240}$Pu	400	1,330	0.012	0.2
^{238}Pu	188*	630	0.005	0.09
^{241}Am	460*	1,530	0.034	0.31

*Estimated from ratio to ^{239,240}Pu using data from Kashparov et al. (2003). d.w. = dry weight.

We have also estimated activity concentrations of transuranium elements in crops grown on hypothetical agricultural land in a zone of 400 kBq m^{-2} $^{239+240}$Pu contamination (Table 3.15). Activity concentration in soil was estimated for cultivated land assuming soil density of 1,500 kg m^{-3} and uniform contamination to a plough depth of 20 cm. CRs used for cereals and potatoes were recommended values from IAEA (1994).

It is clear from Tables 3.14 and 3.15 that activity concentrations of ^{241}Am and isotopes of Pu are expected to be very low in foodstuffs produced on (hypothetical) agricultural land in the most contaminated areas of the 30-km zone. Though these very long-lived isotopes will become relatively more important than ^{90}Sr and ^{137}Cs in several hundred years time, they are not expected to present a significant ingestion dose risk from hypothetical agricultural activity. This assumes that potatoes and other root crops are thoroughly washed of soil prior to cooking and consumption.

3.3 FOREST ECOSYSTEMS

3.3.1 Cycling of radioactivity in the forest ecosystem

Tree canopies efficiently intercept rainfall and atmospheric pollutants (e.g., Ould-Dada et al., 2002): 60–90% of the initial radioactive fallout was intercepted

Table 3.16. ^{137}Cs in various components of a pine forest, Bourakovka, Chernobyl in 1990.*
^{90}Sr (from weapons tests)† in different components of a pine forest in Sweden in 1990.**
*From Thiry et al. (1990), quoted in SCOPE (1993). **From Melin et al. (1994).

Compartment	^{137}Cs activity concentration (kBq kg^{-1}) (d.w.)	Compartment	^{90}Sr activity concentration (Bq kg^{-1}) (d.w.)
Tree (pine)		**Tree (Scots pine)**	
Needles/leaves	10.1	Current needles	5
Twigs (wood)	7.6	Older needles	8
Bark	25.1	Current branches	9
Trunk (wood)	1.6	Older branches	11
		Stem/bark	13
		Stem/wood	4
		Roots	5
Understorey			
Mosses	270–515	*Vaccinum myrtillus*	15
Lichens	275–510	*Deschampsia flexuosa*	10
Heath	500	*Pteridum aquilinum*	38
Fungi	380	Herbs	76
		Litter/humus	64

†Note that the ^{137}Cs measurements are in kBq kg^{-1} whereas the ^{90}Sr measurements are in Bq kg^{-1}.
d.w. = dry weight.

by the tree canopy (Tikhomirov and Shcheglov, 1994). In studies on Swedish forests, Melin et al. (1994) stated that 'a rough estimate indicates that the [evergreen] spruce stand will intercept about 90% of the deposited radionuclides while the unfoliated deciduous stands of beech, birch and alder will intercept less than 35% of dry deposited radionuclides.' The Chernobyl accident occurred at a time of the year when deciduous trees had low leaf cover, and hence low ability to intercept radionuclides. Initial high doses of radiation to leaves of (evergreen) pine trees had a lethal effect on some areas of pine forest (the so-called Red Forest) close to the Chernobyl reactor (see Chapter 8). During the first year following fallout, the majority of radioactivity migrated (by rainfall washout and leaf fall) from the leaf canopy to the forest soil. From 1987 onwards, more than 90% of the total ^{137}Cs in forest systems in the Chernobyl zone was in the soil (Tikhomirov and Shcheglov, 1994). Table 3.16 illustrates activity concentrations in various components of forest vegetation. Activity concentrations of both ^{90}Sr and ^{137}Cs in understorey vegetation are significantly higher than in tree tissues.

Storages and fluxes of raionuclides in the forest ecosystem are illustrated in Figure 3.10. Radiocaesium is efficiently retained in the surface layers of soil and yet remains available for bioaccumulation for long periods of time. Rates of erosional loss of radiocaesium from the forest system are low – less than 1% per year (Tikhomirov and Shcheglov, 1994; Nylén, 1996). Rates of vertical migration in forest soils are also low. For example, forty years after the nuclear weapons test era, the maximum activity concentrations of weapons-test derived ^{137}Cs, ^{90}Sr and

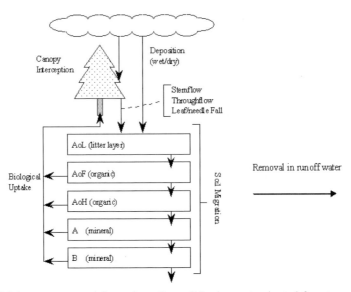

Figure 3.10. Major storages and fluxes in radionuclides in contaminated forest ecosystems. From Shaw et al. (2003).

^{239,240}Pu were still found in the surface layer in undisturbed forest soils in Korea (Lee and Lee, 2000). Low rates of migration of radiocaesium were observed in forests contaminated by Chernobyl fallout (Tikhomirov and Shcheglov, 1994), as illustrated in Figure 3.11.

As with grassland systems (see Chapter 2), radiocaesium from Chernobyl appears to have migrated less far down the soil profile than weapons-test radiocaesium (Figure 3.11b). As discussed in Chapter 2, differences may be due to different times since deposition or to initially different environmental conditions and physical-chemical forms of deposited radionuclides. Both ^{137}Cs and ^{239,240}Pu are efficiently bound in the upper organic horizons of forest soils, although it has been observed that a large proportion of the ^{137}Cs is bound to mineral fractions in the organic horizons (Bunzl et al., 1998). Due to the strong attachment of radiocaesium to the solid phase, a significant proportion of the vertical mixing is likely to be caused by bioturbation (Bunzl, 2002).

The distribution of a number of radionuclides (^{60}Co, ^{106}Ru, ^{125}Sb, ^{144}Ce and ^{154}Eu) were studied at Kopachi and Dityatky (7 and 26 km from Chernobyl, respectively) by Rühm et al. (1996). This study showed lower mobility of these radionuclides than was observed for ^{137}Cs. This low mobility may have been influenced (particularly at the Kopachi site close to Chernobyl) by deposition of radionuclides in fuel particles (see Chapter 2).

3.3.2 Transfer of radionuclides to fungi, berries and understorey vegetation

Understorey vegetation efficiently accumulates radiocaesium in the forest ecosystem (Table 3.17). This contamination is an important transfer pathway to humans since

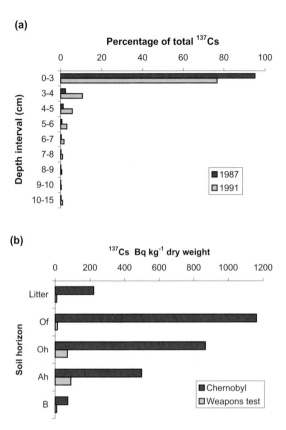

Figure 3.11. Radiocaesium profiles in forest soils in (a) the Chernobyl 30-km zone at two different times after the accident and (b) in a forest soil in Germany (in 1996) contaminated by Chernobyl and weapons test fallout (data is decay corrected to 1 May, 1986 and a ^{134}Cs : ^{137}Cs ratio of 0.57 was used to distinguish weapons-test and Chernobyl ^{137}Cs).
(a) From data in Tikhomirov and Shcheglov (1994). (b) From data in Rühm *et al.* (1999).

many species of fungi and berries are commonly eaten, and understorey vegetation is eaten by game animals such as boar and deer. Accumulation of radiocaesium by forest products is determined by its vertical distribution and bioavailability in the soil, and the rooting behaviour of the particular species (Fesenko *et al.*, 2001).

Transfer of radionuclides to fungi

Fungi were found to strongly accumulate radiocaesium, with aggregated transfer factors (fresh weight basis) in the range $5-10 \times 10^{-3}$ m^2 kg^{-1} for species such as *Boletus edulis* and *Cantharella cibarius* and up to ca. 100×10^{-3} m^2 kg^{-1} or more for species such as *Russula*, *Xerocomus badius* and *Suillus luteus* (Kenigsberg *et al.*, 1996; see Table 3.17). Uptake in different species is strongly influenced by their method of nutrition: in general, the saprotrophs and wood degrading fungi have

Table 3.17. Aggregated transfer factors (fresh weight) of ^{137}Cs in various species of edible fungi collected in Belarus.*
From Kenigsberg et al. (1996).

Species	T_{ag} 1989–1990 (m² kg⁻¹)		T_{ag} 1994 (m² kg⁻¹)	
	Mean	Standard deviation	Mean	Standard deviation
Boletus edulis	7.8×10^{-3}	6.0×10^{-3}	13.9×10^{-3}	11.0×10^{-3}
Cantharellus cibarius	8.6×10^{-3}	10.4×10^{-3}	11.7×10^{-3}	14.6×10^{-3}
Xerocomus badius	110.6×10^{-3}	50.9×10^{-3}	83.6×10^{-3}	74.4×10^{-3}
Russula	28.3×10^{-3}	42.7×10^{-3}	50.3×10^{-3}	74.8×10^{-3}
Tricholoma flavovirens	8.4×10^{-3}	7.3×10^{-3}	43.7×10^{-3}	25.0×10^{-3}
Suillus luteus	98.3×10^{-3}	68.3×10^{-3}	41.7×10^{-3}	33.4×10^{-3}
Armillaria mellea	7.4×10^{-3}	4.2×10^{-3}	4.8×10^{-3}	4.6×10^{-3}
Leccinum scabrum	46.4×10^{-3}	57.2×10^{-3}	48.9×10^{-3}	72.9×10^{-3}

*Variation around these averages is very high, such that other authors have suggested that significant differences cannot be determined between species of the same nutritional type (Barnett et al., 1999).

relatively lower contamination while those fungi forming symbioses with tree roots (mycorrhizal fungi such as *Xerocomus* and *Lactarius*) have a relatively high degree of uptake. The soil type and the depth at which the mycelium sources its nutrients are also important factors causing variation in uptake rates (Kenigsberg et al., 1996). As with pasture vegetation (Table 3.4), ^{137}Cs uptake to fungi on hydromorphic soils was significantly greater than on automorphic soils. Automorphic soils are those formed from normal parent materials and under normal drainage conditions, hydromorphic soils are formed under conditions of periodic or permanent saturation.

Radiocaesium activity concentrations in fungi often showed relatively little change over time during the first decade after the accident (see Table 3.17 and Figures 3.8 and 3.12). No significant decline in activity concentration of fungi was observed during the period 1986–1994, though there was some evidence of a slow reduction (T_{eff} = 5–8 years) in some species (Kenigsberg et al., 1996; IAEA, 2002). At a site in Sweden there was a reduction of 60–80% in ^{137}Cs in many different species over the period 1991–1998 (Moberg et al., 1999). Though rates of decline varied at different sites and among different species, the generally slow reduction in activity concentration in fungi is consistent with slow loss of radiocaesium from the system (by vertical migration and runoff) and low 'fixation' ability of highly organic forest soils.

Reviews of transfer of radiocaesium to fungi in the UK (Barnett et al., 1999) and over Europe (Gillett and Crout, 2000) showed variation in transfer factors of three orders of magnitude within an individual species. For all species, Gillett and Crout (2000) observed variation in the transfer ratio (defined as activity concentration in mushroom divided by the original ^{137}Cs deposition decay corrected to the time of sampling and approximately equal to the T_{ag}) in the range <0.001 to >10 m² kg⁻¹ on a dry weight basis. Transfer ratios were significantly related to nutritional

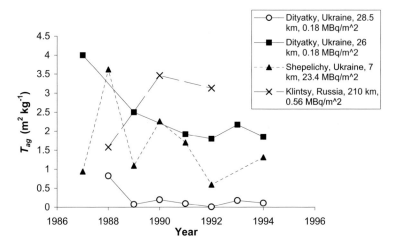

Figure 3.12. Aggregated transfer factor for ^{137}Cs (dry mass basis) in a very highly accumulating mushroom species (*Paxillus involutus*). Note that T_{ag} calculated on a dry mass basis is approximately 10 times higher than that calculated on a fresh mass basis. Variation is very high, even for this single species, though there is evidence of a decline over time in some sites. The legend shows the sampling site, distance from Chernobyl and average ^{137}Cs fallout.
From data presented in Belli and Tikhomirov (1996).

type: saprophytic and parasitic species had approximately similar transfer ratios, but uptake in microrrhyzal species was significantly higher than in other species (Gillett and Crout, 2000), as observed in an fSU study (Kenigsberg *et al.*, 1996). Apart from radioactive decay, these workers observed no significant decline in activity concentrations in fungi during a 10-year period after the Chernobyl accident.

Activity concentrations of ^{90}Sr in fungi were much lower than were observed for ^{137}Cs. This was primarily due to much lower fallout of ^{90}Sr in most areas, but was also due to the generally lower bioaccumulation of ^{90}Sr in fungi (Table 3.18).

Table 3.18. Comparison of mean T_{ag} values for ^{137}Cs and ^{90}Sr in fungi in the Bragin district of Belarus, 1994–1995. Mean activity concentrations in different species were of the order of 10^4–10^5 Bq kg^{-1} for ^{137}Cs and 500 Bq kg^{-1} for ^{90}Sr.
From data in Aslanoglou *et al.* (1996).

Species	Aggregated transfer factor (m² kg⁻¹)	
	^{137}Cs (f.w.)	^{90}Sr (d.w.)
Cantharellus cibarius	120×10^{-3}	5.9×10^{-3}
Boletus edulis	110×10^{-3}	6.0×10^{-3}
Russula app.	570×10^{-3}	4.8×10^{-3}

d.w. = dry weight; f.w. = fresh weight.

Berries and other understorey vegetation

As with fungi, uptake of radiocaesium to forest berries is high in comparison with foodstuffs grown in agricultural systems. Aggregated transfer factors (fresh weight basis) of around $1-10 \times 10^{-3}\,\mathrm{m^2\,kg^{-1}}$ were typically observed for various species of forest berries, though values up to around $40-50 \times 10^{-3}\,\mathrm{m^2\,kg^{-1}}$ were observed in various countries during 1987–1994 (Rantavaara, 1990; Kenigsberg et al., 1996; Zibold et al., 2001). Note that the values presented in Zibold et al. were here recalculated from T_{ag} values originally presented on a dry mass basis using a f.w. : d.w. ratio of 10 (Carini, 2001).

Aggregated transfer coefficient values for radiocaesium to different berry species have recently been reviewed (Beresford et al., 2001). The different species appear to fall into one of three categories on the basis of their T_{ag} values (although there is considerable within-species variability):

- Predominantly non-forest berries – *F. vesca*, *R. idaeus* and *R. fruticosus* – with a mean T_{ag} value of $0.75 \times 10^{-3}\,\mathrm{m^2\,kg^{-1}}$ (f.w.).
- Berries which grow in wet (and forest) ecosystems – *R. chamaemorus* and *V. oxycoccus* – with a mean T_{ag} value of $10 \times 10^{-3}\,\mathrm{m^2\,kg^{-1}}$ (f.w.).
- Other *Vaccinium* species growing within forests – *V. myrtillus* and *V. vitis-idaea* – with a mean T_{ag} value of $5.5 \times 10^{-3}\,\mathrm{m^2\,kg^{-1}}$ (f.w.).

Activity concentrations of Pu isotopes were generally low in fungi and forest berries (Lux et al., 1995). This was partly due to generally lower transfer factors than radiocaesium, but was also due to the much lower release of Pu isotopes in the accident (see Table 1.2). Various other radionuclides were measured in different understorey vegetation by Lux et al. (1995). Using these data, we have calculated mean transfer factors relative to ^{137}Cs (Table 3.19) showing that mean bioaccumula-

Table 3.19. Comparison of concentration ratios[†] of ^{137}Cs with other radionuclides (RN) in understorey vegetation (here defined as activity concentration in plant dry mass divided by activity concentration in the Of – horizon of soil).

Radionuclide	No. of samples	CR Geometric mean (Range)	Ratio of CR values: CR(^{137}Cs) : CR(RN)
Co-60	11	0.014 (0.005–0.16)	14
Sr-90	9	0.14 (0.03–0.43)	1.4
Ru-106	12	0.0022 (<0.002–0.005)	90
Sb-125	12	0.0053 (0.002–0.02)	36
Cs-137	18	0.19 (0.05–0.93)	1.0
Ce-144	14	0.0073 (<0.002–0.43)	26
Eu-154	12	0.0049 (<0.001–0.16)	39
Pu-239	17	0.012 (0.002–0.17)	16

[†] Calculated from measurements at three forest sites 6–28.5 km from Chernobyl during 1992, as presented in Lux et al. (1995). Geometric means were calculated, 'less than' values being included assuming a value equal to half the 'less than' value.

tion for most radionuclides was at least a factor of 10 less than observed for ^{137}Cs, though bioaccumulation of ^{90}Sr was close to that of ^{137}Cs. Note, however, the wide variation in measured values: Lux et al. (1995) noted high transfer factors of ^{144}Ce in a few cases.

Due to their generally lower radiocaesium transfer factors and lower consumption rates in the Ukraine, Belarus and Russia, forest berries posed a smaller radiological hazard to humans than fungi. But fungi and other forest plants contribute significantly to the diet of grazing animals and are therefore the most important pathways for radiocaesium transfer to game animals.

Measurements of ^{99}Tc in a forest soil and strawberry (*Fragaria vesca*) leaves led Uchida et al. (2000) to conclude that their results 'indicate that soil-to-plant transfer factors for Tc are similar to those for Cs'. At 6 sites in the 30-km zone, the CR (dry weight basis) of ^{99}Tc in strawberry leaves were in the range 0.086–0.448 compared with 0.062–0.389 for ^{137}Cs (Uchida et al., 2000). Maximum ^{99}Tc activity concentration in strawberry leaves was 6.0 Bq kg^{-1} dry weight (Uchida et al., 2000).

Plutonium in fungi and berries

The CR values in Table 3.19 may indicate that Pu contamination of berries could potentially be significant in the most contaminated parts of the 30-km zone, though more empirical data of activity concentrations in berries would be required before firm conclusions could be made. The CR estimates for ^{239}Pu (Table 3.19) are somewhat higher than those observed in other studies. The range of CR in Table 3.19 (0.002–0.17 d.w. $\approx$ (0.2–17) $\times 10^{-3}$ f.w.) compares with (0.027–0.74) $\times 10^{-3}$ (f.w.) in strawberry plants (*Fragaria vesca*) from a review by Carini (2001) and (0.3–3) $\times 10^{-3}$ in bilberry plants (*Vaccinium myrtillus*) from accumulation of weapons test fallout (Riekkinen and Jaakkola, 2001, re-calculated to f.w. basis). Note that the latter values may have been influenced by industrial pollution which was found to increase Pu uptake (Riekkinen and Jaakkola, 2001).

It may be that the measurements in Table 3.19 were influenced by adsorption of soil to plant surfaces. In field experiments, Pu contamination of plants was found to be dominated by absorption of deposited and resuspended materials to surfaces and only a small proportion was due to root uptake (Hinton and Pinder, 2001). At two sites in the 30-km zone, Hinton et al. (1996) observed 0.002 and 0.034 kg kg^{-1} of soil loaded on (dried) plant (cabbage and kholrabi) surfaces. If adsorption of soil to leaf surfaces was the major contamination pathway and assuming that the adhered soil activity concentration was the same as the average for the surface soil, this would give CRs (d.w. basis) of 0.002 and 0.034. These values are of the same order as those given for Pu in understorey vegetation in Table 3.19, though note that the plant types are very different. Though no certain conclusions can be made, it is possible that the relatively high bioaccumulation of Pu observed by Lux et al. (1995) is due to surface adsorption of soil particles. A study of Pu uptake to meadow vegetation in Belarus (Sokolik et al., 2004), however, showed a strong relationship between plant uptake of Pu and its solids–aqueous distribution coefficient in the soil, implying a root uptake mechanism in this case.

Uptake of Pu to fungi could potentially be significant in some areas. In Poland, mean T_{ag} (d.w. basis) for various mushroom species were in the range 0.09–$3.1 \times 10^{-3}\,\text{m}^2\,\text{kg}^{-1}$, but the maximum observed value was $38 \times 10^{-3}\,\text{m}^2\,\text{kg}^{-1}$ (Mietelski et al., 2002). In the same study, samples in the Ukraine gave T_{ag} (d.w.) values in the range $0.75\text{–}0.91 \times 10^{-3}\,\text{m}^2\,\text{kg}^{-1}$ for three different species, but $14.6 \times 10^{-3}\,\text{m}^2\,\text{kg}^{-1}$ for one other species. Only one mushroom of each species was measured in the Ukrainian samples. These transfer factors may imply estimated activity concentrations of up to several hundred becquerels per kilogram $^{239+240}$Pu in fresh fungi in the most contaminated areas of the exclusion zone, though this tentative estimate has a degree of uncertainty. By washing one of the samples, Mietelski et al. (2002) concluded that external contamination by soil particles was not the most important uptake mechanism.

3.3.3 Transfer of radionuclides to game and semi-domestic animals

Game and semi-domestic animals such as deer, reindeer, moose and wild boar accumulate significant quantities of radiocaesium through grazing forest vegetation and other semi-natural ecosystems.

After the Chernobyl accident, reindeer and roe deer in areas of Scandinavia, Germany and Austria had activity concentrations substantially over $1,000\,\text{Bq}\,\text{kg}^{-1}$ and exceeded intervention levels (Åhman and Åhman, 1994; Strebl et al., 1996; Zibold et al., 2001). In 1987, 29% of the slaughtered reindeer in the reindeer herding district of Sweden were rejected as food because their radiocaesium activity concentrations were above the intervention level of $1,500\,\text{Bq}\,\text{kg}^{-1}$ (Åhman and Åhman, 1994). In 2004, less than 1% was rejected (partly due to countermeasures still in force).

The aggregated transfer factor of radiocaesium in game animals varies significantly over time and according to their food source. Studies in Germany showed that the activity concentration of ^{137}Cs in roe deer varied according to changes in the diet from green plants to fungi in autumn (Figure 3.13; Zibold et al., 2001). Similar observations were made for reindeer in Sweden where lichen and mushroom consumption in autumn and winter caused significant temporary increases in ^{137}Cs in reindeer meat (Åhman and Åhman, 1994).

Aggregated transfer factors of ^{137}Cs in game and semi-domestic animals are summarised in Table 3.20. This table also illustrates the significant variation in effective ecological half-life observed in game animals over the period after the accident. Effective ecological half-lives in products from semi-natural ecosystems were generally longer than in agricultural and aquatic systems (cf. Chapter 2).

^{90}Sr accumulation in meat of game animals was significantly lower than ^{137}Cs, even after accounting for the approximately 10 times greater deposition of ^{137}Cs in the district studied (Table 3.21). Activity concentrations of ^{90}Sr in bone, however, were similar to those of ^{137}Cs in meat, despite the lower ^{90}Sr fallout. This is due to a substantially higher uptake of ^{90}Sr in bone due to its chemical similarity to calcium.

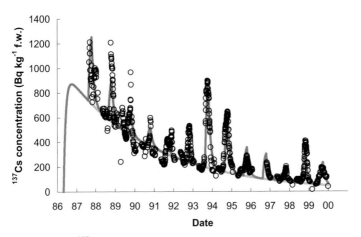

Figure 3.13. Change in ^{137}Cs activity concentration in roe deer meat in a spruce forest, Ochsenhausen, Germany. The steady exponential decline ($T_{eff} = 3.5$ yrs) was due to reductions in contamination of green plants on which reindeer feed for most of the year. The sharp peaks are due to consumption of more contaminated fungi in autumn.
From Zibold *et al.* (2001); reproduced from data kindly supplied by Gregor Zibold, Fachhochschule Weingarten.

Table 3.20. Range in transfer factors and effective ecological half-lives observed in game during the first few years after Chernobyl.

Study	Study period	T_{ag} (m^2 kg^{-1}) (f.w.)	Effective ecological half-life
Roe deer, Spruce forest, Germany[1]	1987–1994	$(6–60) \times 10^{-3}$	$T_{eff} = 3.5$ yr
Roe deer, mixed forest on peat bog, Germany[2]	1988–1999	$(5–70) \times 10^{-3}$	$T_{eff} = 17$ yr, due to low fixation in peat bog soil.
Various game animals, Russia[3]	Several years post Chernobyl	$(0.5–28) \times 10^{-3}$	
Roe deer, Russia[4]	1992–1994	$(0.86–18.5) \times 10^{-3}$	
Wild boar, Russia[4]	1992–1994	$(0.13–69.0) \times 10^{-3}$	
Reindeer, winter feeding on lichen and fungi, Sweden[5]	1986–1992	$(120–760) \times 10^{-3}$ (during winter 86–87)	$T_{eff} = 2.3–5.0$ yr
Reindeer, Norway[6]	1987–1990		$T_{eff} = 2$ yr
Reindeer, Norway[7]	1987–1991		$T_{eff} = 3.5$ yr
Moose, Sweden[8]	1986–1990	$(6–20) \times 10^{-3}$	T_{eff} up to 30 yr
Moose, Finland[9]	1986–1989	$(5–45) \times 10^{-3}$	
Small game (hare, terrestrial birds, waterfowl), Finland[9]	1986–1989	$(0–149) \times 10^{-3}$	

[1] Klemt *et al.* (1998); [2] from data in Zibold *et al.* (2001); [3] Kenigsberg *et al.* (1996); [4] Eriksson *et al.* (1996); [5] Åhman and Åhman (1994); [6] Bretten *et al.* (1992); [7] Hove *et al.* (1992), quoted in Åhman and Åhman (1994); [8] Bergman *et al.* (1991); [9] Rantavaara (1990).

Table 3.21. ^{137}Cs and ^{90}Sr in game animals in the Bragin district of Belarus, 1994–1995.
From Eriksson et al. (1996).

Animal	Activity concentration (Bq kg^{-1})		
	^{137}Cs (muscle, f.w.)	^{90}Sr (muscle, d.w.)	^{90}Sr (bone, d.w.)
Moose *Alces alces*	200–12,000	4–95	11,000–95,000
Roe deer *Capreolus capreolus*	3,600–10,700	4–23	4,200–61,000
Wild boar *Sus scrofa*	4,400–62,000	3–10	4,200–30,900

d.w. = dry weight; f.w. = fresh weight.

3.3.4 Radionuclides in trees

Initial contamination of trees was due to foliar (i.e., leaf) absorption and adsorption of radionuclides deposited to the canopy and external surfaces including bark. For a period of up to 2–4 years, this initial absorption was the dominant pathway of radionuclide contamination of the tree (Tikhomirov and Shcheglov, 1994). During this period, contamination was independent of variations in soil type and most contamination was observed in the outer tree surfaces (leaves, bark). Over the longer term, the primary route of contamination is uptake from the soil via the root system and is therefore dependent on soil properties. Because foliar absorption was the dominant pathway in the early period (first 2–4 years), root uptake only became important for relatively long-lived radionuclides such as ^{90}Sr and ^{137}Cs.

In general, radiocaesium and radiostrontium activity concentrations in trees are significantly lower than in soils, leaf litter and understorey vegetation (Tables 3.16 and 3.22). Predictions of long-term changes in radionuclide activity concentration in

Table 3.22. Radiocaesium transfer factors in different parts of trees at Dityatki, 28 km south of Chernobyl during 1987.
From Belli and Tikhomirov (1996).*

Compartment	Aggregated transfer factor T_{ag} ($\times 10^{-3}$ m^2 kg^{-1}) (d.w.)		
	Pine	Birch	Oak
External bark	29.7	28.2	21.3
Internal bark	4	3.2	2
Wood	0.83	1.2	0.66
Branches	7.7	10	8.8
Leaves	7.2 (young needles) 36.2 (old needles)	11.2	5.2

* Note that values presented are multiplied by 10^{-3} to give the T_{ag}. d.w. = dry weight.

trees are dependent on soil type, species and age of tree. Whilst contamination of the external parts of the tree tended to decrease over time due to washoff of deposited radioactivity, internal contamination of the wood tended to increase due to root uptake. The degree of root uptake is dependent on both soil type and the physico-chemical form of the initial radioactive deposition. During the period 1986–1994, radiocaesium in the wood of birch and pine increased over time (by a factor 2–3) in trees grown on peaty soil, but decreased over time (by a factor 5–10) in trees grown on soddy–podzolic sandy soil (Belli and Tikhomirov, 1996). This is believed to have been due to the greater long-term availability for root uptake of radiocaesium in the peaty soil compared to the soddy–podzolic sandy soil since the latter more efficiently 'fixes' radiocaesium.

Trees of the same species (pine) growing on similar soils (podzols) were also found by Belli *et al.* (2000) to accumulate radiocaesium more readily at sites which did not receive 'hot particles' in the initial deposit, compared with those growing within 10 km of the Chernobyl reactor where hot particles were abundant. Furthermore, because of their higher growth rates, young trees tended to accumulate more radiocaesium via root uptake than old trees. In 1990, needles, wood, internal bark and branches of young trees (15–20 years) were more contaminated with ^{137}Cs than mature (50–55 years) trees by a factor of 2.5–3.5. The difference between young and old trees was not so great for external bark because the external contamination masked root uptake (Mamikhin *et al.*, 1997).

As in the case of its nutrient analogue, potassium, the rate of radiocaesium cycling within forests is rapid and a quasi-equilibrium of its distribution is believed to be reached a few years after atmospheric fallout (Shcheglov *et al.*, 2001) which is supported by post-Chernobyl observations (Soukhova *et al.*, 2003). Thus, it may be that ^{137}Cs activity concentrations in trees will change only slowly over time in the coming decades, with a decline likely to be at a rate close to its physical decay rate.

Within the tree trunk, the highest concentrations of radiocaesium are generally found in the bark and outer tree rings (Table 3.22). Studies on tree rings in coniferous (pine, *P. sylvestris*) and deciduous (birch, *B. pendula*) trees in 1996 showed maximum activity concentrations in the outer rings, with lower activity concentrations observed in the ring formed in 1986 (Soukhova *et al.*, 2003). In pine, activity concentrations in the centre (the pith) were approximately the same as in the ring formed in 1986, whilst in birch, there was an increase again inwards from the 1986 ring to the centre of the tree.

3.4 RADIATION EXPOSURES FROM INGESTION OF TERRESTRIAL FOODS

3.4.1 Reference levels of radioactivity in foodstuffs

A reference or intervention level is a concentration of radioactivity in a foodstuff at which some action must be taken by regulatory authorities. If the reference level is

Table 3.23. Agreed CFILs of radionuclides in foods in place in the EC.
Adapted from Smith and Beresford (2003).

Radioactivity in food ($Bq\,kg^{-1}$ or $Bq\,l^{-1}$)	Baby foods	Dairy produce	Other foods (except minor foodstuffs)	Liquid foodstuffs
Isotopes of Strontium notably ^{90}Sr	75	125	750	125
Isotopes of Iodine notably ^{131}I	150	500	2,000	500
α-emitting Isotopes of Pu and trans-Pu elements (e.g., ^{239}Pu, ^{241}Am)	1	20	80	20
All other nuclides of half-life >10 days, notably ^{134}Cs, ^{137}Cs	400	1,000	1,250	1,000

exceeded it can result in a ban, or limit on consumption of the product, but (where appropriate) may simply indicate the need for further investigation. Reference levels vary between countries and according to circumstance. In part, reference levels are set such that the overall ingested dose resulting from consumption of foodstuffs at this level is below a certain limit. For example, they may be higher in the event of a short-term, temporary increase in levels of radioactivity in food than in the case of long-term contamination. They may also vary according to the number of foodstuffs contaminated: if a number of foodstuffs, forming an important part of the diet, are being considered, the reference level may be lower than that for the case in which a single, less important product is considered.

An example of a reference level used in the European Community (EC) (EURATOM Council Regulations No. 3958/87, No. 994/89, No. 2218/89, No. 770/90) are the Council Food Intervention Limits (CFILs), as shown in Table 3.23. These CFILs are not in force today but will enter into force after a decision taken by the Commission in the case of an accidental situation. They would be in force for at most 3 months, after which they may be adjusted to better fit the particular accidental situation. Reference levels are usually determined by calculating the mean activity concentration in foodstuffs which, assuming consumption over a one-year period, would lead to a radiation dose which is deemed to present a negligible risk. The dose is determined by the radioactivity in a product and its consumption rate, thus CFILs may be affected by the rate of consumption of a foodstuff. For example, minor foodstuffs such as herbs and spices have higher CFILs (by a factor of 10) than the values shown in Table 3.23, which assume high consumption rates and apply to major foodstuffs such as dairy produce, potatoes and beef. The CFILs in the EC assume that the average activity concentration of a person's total diet is a fraction (10%) of the contamination of certain individual foodstuffs. This assumption is considered to be appropriate since the total diet, particularly in western Europe, combines foodstuffs of different types originating from very disperse areas.

The intervention limits for ^{137}Cs activity concentrations in foodstuffs in the Ukraine, Belarus and Russia are shown in Table 3.24. These are generally lower than the EC levels partly because there are multiple exposure pathways in the

Table 3.24. Intervention limits for the ^{137}Cs activity concentration in foodstuffs within Belarus, Russia and the Ukraine as in place in 1999. Permissible levels for milk and meat in all three countries have since dropped to approximately 100 Bq l^{-1} for milk and 500 (200 in the Ukraine) Bq kg^{-1} for meat.
Adapted from Smith and Beresford (2003).

	Belarus	Russia (Bq kg^{-1})	Ukraine
Fresh milk	111	370	370
Butter	185	370	
Cheese		370	
Beef	600	740	740
Pork	370	740	740
Chicken	370	740	740
Bread	74	370	370
Potatoes	100	370	590
Other vegetables	100	370	590
Cereals and legumes	100	370	
Cultivated berries	100	370	
Other fresh fruit	100	370	590
Baby food	37	185	185

Chernobyl affected areas and diets tend to be much more localised in these countries than in the EC. In addition, limits in the most affected areas have generally decreased with time after the accident. In the first year after the accident, for pragmatic reasons, limits were higher than at present. It has been argued that current limits in the Ukraine, Belarus and Russia are too restrictive, leading to the implementation of food bans and expensive remediation measures where actual radiation risks are low.

3.4.2 Radiation exposures from agricultural foodstuffs

Studies of inhabitants of the Bryansk region of Russia (Travnikova et al., 1999) showed that milk formed a major part of their diet. Table 3.25 summarises the consumption of different food groups before and after the Chernobyl accident. After the accident, the consumption of milk and meat from private farms and markets dropped considerably (note also that private cows were removed from many areas). To some extent this was offset by an increase in consumption from state shops, but overall the consumption of milk declined to approximately 40% of the pre-accident level (Travnikova et al., 1999). For the Bryansk population, the study found that animal products (milk and meat) formed an average 60 and 85% of the ^{137}Cs intake for urban and rural populations respectively.

The intake of ^{137}Cs from consumption of agricultural and natural foodstuffs in a village in the Ukraine is shown in Table 3.26, clearly showing that ingestion doses

Table 3.25. Average annual consumption of foodstuffs by the population of a village in Bryansk, Russia, before and after the Chernobyl accident (kg yr^{-1}).
From Travnikova *et al.* (1999).

Source[†]	Milk	Meat	Bread	Potatoes	Vegetables	Fruit	Eggs	Fish
1985								
Private	285	66	–	241	117	69	33	11
Shops	–	–	120	–	–	–	–	37
Total	*285*	*66*	*120*	*241*	*117*	*69*	*33*	*48*
1990								
Private	44	18		179	55	26	7	4
Shops	40	40	117	–	15	44	–	15
Total	*84*	*58*	*117*	*179*	*69*	*69*	*7*	*18*
1996								
Private	186	58	–	212	80	58	15	11
Shops	–	7	113	–	4	4	–	4
Total	*186*	*66*	*113*	*212*	*84*	*62*	*15*	*15*

[†] Consumption of private produce (including food sold in local markets) is distinguished from produce bought from shops (mainly collective farm produce which may not always be sourced locally).

varied considerably, largely as a result of variation in consumption of milk and forest products.

In recent years, intervention limits have generally not been exceeded in intensively produced foodstuffs (i.e., the output of former collective farms) (Firsakova *et al.*, 1996; Prister *et al.*, 1998) (Table 5.4). In 1998, the estimated annual average dose to the populations living in the most affected areas of the fSU (^{137}Cs deposition 1 MBq m^{-2}), assuming no countermeasures, was 1–2 mSv (De Cort *et al.*, 1998). For the vast majority of the population in the most affected areas, dose rates are currently expected to be less than this. Those individuals consuming significant quantities of forest products, however, may have much higher ingestion dose rates.

3.4.3 People now living in the abandoned areas

There are people who are living within settlements which are officially abandoned in all three affected fSU countries. People now living within abandoned areas produce most of their own foodstuffs (including potatoes and other vegetables, fruit, milk (cow and goat), meat (pork, beef and chicken) and eggs) and also collect wild foods such as edible fungi and berries (Beresford and Wright, 1999). They do not have access to radiocaesium binders to reduce transfer to animals and it is unlikely that they have access to mineral fertilisers that have the radiological benefit of reducing

Table 3.26. Example of consumption rates of different foodstuffs (on a fresh weight basis) and the contribution of each foodstuff to the daily ^{137}Cs intake, as determined during June/July 1997 in Milyach, the Ukraine.* For comparison, a radiocaesium intake rate of 77 kBq y^{-1} leads to an ingestion dose of 1 mSv y^{-1}.†

* Adapted from Beresford and Wright (1999). † From ICRP (1993).

Foodstuff	^{137}Cs activity concentration (kBq kg^{-1})	Consumption rate (kg y^{-1}) (f.w.)	^{137}Cs intake rate‡ (kBq y^{-1})
Milk	0.087	370	32
			8.0–330
Meat	0.058	37	2.1
			0.66–8.0
Potatoes	0.027	260	6.9
			2.4–16
Other vegetables	0.023	73	1.6
			0.47—6.6
Fungi	0.215	7	15.7
			0–280
Berries	0.456	3	1.5
			0–5.1
Fish	0.055	7	0.4
			0.11–5.5
Bread	0.022	180	4.0
			2.6–10
Total intake	0.079	990	79
			19–510

‡ Results are presented as median values and the range in intake rates is indicated in italics.
f.w. = fresh weight.

the root uptake of radiocaesium. Figure 3.14 presents the ^{137}Cs activity concentration in milk from cows owned by people living within the 30-km zone; care should be taken in interpreting any apparent temporal trends as the locations for which data are available differs between years. For some of the milk samples presented within Figure 3.14, corresponding ^{90}Sr values were also available; ^{90}Sr concentrations in milk ranged from 5–210 Bq kg^{-1}.

In areas which were abandoned during 1987–1991, countermeasures are likely to have been applied to reduce the transfer of radiocaesium to foodstuffs and possibly to reduce external doses during the period prior to abandonment. Some of these measures, such as ploughing or methods of reducing external dose, will result in a lower radiation exposure to people currently living within abandoned areas than would otherwise be the case. However, increased fertilisation and liming rates, applied as countermeasures prior to abandonment, are unlikely to be currently resulting in lower ^{137}Cs transfers to food products.

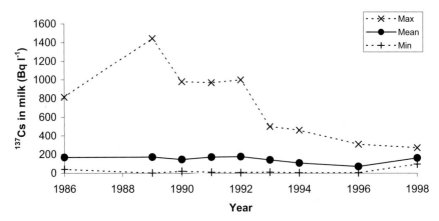

Figure 3.14. A summary of the ^{137}Cs activity concentration measured in the milk of cattle owned by people living within the 30-km exclusion zone. Total number of samples is 372 from 21 settlements; sampled settlements vary between years.
From Beresford and Wright (1999).

3.4.4 Radiation exposures via the forest pathway

Gathering fungi and berries (so called 'forest gifts') is an important cultural activity and forms a key part of the diet in the Ukraine, Belarus and Russia. External doses to both recreational forest users and forestry workers can also be significant. In a study in the Bryansk region of Russia, Fesenko *et al.* (2000) observed that external doses, due to recreational forest use, only added 3–10% to the average annual external dose. External doses to people working in the forest, however, were 1.7–2.5 times higher than the average external dose for the whole population (Fesenko *et al.*, 2000). This study also showed that, for forestry workers and people regularly consuming forest products (including milk from cows grazing in the forest), internal doses from consumption of forest products exceeded annual external doses as shown in Table 3.27.

In a survey of rural settlements within Belarus, Russia and the Ukraine, 40–75% of interviewees gathered and consumed wild fungi, 60–70% forest berries and 20–40% fish from local lakes (Strand *et al.*, 1996). Shutov *et al.* (1996) estimated that fungi and berries could contribute up to 60–70% of dietary ^{137}Cs intake of those adults within Russia. Indeed, within the rural population in the affected parts of Russia a mean increase in the whole body radiocaesium activity of 60–70% in autumn as a result of fungi consumption has been noted (Skuterud *et al.*, 1997). Consumption of fungi was positively correlated with ^{137}Cs in the body (Figure 3.15). Similar correlations were reported between the consumption of both forest berries and freshwater fish caught in local lakes and ^{137}Cs whole body measurements in two Russian settlements (Strand *et al.*, 1996). Forest edges may also be used to graze domestic animals such as cows and sheep which can lead to transfers of radioactivity to their milk.

Table 3.27. Mean effective dose[†] in 15 forest units in the Novozybokov district, Bryansk region, Russia.
Calculated from data presented in Fesenko et al. (2000).

Group	Effective dose (1987)			Effective dose (1996)		
	Total dose (mSv yr^{-1})	% Internal	% External	Total dose (mSv yr^{-1})	% Internal	% External
Population	15.3	72	28	2.6	67	33
Foresters	27.6	69	31	5.6	71	29

[†] These estimates assume that no countermeasures were in place: actual doses may be lower than these estimates, particularly in 1987 when restrictions were better adhered to by the population. Mean ^{137}Cs in the district was approximately 750,000 Bq m^{-2} (Fesenko et al., 2000).

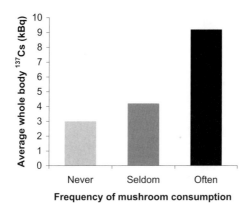

Figure 3.15. A comparison of the consumption rate of fungi (expressed qualitatively as frequency eaten) and the whole body ^{137}Cs burden determined in people ($n = 102$) living in an urban area of Russia (town of Klincy, population = 70,000–80,000) during July 1997.
From Mehli and Strand (1998).

Restrictions have been placed on hunting and the sale of meat from game and semi-domestic animals in some parts of western Europe (Åhman and Åhman, 1994) and in the Ukraine, Belarus and Russia. Restrictions varied between different countries. For example, the current limit for sale of reindeer meat in Norway is 3,000 Bq kg^{-1} whilst in Sweden it is 1,500 Bq kg^{-1}. In addition to limits for sale of meat, dietary advice has been given to groups whose diet includes a large proportion of reindeer meat. Despite restrictions in the Ukraine, Belarus and Russia, anecdotal evidence suggests that some illegal hunting continued in the Chernobyl exclusion zone after the accident.

Exposures from the collection and use of timber products can be significant (Ravila and Holm, 1994), but are believed to be less important than consumption of forest products and external exposures to people in the forest (Fesenko et al., 2000). Use of wood as a domestic and industrial heat (energy) source, and concentration of radioactivity in the resulting ash can lead to exposures: the concentration of radiocaesium in wood ash is 50–100 times greater than in the original wood. In some places, wood ash is used as a fertiliser for gardens, leading to a potential exposure pathway from transfer to vegetables. Radiocaesium is relatively weakly bioaccumulated in wood, so use of timber products for building is not likely to give rise to significant doses (IAEA, 2003). Timber is however a potential route for transport of radioactivity over long distances away from contaminated areas.

3.4.5 Time dependence of exposures

After the initial period (during which exposures were determined primarily by short-lived radionuclides, particularly ^{131}I in milk), long-lived radiocaesium and (to a much lesser extent) radiostrontium formed the vast majority of the dose. As discussed in Chapter 2, in intensive agricultural systems, activity concentrations of radiocaesium declined quite rapidly during the first five or so years after the accident due to redistribution within the soil column (by ploughing and vertical migration) and 'fixation' to the soil. Effective ecological half-times were typically of order 1–4 years during the period of approximately 0.5–5 years after fallout, leading to declines in activity concentrations in foodstuffs of approximately one order of magnitude. After this period (from, say, the mid-1990s onwards) these rates of decline reduced significantly so that effective ecological half-lives increased to between 6 and 30 years (the maximum being the physical half-life of ^{137}Cs) (Smith et al., 2000a).

Effective ecological half-times (T_{eff}) for radiocaesium were highly variable in forest and other semi-natural ecosystems (see, e.g., Table 3.20), but were generally longer than those observed in intensive agricultural and freshwater systems. In the period 1987–1996, effective half-lives for fungi, berries and game animals were of order of 10–30 years in some cases, though more rapid declines (of the order of 2–6 years) in radiocaesium content were also observed (Åhman and Åhman, 1994; Fesenko et al., 2000; Zibold et al., 2001).

The relatively slower reduction in ^{137}Cs content in forest and other semi-natural products compared to intensive agricultural systems meant that their importance in the formation of internal dose increased over time. Fesenko et al. (2000) stated that 'the relative contribution of fungi and berries to the internal doses of the population [in forest areas of the Bryansk region, Russia] varies from 10–15% in 1987 to 40–45% in 1996'. Overall, however, this study (Fesenko et al., 2000) showed that both external and internal doses declined significantly from 1987–1996 (Table 3.27), due mainly to decay of ^{134}Cs, redistribution within the soil column and reduction in ^{137}Cs contents in both agricultural and semi-natural produce. In some areas of the fSU it has been estimated that radiation doses in excess of $0.5\,\mathrm{mSv\,y^{-1}}$ could be received by average rate consumers of forest berries until 2050 and of mushrooms until 2150 (Beresford et al., 2001).

In the very long term (hundreds to thousands of years), it is possible that contamination of fungi and forest berries by Pu isotopes could be important in

Table 3.28. ^{137}Cs transfer factors (f.w. unless stated) and illustrations of activity concentrations of different foodstuffs from measurements made in the early 1990s. Note that measured T_{ag} values can vary considerably, and the values given here are for illustrative purposes only. The higher contamination density is illustrative of land in the 0.55–1.5 MBq m^{-2} contaminated zones in the Ukraine, Belarus and Russia, the lower is illustrative of the most contaminated areas in western Europe.

Foodstuff and site of T_{ag} estimate	Data source	Transfer factor (T_{ag}) m^2 kg^{-1}	Predicted ^{137}Cs in foodstuffs at different contamination densities	
			2×10^4 Bq m^{-2}	1×10^6 Bq m^{-2}
Grains, collective farm	(1)	0.1×10^{-3}	2 Bq kg^{-1}	100 Bq kg^{-1}
Potato, collective farm	(1)	0.05×10^{-3}	1 Bq kg^{-1}	50 Bq kg^{-1}
Milk, Dubrovitsa, Ukraine 1994–1995	(2)	0.15×10^{-3} Gley soil 3.7×10^{-3} Peat soil	3 Bq kg^{-1} 74 Bq kg^{-1}	150 Bq kg^{-1} 3,700 Bq kg^{-1}
Sheep meat, Norway, 1993	(3)	42.7×10^{-3}	854 Bq kg^{-1}	42,700 Bq kg^{-1}
Beef, intensive agriculture	(1)	0.88×10^{-3}	18 Bq kg^{-1}	880 Bq kg^{-1}
Beef, recommended T_{ag}, semi-natural system	(4)	6×10^{-3}	120 Bq kg^{-1}	6,000 Bq kg^{-1}
Wild mushroom *Suillus luteus*, 1994, Belarus	(5)	41.7×10^{-3}	834 Bq kg^{-1}	41,700 Bq kg^{-1}
Berries, *Vaccinium myrtillus* 1989–1994, Belarus	(5)	7.7×10^{-3}	154 Bq kg^{-1}	7,700 Bq kg^{-1}
Game animals	(6)	10×10^{-3}	200 Bq kg^{-1}	10,000 Bq kg^{-1}
Predatory fish, range of lakes of varying type and potassium conc., 1993–1997	(7)	1.0×10^{-3} Low 50×10^{-3} High	20 Bq kg^{-1} 1,000 Bq kg^{-1}	500 Bq kg^{-1} 50,000 Bq kg^{-1}
Birch, pine wood, 1992–1994 (d.w.)	(8)	0.25×10^{-3} Soddy–podzolic sandy soil 5×10^{-3} Peat soil	5 Bq kg^{-1} 100 Bq kg^{-1}	250 Bq kg^{-1} 5,000 Bq kg^{-1}
Tree leaves, 1991–1992 (d.w.)	(8)	2×10^{-3} Soddy–podzolic sandy soil 60×10^{-3} Peat soil	40 Bq kg^{-1} 1,200 Bq kg^{-1}	2,000 Bq kg^{-1} 60,000 Bq kg^{-1}

(1) Kryshev and Ryazantsev (2000); (2) Howard *et al*. (1996); (3) Dahlgaard (1994); (4) IAEA (1994); (5) Kenigsberg *et al*. (1996); (6) Representative value from Table 3.20; (7) Smith *et al*. (2000b); (8) Belli and Tikhomirov (1996).

some small areas of the 30-km zone. Realistically, though, it seems unlikely that Pu ingestion doses will ever be significant. By the time that ^{90}Sr and ^{137}Cs no longer dominate the dose (in one to two hundred years or so), erosion, vertical migration and soil fixation processes are likely to have reduced uptake of Pu to plants significantly.

3.4.6 Comparison of radiocaesium transfers to various products

Because of the wide variation in radiocaesium transfer factors and fallout levels, it is difficult accurately to summarise differences in activity concentrations in foodstuffs after Chernobyl. Table 3.28 illustrates this wide variation. Data was chosen to reflect transfer factors in the early-mid 1990's in order to minimise variation over time. The data highlight the importance of management practices (contrast between produce from collective and private farms), relatively high activity concentrations in semi-natural products, and the importance of factors such as soil type and potassium concentration.

3.5 REFERENCES

Åhman, B. and Åhman, G. (1994) Radiocesium in Swedish reindeer after the Chernobyl fallout: Seasonal variations and long-term decline. *Health Physics*, **66**, 503–512.

Alexakhin, R., Firsakova, S., Rauret, G., Arkhipov, N., Vandecasteele, C.M., Ivanov, Y., Fesenko, S. and Sanzharova, N. (1996) Fluxes of radionuclides in agricultural environments. In: Karaoglou, A., Desmet, G., Kelly, G.N. and Menzel, H.G. (eds). *The radiological Consequences of the Chernobyl Accident*, pp. 39–47. European Commission, Brussels.

Anspaugh, L.R., Simon, S.L., Gordeev, K.I., Likhtarev, I.A., Maxwell, R.M. and Shinkarev, S.M. (2002) Movement of radionuclides in terrestrial ecosystems by physical processes. *Health Physics*, **82**, 669–679.

Aslanoglou, X., Assimakopoulos, P.A., Averin, V., Howard, B.J., Howard, D.C., Karamanis, D.T. and Stamoulis, K. (1996) Impact of the Chernobyl accident on a rural population in Belarus. In: Karaoglou, A., Desmet, G., Kelly, G.N. and Menzel, H.G. (eds), *The Radiological Consequences of the Chernobyl Accident*, pp. 363–378. European Commission, Brussels.

Assimakopoulos, P.A., Ioannides, K.G., Karamanis, D., Lagoyannis, A., Pakou, A.A., Koutsotolis, K., Nikolaou, E., Arkhipov, A., Arkhipov, N., Gaschak, S., *et al.* (1995) Ratios of transfer coefficients for radiocesium transport in ruminants. *Health Physics*, **69**, 410–414.

Balonov, M., Anisimova, L.I. and Perminova, G.S. (1999) Strategy of population protection and area rehabilitation in Russia in the remote period after the Chernobyl accident. *Journal of Radiological Protection*, **19**, 261–269.

Barnett, C.L., Beresford, N.A., Self, P.L., Howard, B.J., Frankland, J.C., Fulker, M.J., Dodd, B.A. and Marriott, J.V.R. (1999) Radiocaesium activity concentrations in the fruit-bodies of macrofungi in Great Britain and an assessment of dietary intake habits. *The Science of the Total Environment*, **231**, 67–83.

Belli, M. and Tikhomirov, F. (eds) (1996) *Behaviour of radionuclides in natural and semi-natural environments*. Report EUR16531, European Commission, Luxembourg.

Belli, M., Bunzl, K., Delvank, B., Gerzabeck, M., Rafferty, B., Shaw, G. and Wirth, E. (2000) *SEMINAT – Long-term dynamics of radionuclides in semi-natural environment: derivation of parameters and modelling*. Agenzia Nazionale per la protezione dell'Ambiente, Rome. ISBN 88-448-0286-4.

Beresford, N.A. (1989a) The transfer of Ag-110m to sheep tissues. *Science of the Total Environment*, **85**, 81–90.

Beresford, N.A. (1989b) Field observations of ^{110m}Ag, originating from the Chernobyl accident, in west Cumbrian vegetation and soil samples. *Journal of Radiological Protection*, **9**, 281–3.

Beresford, N.A., Lamb, C.S., Mayes, R.W., Howard B.J. and Colgrove, P.M. (1989) The effect of treating pastures with bentonite on the transfer of Cs-137 from grazed herbage to sheep. *Journal of Environmental Radioactivity*, **9**, 251–264.

Beresford, N.A. and Howard, B.J. (1991) The importance of soil adhered to vegetation as a source of radionuclides ingested by grazing animals. *Science of the Total Environment*, **107**, 237–254.

Beresford, N.A., Mayes, R.W., Howard, B.J., Eayres, H.E., Lamb, C.S., Barnett, C.L. and Segal, M.G. (1992) The bioavailability of different forms of radiocaesium for transfer across the gut of ruminants. *Radiation Protection Dosimetry*, **41**, 87–91.

Beresford, N.A., Barnett, C.L., Crout, N.M.J. and Morris, C. (1996) Radiocaesium variability within sheep flocks: Relationships between the ^{137}Cs activity concentrations of individual ewes within a flock and between ewes and their progeny. *Science of the Total Environment*, **177**, 85–96.

Beresford, N.A., Mayes, R.W., Hansen, H.S., Crout, N.M.J., Hove, K. and Howard, B.J. (1998) Generic relationship between calcium intake and radiostrontium transfer to milk of dairy ruminants. *Radiation and Environmental Biophysics*, **37**, 129–131.

Beresford, N.A. and Wright S.M. (eds) (1999) *Self-help Countermeasure Strategies for Populations Living within Contaminated Areas of the Former Soviet Union and an Assessment of Land Currently Removed from Agricultural Usage*. Institute of Terrestrial Ecology: Grange-over-Sands, U.K., 82 pp.

Beresford, N.A., Mayes, R.W., Colgrove, P.M., Barnett, C.L., Bryce, L., Dodd, B.A. and Lamb, C.S. (2000a). A comparative assessment of the potential use of alginates and dietary calcium manipulation as countermeasures to reduce the transfer of radiostrontium to the milk of dairy animals. *Journal of Environmental Radioactivity*, **51**, 321–342.

Beresford, N.A., Mayes, R.W., Cooke, A.I., Barnett, C.L., Howard, B.J., Lamb, C.S. and Naylor, G.P.L. (2000b) The importance of source dependent bioavailability in determining the transfer of ingested radionuclides to ruminant derived food products. *Environmental Science and Technology*, **34**, 4455–4462.

Beresford, N.A., Gashchak, S., Lasarev, N., Arkhipov, A., Chyorny, Y., Astasheva, N., Arkhipov, N., Mayes, R.W., Howard, B.J., Baglay, G., Loginova, L. and Burov, N. (2000c) The transfer of ^{137}Cs and ^{90}Sr to dairy cattle fed fresh herbage collected 3.5 km from the Chernobyl nuclear power plant. *Journal of Environmental Radioactivity*, **47**, 157–170.

Beresford, N.A., Voigt, G., Wright, S.M., Howard, B.J., Barnett, C.L., Prister, B., Balonov, M., Ratnikov, A., Travnikova, I., Gillett, A.G., *et al.* (2001) Self-help countermeasure strategies for populations living within contaminated areas of Belarus, Russia and the Ukraine. *Journal of Environmental Radioacititvty*, **56**, 215–239.

Beresford, N. A., Mayes, R.W., Barnett, C.L. and Lamb, C.S. (2002a) Dry Matter Intake – A generic approach to predict the transfer of radiocaesium to ruminants? *Radioprotection – colloques*, **37**, 373–378.

Beresford, N.A., Barnett, C.L., Coward, P.A., Howard, B.J. and Mayes, R.W. (2002b) A simple method for the estimation of the bioavailability of radiocaesium from herbage contaminated by adherent soil. *Journal of Environmental Radioactivity*, **63**, 77–84.

Beresford, N.A. (2003) Does size matter? *Proceedings of the international conference on the protection of the environment from the effects of ionizing radiation*, Stockholm, Sweden, pp. 182–185. International Atomic Energy Agency, IAEA-CN-109, Vienna.

Bergman R., Nylén T., Palo, T. and Lidström, K. (1991) The behaviour of radioactive caesium in a boreal ecosystem. In: Moberg, L. (ed.), *The Chernobyl Fallout in Sweden*. Arprint, Stockholm, ISBN 91-630-0721-5.

Boccolini, A., Gentili, A., Guidi, P., Sabbatini, V. and Toso, A. (1988) Observations of silver-110 m in the marine mollusc Pinna noblis. *Journal of Environmental Radioactivity*, **6**, 191–193.

Bretten, S., Gaare, E., Skogland, T. and Steinnes, E. (1992) Investigations of radiocaesium in the natural terrestrial environment in Norway following the Chernobyl accident. *Analyst*, **117**, 501–503.

Bunzl, K., Kracke, W., Schimmack, W. and Zelles, L. (1998) Forms of fallout ^{137}Cs and $^{239+240}$Pu in successive horizons of a forest soil. *Journal of Environmental Radioactivity*, **39**, 55–68.

Bunzl, K. (2002) Transport of fallout radiocaesium in the soil by bioturbation: A random walk model and application to a forest soil with a high abundance of earthworms. *Science of the Total Environment*, **293**, 191–200.

Byrne, A.R. (1988) Radioactivity in fungi in Slovenia, Yugoslavia, following the Chernobyl accident. *Journal of Environmental Radioactivity*, **6**, 177–183.

Cambray, R.S., Cawse, P.A., Garland, J.A., Gibson, J.A.B., Johnson, P., Lewis, G.N.J., Newton, D., Salmon, L. and Wade, B.O. (1987) Observations on radioactivity from the Chernobyl accident. *Nuclear Energy*, **26**, 77–101.

Carini, F. (2001) Radionuclide transfer from soil to fruit. *Journal of Environmental Radioactivity*, **52**, 237–279.

Chamberlain, A.C. and Chadwick, R.C. (1966) Transport of iodine from atmosphere to ground. *Tellus*, **18**, 226–237.

Chamberlain, A.C. (1970) Interception and retention of radioactive aerosols by vegetation. *Atmospheric Environment*, **4**, 57–58.

Colgan, T. (1992) *Chernobyl Radioactivity in Irish Mountain Sheep*. Irish Food Science Review. January 1992, pp. 5–9.

Comar, C.L., Wasserman, R.H. and Lengemann, F.W. (1966) Effect of dietary calcium on secretion of strontium into milk. *Health Physics*, **12**, 1–6.

Coughtrey, P.J. and Thorne, M.C. (1983) *Radionuclide Distribution and Transport in Terrestrial and Aquatic Ecosystems. A critical review of data* (Volume 1). A.A. Balkema, Rotterdam.

Cremers, A., Elsen A., De Preter, P. and Maes, A. (1988) Quantitative analysis of radiocaesium retention in soils. *Nature*, **335**, 247–249.

Cross, M.A. (2001) Mathematical modelling of ^{90}Sr and ^{137}Cs in terrestrial and freshwater environments. Ph.D. thesis, University of Portsmouth, UK.

Crout, N.M.J., Beresford, N.A., Mayes, R.W., MacEachern, P.J., Barnett, C.L., Lamb, C.S. and Howard, B.J. (2000) A model for radioiodine transfer to goat milk incorporating the influence of stable iodine. *Radiation and Environmental Biophysics*, **39**, 59–65.

Dahlgaard, H. (1994) *Nordic radioecology: The transfer of radionuclides through Nordic ecosystems to man.* Elsevier, Amsterdam.

De Cort, M., Fridman, Sh.D., Izrael, Yu.A., Jones, A.R., Kelly, G.N., Kvanikova, E.V., Matveenko, I.I., Nazarov, I.M., Stukin, E.D., Tabachny, L.Ya. and Tsaturov, Yu.S. (1998) *Atlas of Caesium Deposition on Europe after the Chernobyl Accident.* EUR 16733, EC, Office for Official Publications of the European Communities, Luxembourg.

Desmet, G., Nassimbeni, P. and Belli, M. (eds) (1990) *Transfer of radionuclides in natural and semi-natural ecosystems.* Elsevier Applied Science, London.

Drozdovitch, V.V., Goulko, G.M., Minenko, V.F., Paretzke, H.G., Voigt, G. and Kenigsberg, J.I. (1997) Thyroid dose reconstruction for the population of Belarus after the Chernobyl accident. *Radiation and Environmental Biophysics,* **36**, 17–23.

Eisler, R. (2003) The Chernobyl Nuclear Power Plant reactor accident: Ecotoxicological update. In: Hoffman, D.J., Rattner, B.A., Burton, G.A. Jr. and Cairns, J. Jr. (eds). *Handbook of Ecotoxicology* (2nd edition), pp. 702–736. CRC Press, Boca Raton.

Eriksson, O., Gaichenko, V., Gashchak, S., Jones, B., Jungskär, W., Chizevsky, I., Kurman, A., Panov, G., Ryabtsev, I., Shcherbatchenko, A., et al. (1996) Evolution of the contamination rate in game. In: Karaoglou, A., Desmet, G., Kelly, G.N. and Menzel, H.G. (eds), *The Radiological Consequences of the Chernobyl Accident,* pp. 147–154. European Commission, Brussels.

Fabbri, S., Piva, G., Sogni, R., Fusconi, G., Lusardi, E. and Borasi, G. (1994) Transfer kinetics and coefficients of ^{90}Sr, ^{134}Cs and ^{137}Cs from forage contaminated by Chernobyl fallout to milk of cows. *Health Physics,* **66**, 375–379.

Fesenko, S.V., Voigt, G., Spiridonov, S.I., Sanzharova, N.I., Gontarenko, I.A., Belli, M. and Sansone, U. (2000) Analysis of the contribution of forest pathways to the radiation exposure of different population groups in the Bryansk region of Russia. *Radiation and Environmental Biophysics,* **39**, 291–300.

Fesenko, S.V., Soukhova, N.V., Sanzharova, N.I., Avila, R., Spiridonov, S.I., Klein, D. and Badot, P-M. (2001) ^{137}Cs availability for soil to understorey transfer in different types of forest ecosystems. *The Science of the Total Environment,* **269**, 87–103.

Firsakova, S., Hove K., Alexakhin, R., Prister, B., Arkhipov, N. and Bogdanov, G. (1996). Countermeasures implemented in intensive agriculture. In: Karaoglou, A., Desmet, G., Kelly, G.N. and Menzel, H.G. (eds). *The Radiological Consequences of the Chernobyl Accident,* pp. 379–387. European Commission, Brussels.

Frissel, M.J., Noordijk, H., and Van Bergeijk, K.E. (1990) The impact of extreme environmental conditions, as ocurring in natural ecosystems, on the soil-to-plant transfer of radionuclides. In: Desmet, G., Nassimbeni, P. and Belli, M. (eds). *Transfer of Radionuclides in Natural and Semi-natural Ecosystems,* pp. 40–47. Elsevier Applied Science, London.

Frissel, M.J. (2001) *The Classification of Soil Systems on the Basis of Transfer Factors of Radionuclides from Soil to Reference Plants.* Report of the Second FAO/IAEA Research Coordination Meeting (RCM) held in Vienna, Austria from 12–16 March 2001. IAEA-319-D6-RC-727.2, IAEA, Vienna.

Frissel, M.J., Deb, D.L., Fathony, M., Lin, Y.M., Mollah, A.S., Ngo, N.T., Othman, I., Robison, W.L., Skarlou-Alexiou, V., Topcuoğlou, S., Twining, J.R., et al. (2002) Generic values for soil-to-plant transfer factors of radiocaesium. *Journal of Environmental Radioactivity,* **58**, 113–128.

Gillett, A.G. and Crout, N.M.J. (2000) A review of ^{137}Cs transfer to fungi and consequences for modelling environmental transfer. *Journal of Environmental Radioactivity,* **48**, 95–121.

Hansen, H.S. and Hove, K. (1991) Radiocesium bioavailability: Transfer of Chernobyl and tracer radiocesium to goat milk. *Health Physics*, **60**, 665–673.

Hinton, T.G., Kopp, P., Ibrahim, S., Bubryak, I., Syomov, A., Tobler, L. and Bell, C. (1995) A comparison of techniques used to estimate the amount of resuspended soil on plant surfaces. *Health Physics*, **68**, 523–531.

Hinton, T.G., McDonald, M., Ivanov, Y., Arkhipov, N. and Arkhipov, A. (1996) Foliar absorption of resuspended ^{137}Cs relative to other pathways of plant contamination. *Journal of Environmental Radioactivity*, **30**, 15–30.

Hinton, T.G. and Pinder, J.E. (2001) A review of plutonium releases from the Savannah River Site. In: Kudo, A. (ed.), *Plutonium in the Environment*, pp. 413–435. Elsevier, Amsterdam.

Horrill, A.D. and Howard, D.M. (1991) Chernobyl fallout in three areas of upland pasture in West Cumbria. *Journal of Radiological Protection*, **11**, 249–257.

Hove, K., Garmo, T.H., Hansen, H.S., Pedersen, Ø., Staaland, H. and Strand, P. (1992) Duration and variation in radiocaesium content of products from domestic animals grazing natural pasture. *Information from the Norwegian Agricultural Advisory Service*, **13**, 256–268.

Howard, B.J., Beresford, N.A., Burrow, L., Shaw, P.V. and Curtis, E.J.C. (1987) A comparison of caesium 137 and 134 activity in sheep remaining on upland areas contaminated by Chernobyl fallout with those removed to less active lowland pasture. *Journal of the Society of Radiological Protection*, **7**, 71–73.

Howard, B.J., Beresford, N.A. and Hove, K. (1991) Transfer of radiocaesium to ruminants in natural and semi-natural ecosystems and appropriate countermeasures. *Health Physics*, **61**, 715–725.

Howard, B.J., Beresford, N.A., Mayes, R.W. and Lamb, C.S. (1993) Transfer of ^{131}I to sheep milk from vegetation contaminated by the Chernobyl fallout. *Journal of Environmental Radioactivity*, **19**, 155–161.

Howard, B.J. and Beresford, N.A. (1994) Radiocaesium contamination of sheep in the United Kingdom after the Chernobyl accident. In: Ap Dewi, I., Axford, R.F.E., Fayez, M., Marai, I. and Omed, H. (eds), *Pollution in Livestock Production Systems*, pp. 97–118. CAB International, Wallingford, UK.

Howard, B.J., Hove, K., Prister, B., Ratnikov, A., Travnikova, I., Averin, V., Pronevitch, V., Strand, P., Bogdanov, G. and Sobolev, A. (1996) Fluxes of radiocaesium to milk and appropriate countermeasures. In: Karaoglou, A., Desmet, G., Kelly, G.N. and Menzel, H.G. (eds), *The Radiological Consequences of the Chernobyl Accident*, pp. 349–362. European Commission, Brussels.

Howard, B.J., Beresford, N.A., Mayes, R.W., Hansen, H.S., Crout, N.M.J. and Hove, K. (1997) The use of dietary calcium intake of dairy ruminants to predict the transfer coefficient of radiostrontium to milk. *Radiation and Environmental Biophysics*, **36**, 39–43.

Howard, B.J., Beresford, N.A. and Voigt, G. (2001). Countermeasures for animal products: A review of effectiveness and potential usefulness after an accident. *Journal of Environmental Radioactivity*, **56**, 115–137.

IAEA (1994) *Handbook of Transfer Parameter Values for the Prediction of Radionuclide Transfer in Temperate Environments*. Technical Report Series No. 364. International Atomic Energy Agency, Vienna.

IAEA (2002) *Report of the Forest Working Group of the BIOMASS Programme*. IAEA, Vienna.

IAEA (2003) *Assessing Radiation Doses to the Public from Radionuclides in Tmber and Wood Products.* IAEA-TECDOC-1376, IAEA, Vienna. ISBN 92-0-110903-2.
ICRP (1993). *Age-dependent Doses to Members of the Public from Intake of Radionuclides. Part 2: Ingestion Dose Coefficients.* International Commission on Radiological Protection: Publication 67. Pergamon, Oxford.
Jacobs, M.B. (1958) *Chemical Analyses of Foods and Food Products* (3rd Edition). D. van Nostrand Company Inc., Princeton.
Karlén, G., Johanson, K.J. and Bertilsson, J. (1995) Transfer of ^{137}Cs to cow's milk: Investigations on dairy farms in Sweden. *Journal of Environmental Radioactivity*, **28**, 1–15.
Kashparov, V.A., Lundin, S.M., Zvarych, S.I., Yoshchenko, V.I., Levchuk, S.E., Khomutinin, Y.V., Maloshtan, I.M. and Protsak, V.P. (2003) Territory contamination with the radionuclides representing the fuel component of Chernobyl fallout. *Science of the Total Environment*, **317**, 105–119.
Kenigsberg, J., Belli, M., Tikhomirov, F., Buglova, E., Shevchuk, V., Renaud, Ph., Maubert, H., Bruk, G. and Shutov, V. (1996) Exposures from consumption of forest produce. In: Karaoglou, A., Desmet, G., Kelly, G.N. and Menzel, H.G. (eds). *The Radiological Consequences of the Chernobyl Accident*, pp. 271–281. European Commission, Brussels.
Kenik, I., Rolevich, I., Ageets, V., Gurachesky, V. and Poplyko, I. (1999) Long-term strategy of rehabilitation of Belarusian territories contaminated by radionuclides. *Radioprotection*, **34**, 13–24.
Kirchner, G. (1994) Transport of iodine and cesium via the grass–cow–milk pathway after the Chernobyl accident. *Health Physics*, **66**, 653–665.
Klemt, E., Drissner, J., Kaminski, S., Miller, R., and Zibold, G. (1998) Time dependency of the bioavailability of radiocaesium in lakes and forests. In: Linkov, I. and Schell, W.R. (eds), *Contaminated Forests*, pp. 95–101. Kluwer, Dordrecht.
Korobova, E., Ermakov, A. and Linnik, V. (1998) ^{137}Cs and ^{90}Sr mobility in soils and transfer in soil–plant systems in the Novozybkov district affected by the Chernobyl accident. *Applied Geochemistry*, **13**, 803–814.
Kryshev, I.I. and Ryazantsev, E.P. (2000) *Ecological Safety of Nuclear Energy Complexes of Russia*, 383 pp. Izdat, Moscow (in Russian).
Lee, M.H and Lee, C.W. (2000) Association of fallout-derived ^{137}Cs, ^{90}Sr and ^{239,240}Pu with natural organic substances in soils. *Journal of Environmental Radioactivity*, **47**, 253–262.
Long, S., Pollard, D., Cunningham, J.D., Astasheva, N.P., Donskaya, G.A. and Labetsky, E.V. (1995) The effects of food processing and direct decontamination techniques on the radionuclide content of foodstuffs: A literature review. *Journal of Radioecology*, **3**, 15–30.
Lux, D., Kammerer, L., Rühm, W. and Wirth, E. (1995) Cycling of Pu, Sr, Cs, and other long-living radionuclides in forest ecosystems of the 30 km zone around Chernobyl. *Science of the Total Environment*, **173/4**, 375–384.
Malek, M.A., Hinton, T.G. and Webb, S.B. (2002) A comparison of ^{90}Sr and ^{137}Cs uptake in plants via three pathways at two Chernobyl-contaminated sites. *Journal of Environmental Radioactivity*, **58**, 129–141.
Mamikhin, S.V., Tikhomirov, F.A. and Shcheglov, A.I. (1997). Dynamics of ^{137}Cs in the forests of the 30-km zone around the Chernobyl nuclear power plant. *Science of the Total Environment*, **193**, 169–177.
Martin, C.J., Heaton, B. and Robb, J.D. (1988) Studies of ^{131}I, ^{137}Cs and ^{103}Ru in milk, meat and vegetables in North East Scotland following the Chernobyl accident. *Journal of Enviornmental Radioactivity*, **6**, 247–259.

Martin, C.J., Heaton, B. and Thompson, J. (1989) Cesium-137, ^{134}Cs and ^{110m}Ag in lambs grazing pasture in NE Scotland contaminated by Chernobyl fallout. *Health Physics*, **56**, 459–464.

Mehli, H. and Strand, P. (eds) (1998) *RECLAIM – Time-dependent Optimalisation of Strategies for Countermeasures Use to Reduce Population Radiation Dose and Reclaim Abandoned Land.* First progress report: EU contract no. ERBIC15-CT96-0209. Norwegian Radiation Protection Authority, Østerås.

Melin, J., Wallberg, L. and Suomela, J. (1994) Distribution and retention of cesium and strontium in Swedish boreal forest ecosystems. *Science of the Total Environment*, **157**, 93–105.

Mietelski, J.W., Baeza, A.S., Guillen, J., Buzinny, M., Tsigankov, N., Gaca, P., Jasinska, M. and Tomankiewicz, E. (2002) Plutonium and other alpha emitters in fungi from Poland, Spain and Ukraine. *Applied Radiation and Isotopes*, **56**, 717–729.

Moberg, L., Hubbard, L., Avila, R., Wallberg, L., Feoli, E., Scimone, M., Milesi, C., Mayes, B., Iason, G., Rantavaara, A., et al. (1999) *An Integrated Approach to Radionuclide Flow in Semi-natural Ecosystems Underlying Exposure Pathways to Man.* Final report of the LANDSCAPE project. SSI Report 99:19, Swedish Radiation Protection Authority, Stockholm.

Mück, K., Sinojmeri, M., Whilidal, H. and Steger, F. (2001) The long-term decrease of ^{90}Sr availability in the environment and its transfer to man after a nuclear fallout. *Radiation Protection Dosimetry*, **94**, 251–259.

Mück, K. (2003) Sustainability of radiologically contaminated territories. *Journal of Environmental Radioactivity*, **65**, 109–130.

Nisbet, A.F., Mercer, J.A., Hesketh, N., Liland, A., Thorring, H., Bergan, T., Beresford, N.A., Howard, B.J., Hunt, J. and Oughton, D. H. (2004) *Datasheets on countermeasures and waste disposal options for the management of food production systems contaminated following a nuclear accident.* NRPB-W58, National Radiological Protection Board, Didcot.

Nylén, T. (1996) Uptake, turnover and transport of radiocaesium in boreal forest ecosystems. Ph.D. thesis, Swedish University of Agricultural Sciences, Uppsala Sweden, ISBN 91-576-5149-3.

Ould-Dada, Z., Copplestone, D., Toal, M. and Shaw, G. (2002) Effect of forest edges on deposition of radioactive aerosols. *Atmospheric Environment*, **36**, 5595–5606.

Paasikallio, A., Rantavaara, A. and Sippola, J. (1994) The transfer of cesium-137 and strontium-90 from soil to food crops after the Chernobyl accident. *Science of the Total Environment*, **155**, 109–124.

Papastefanou, C., Manolopoulou, M., Stoulos, S. and Ioannidou, A. (1991) Seasonal variations of ^{137}Cs content of milk after the Chernobyl accident. *Health Physics*, **61**, 889–891.

Pfau, A.A., Fischer, R., Heinrich, H.C. and Handl, J. (1989) Radiosilver ^{110m}Ag from Chernobyl and its transfer from plant to ruminants. XIXth ENSA-Annual Meeting. Vienna, August 29–September 2, 1988. IAEA, Vienna.

Pietrzak-Flis, Z., Krajewski, P., Krajewska, G. and Sunderland, N.R. (1994) Transfer of radiocesium from uncultivated soils to grass after the Chernobyl accident. *Science of the Total Environment*, **141**, 147–153.

Pomeroy, I.R., Nicholson, K.W. and Branson, J.R. (1996) *Monitoring Chernobyl radiocaesium in Cumbrian grass, 1992–1995.* AEA Technology report AEAT/18142002/REMA-197. AEA Technology, Abingdon.

Prister, B.S., Sobolev, A.S., Bogdanov, G.A., Los, I.P., Howard, B.J. and Strand, P. (1996) Estimation of the balance of radiocaesium in the private farms of Chernobyl zone and countermeasures with regard to the reduction of health risk for rural inhabitants. In: Karaoglou, A., Desmet, G., Kelly, G.N. and Menzel, H.G. (eds), *The Radiological Consequences of the Chernobyl Accident*, pp. 319–321. European Commission, Brussels.

Prister, B., Alexakhin, R., Firsakova, S. and Howard, B.J. (1998) Short and long term environmental assessment. *Proceedings of the EC/CIS workshop on restoration of contaminated territories resulting from the Chernobyl accident 29–30 June 1998*, Brussels.

Pröhl, G. and Hoffman, F.O. (1996) Radionuclide interception and loss processes in vegetation. In: *Modelling of Radionuclide Interception and Loss Processes in Vegetation and Transfer in Semi-natural Ecosystems*. IAEA TECDOC-857. International Atomic Energy Agency, Vienna.

Pröhl, G., Mück, K., Likhtarev, I., Kovgan, L. and Golikov, V. (2002) Reconstruction of the ingestion doses received by the population evacuated from the settlements in the 30-km zone around the Chernobyl reactor. *Health Physics*, **82**, 173–181.

Rantavaara, A.H. (1990) Transfer of radiocaesium through natural ecosystems to foodstuffs of terrestrial origin in Finland. In: Desmet, G., Nassimbeni, P. and Belli, M. (eds), *Transfer of Radionuclides in Natural and Semi-natural Environments*, pp. 202–209. Elsevier, London.

Ravila, A. and Holm, E. (1994) Radioactive elements in the forest industry. *Science of the Total Environment*, **32**, 339–356.

Riekkinen, I. and Jaakkola, T. (2001) Effect of industrial pollution on soil-to-plant transfer of plutonium in a Boreal forest. *Science of the Total Environment*, **278**, 161–170.

Rosén, K., Andersson, I. and Lönsjö, H. (1995) Transfer of radiocaesium from soil to vegetation and to grazing lambs in a mountain area in Northern Sweden. *Journal of Environmental Radioactivity*, **26**, 237–257.

Rosén, K., Eriksson, Å. and Haak, E. (1996) Transfer of radiocaesium in sensitive agricultural environments after the Chernobyl fallout in Sweden. *Science of the Total Environment*, **182**, 117–133.

Rühm, W., Steiner, M., Wirth, E., Dvornik, A., Zhuchenko, T., Kliashtorin, A., Rafferty, B., Shaw, G. and Kuchma, N. (1996) Dynamic of radionuclides behaviour in forest soils. In: Karaoglou, A., Desmet, G., Kelly, G.N. and Menzel, H.G. (eds), *The Radiological Consequences of the Chernobyl Accident*, pp. 225–228. European Commission, Brussels.

Rühm, W., Yoshida, S., Muramatsu, Y., Steiner, M. and Wirth, E. (1999) Distribution patterns for stable 133Cs and their implications with respect to the long-term fate of radioactive ^{134}Cs and ^{137}Cs in a semi-natural ecosystem. *Journal of Environmental Radioactivity*, **45**, 253–270.

Ruiz, X., Jover, L., Lloerente, G.A., Sanchez-Reyes, A.F. and Febrian, M.I. (1987) Song thrushes *Turdus philomelos* wintering in Spain as biological indicators of the Chernobyl accident. *Ornis Scandinavica*, **19**, 132–141.

Sandalls, J. and Bennett, L. (1992) Radiocaesium in upland herbage in Cumbria, UK: A three year field study. *Journal of Environmental Radioactivity*, **16**, 147–165.

Sanzharova, N.I., Fesenko, S.V., Alexakhin, R.M., Anisimov, V.S., Kuznetsov, V.K. and Chernyayeva, L.G. (1994) Changes in the forms of ^{137}Cs and its availability for plants as dependent on properties of fallout after the Chernobyl nuclear power plant accident. *Science of the Total Environment*, **154**, 9–22.

SCOPE (1993) *Radioecology after Chernobyl: Biogeochemical Pathways of Artificial Radionuclides*, 367 pp. Warner, F. and Harrison, R.M. (eds). SCOPE-50. Wiley, Chichester.

Selnaes, T.D. and Strand, P. (1992) Comparison of the uptake of radiocaesium from soil to grass after nuclear weapons tests and the Chernobyl accident. *Analyst*, **117**, 493–496.

Shaw, G., Avila, R., Fesenko, S., Dvornik, A. and Zhuchenko, T. (2003) Modelling the behaviour of radiocaesium in forest ecosystems. In: Scott, E.M. (ed.), *Modelling Radio-Activity in the Environment*, pp. 315–351. Elsevier, Amsterdam.

Sheppard, S.C. and Evenden, W.G. (1997) Variation in transfer factors for stochastic models: Soil-to-plant transfer. *Health Physics*, **72**, 727–733.

Shcheglov, A. I., Tsvetnova, O.B. and Klyashtorin, A.L. (2001) *Biogeochemical Migration of Technogenic Radionuclides in Forest Ecosystems*. Nauka, Moscow (in English). ISBN 5-02-022568-1.

Shutov, V.N., Bruk, G.Ya, Basalaeva, L.N., Vasilevitskiy, V.A., Ivanova, N.P. and Kaplun, I.S. (1996). The role of fungi and berries in the formulation of internal exposure doses to the population of Russia after the Chernobyl accident. *Radiation Protection Dosimetry*, **67**, 55–64.

Skuterud, L., Balanov, M., Travnikova, I., Strand, P. and Howard, B.J. (1997). Contribution of fungi to radiocaesium intake of rural populations in Russia. *Science of the Total Environment*, **193**, 237–242.

Smith, F.B. and Clark, M.J. (1989) *The Transport and Deposition of Airborne Debris from the Chernobyl Nuclear Power Plant Accident*, 56 pp. Meteorological Office Scientific Paper No. 42, HMSO.

Smith, J.T., Fesenko, S.V., Howard, B.J., Horrill, A.D., Sanzharova, N.I., Alexakhin, R.M., Elder, D.G. and Naylor, C. (1999). Temporal change in fallout ^{137}Cs in terrestrial and aquatic systems: a whole-ecosystem approach. *Environmental Science and Technology*, **33**, 49–54.

Smith, J.T., Comans, R.N.J., Beresford, N.A., Wright, S.M., Howard, B.J. and Camplin, W.C. (2000a) Chernobyl's legacy in food and water. *Nature*, **405**, 141.

Smith, J.T., Kudelsky, A.V., Ryabov, I.N. and Hadderingh, R.H. (2000b) Radiocaesium concentration factors of Chernobyl-contaminated fish: a study of the influence of potassium and 'blind' testing of a previously developed model. *Journal of Environmental Radioactivity*, **48**, 359–369.

Smith, J.T. and Beresford, N.A. (2003) Radionuclides in food: The post-Chernobyl evidence. In: D'Mello, J.P.F. (ed.), *Food Safety*, pp. 373–390. CABI Publishing, Wallingford, UK.

Sokolik, G.A., Ovsiannikova, S.V., Ivanova, T.G. and Leinova, S.L. (2004) Soil–plant transfer of plutonium and americium in contaminated regions of Belarus after the Chernobyl catastrophe. *Environment International*, **30**, 939–947.

Soukhova, N.V., Fesenko, S.V., Klein, D., Spiridonov, S.I., Sanzharova, N.I. and Badot, P.M. (2003) ^{137}Cs distribution among annual rings of different tree species contaminated after the Chernobyl accident. *Journal of Environmental Radioactivity*, **65**, 19–28.

Spezzano, P. and Giacomelli, R. (1991) Transport of ^{131}I and ^{137}Cs from air to cows' milk produced in North-Western Italian farms following the Chernobyl accident. *Journal of Environmental Radioactivity*, **13**, 235–250.

Squire, H.M. (1966) Long-term studies of strontium-90 in soils and pastures. *Radiation Botany*, **6**, 49–67.

Squire, H.M. and Middleton, L.J. (1966) Behaviour of ^{137}Cs in soils and pastures. A long term experiment. *Radiation Botany*, **6**, 413–423.

Strand, P., Howard, B.J. and Averin, V. (eds) (1996) *Transfer of Radionuclides to Animals, their Comparative Importance Under Different Agricultural Ecosystems and Appropriate Countermeasures*. Experimental collaboration project No. 9. Final Report EUR 16539EN. European Commission, Luxembourg.

Strebl, F., Gerzabek, M.H., Karg, V. and Tataruch, F. (1996) ^{137}Cs migration in soils and its transfer to roe deer in an Austrian forest stand. *Science of the Total Environment*, **181**, 237–247.

Thiry, Y., Somber, L., Ronneau, C., Myttenaere, C., Kutlahmedov, Y. and Davidchuk, V.S. (1990) Behaviour of Cs-137 in forested polygons of the Chernobyl contamination zone. *Proceedings of the All-Union Conference: Geochemical pathways of artificial radionuclides in the biosphere*, Gomel, USSR.

Tikhomirov, F.A. and Shcheglov, A.I. (1994) Main investigation results on the forest radioecology in the Kyshtym and Chernobyl accident zones. *Science of the Total Environment*, **157**, 45–57.

Tracy, B.L., Walker, W.B. and McGregor, R.G. (1989) Transfer to milk of ^{131}I and ^{137}Cs released during the Chernobyl reactor accident. *Health Physics*, **56**, 239–243.

Travnikova, I.G., Bruk, G.Ya. and Shutov, V.N. (1999) Dietary pattern and content of Cesium radionuclides in foodstuffs and bodies of Bryansk Oblast rural population after the Chernobyl accident. *Radiochemistry*, **41**, 298–301.

Uchida, S., Tagami, K., Rühm, W., Steiner, M. and Wirth, E. (2000) Separation of Tc-99 in soil and plant samples collected around the Chernobyl reactor using a Tc-selective chromatographic resin and determination of the nuclide by ICP-MS. *Applied Radiation and Isotopes*, **53**, 69–73.

USDA (2004) *National Nutrient Database for Standard Reference, Release 16-1 Nutrient Lists*. US Department of Agriculture Report (available online: http://www.nal.usda.gov/fnic/foodcomp/Data/SR16-1/wtrank/wt_rank.html).

Valcke, E. and Cremers, A. (1994) Sorption–desorption dynamics of radiocaesium in organic matter soils. *Science of the Total Environment*, **157**, 275–283.

Vandecasteele, C.M., Van Hees, M., Hardeman, F., Voigt, G. and Howard, B.J. (2000) The true absorption of ^{131}I, and its transfer to milk in cows given different stable iodine diets. *Journal of Environmental Radioactivity*, **47**, 301–317.

Vinogradov, A.P. (1953) *The Elementary Composition of Marine Organisms*. Sears Foundation, New Haven, 647 pp.

Voigt, G., Müller, H., Paretzke, H.G., Bauer, T. and Röhrmoser, G. (1993) ^{137}Cs transfer after Chernobyl from fodder into chicken meat and eggs. *Health Physics*, **65**, 141–146.

Voigt, G., Rauch, F. and Paretzke, H.G. (1996) Long-term behaviour of radiocaesium in dairy herds in the years following the Chernobyl accident. *Health Physics*, **71**, 370–373.

Ward, G.M., Johnson, J.E. and Stewart, H.F. (1965) Deposition of fallout ^{137}Cs of forage and its transfer to cow's milk. In: Klement, A.W. Jr. (ed.), *Proceedings of the 2nd AEC Symposium on fallout*. National Technical Information Center, Oak Ridge.

Ward, G.M., Keszethelyi, Z., Kanyár, B., Kralovansky, U.P. and Johnson, J.E. (1989) Transfer of ^{137}Cs to milk and meat in Hungary from Chernobyl fallout with comparisons of worldwide fallout in the 1960s. *Health Physics*, **57**, 587–592.

Yasuda, H. and Uchida, S. (1993) Statistical approach for the estimation of Strontium distribution coefficient. *Environmental Science and Technology*, **27**, 2462–2465.

Zhuchenko, Y.M. (1998) Mathematical modelling of radionuclide fluxes from agricultural and wild ecosystems with concern to the radioecological remediation of contaminated territories. PhD Thesis, Obninsk.

Zhuchenko, Yu.M., Firsakova, S.K. and Voigt, G. (2002) Modeling radionuclide effluxes from agricultural and natural ecosystems in Belarus. *Health Physics*, **82**, 881–886.

Zibold, G., Drissner J., Kaminski S., Klemt, E. and Miller, R. (2001) Time-dependence of the radiocaesium contamination of roe deer: Measurement and modelling. *Journal of Environmental Radioactivity*, **55**, 5–27.

4

Radioactivity in aquatic systems

Jim T. Smith, Oleg V. Voitsekhovitch, Alexei V. Konoplev and Anatoly V. Kudelsky

4.1 INTRODUCTION

The Chernobyl Nuclear Power Plant (NPP) is situated next to the Pripyat River which is an important component of the Dnieper River–Reservoir system, one of the largest surface water systems in Europe (Figure 4.1). After the accident, radioactive fallout on the Pripyat catchment threatened to wash downriver into the Kiev Reservoir, a major source of drinking water for the city of Kiev. The radioactive contamination of aquatic systems therefore became a major issue in the immediate aftermath of Chernobyl.

Initial radioactivity concentrations in river water were relatively high as a result of direct fallout onto the river surfaces and washoff of contamination from the surrounding catchment area. During the first few weeks after the accident, however, activity concentrations in river waters rapidly declined because of the physical decay of short-lived isotopes and as radionuclide deposits became absorbed to catchment soils. In the longer term, relatively long-lived radiocaesium and radiostrontium formed the major component of river water contamination. Though long-term levels of these isotopes in rivers were low, temporary increases in activity concentrations during flooding of the Pripyat River caused serious concern in Kiev over the safety of the drinking water supply.

Lakes and reservoirs around Europe were contaminated by fallout to lake surfaces and transfers of radionuclides from their surrounding catchments. Radioactivity concentrations in water declined relatively rapidly in reservoirs and in those lakes with significant inflows and outflows of water, as radionuclides were 'flushed' out of the system. In the long term, most lakes and reservoirs showed similar radiocaesium and radiostrontium activity concentrations to those of their inflowing rivers and streams. In the areas around Chernobyl, however, there are many lakes with no inflowing and outflowing streams ('closed' lake systems). Cycling of radiocaesium in these closed systems led to much higher activity

Figure 4.1. Pripyat–Dnieper River–Reservoir system showing Chernobyl and Kiev with the Kiev Reservoir in between.

concentrations in water and aquatic biota than were seen in open lakes and rivers.

Bioaccumulation of radionuclides (particularly radiocaesium) in fish resulted in activity concentrations (both in western Europe and in the former Soviet Union, fSU) which were in many cases significantly above maximum permissible levels for consumption. In some lakes, particularly in the Ukraine, Belarus and Russia, these problems have continued to the present day and evidence suggests that they will continue for the foreseeable future. Freshwater fish provide an important food source for many of the inhabitants of the contaminated regions of the Ukraine,

Russia and Belarus. Prior to the Chernobyl accident, 17% of the population of the Bryansk region of Russia consumed fish from local rivers and lakes (Balonov and Travnikova, 1990).

4.1.1 Distribution of radionuclides between dissolved and particulate phases

The fraction of a radionuclide which is absorbed to suspended particles in surface waters strongly influences both its transport and bioaccumulation. This fraction is expressed as the distribution coefficient (K_d), the radionuclide activity per kg of solid matter divided by the activity per litre of water. Table 4.1 shows a selection of estimated K_d values for some radiologically important radionuclides (^{131}I, ^{90}Sr, ^{134,137}Cs and Pu isotopes). These emphasise measurements made *in situ*, usually with a long contact time between radionuclide and sediment. For short contact times, K_d values may be lower by an order of magnitude or more. It is clear that ranges in reported values are very large, sometimes covering several orders of magnitude.

It is useful also to consider the fraction f_p of radioactivity which is sorbed to the solid phase (e.g., Håkanson, 1997). Defining C_{aq} (Bq l^{-1}) as the aqueous phase activity of a given radionuclide and C_s (Bq kg^{-1}) as the solid phase activity we can write:

$$K_d = \frac{C_s}{C_{aq}} \quad \text{and} \quad f_p = \frac{sK_d}{(1 + sK_d)} \qquad (4.1)$$

where s is the suspended solids concentration (kg l^{-1}), K_d is the solids–aqueous distribution coefficient (l kg^{-1}) and f_p is the (dimensionless) fraction of the total activity which is in the solid phase. The relationship between K_d, s and f_p is illustrated in Figure 4.2.

In freshwaters, values of suspended solids concentrations typically range between 0.5–50 mg l^{-1}. Assuming the 'best estimate' K_d values given in Table 4.1, for example, suspended solids concentrations within this range lead to particulate sorbed fractions (expressed as a percentage) of 0.05–5% for ^{90}Sr, 4–80% for ^{134,137}Cs and 5–83% for Pu. We have not included a 'best estimate' K_d for ^{131}I since values reported are even more variable than for the other radionuclides. Field measurements in an experimental enclosure (Milton *et al.*, 1992) and in water bodies after Chernobyl (Kryshev, 1995), suggest values of f_p in the range 2–16% and 5–36% respectively.

In marine systems, generally lower particle sorption capacities and higher concentrations of competing ions tend to make radionuclide particle sorbed fractions significantly lower than in freshwaters. In the Baltic Sea after Chernobyl, less than 10% of ^{137}Cs was bound to particles (Carlson and Holm, 1992) and estimates put the average particulate sorbed fraction at approximately 1% (Carlson and Holm, 1992; Knapinska-Skiba *et al.*, 2001).

Table 4.1. K_d values for radiostrontium, radioiodine, radiocaesium and plutonium in freshwaters. 'Best estimate' values are given where appropriate. Note that different review estimates may contain some of the same data.

RN	Reported (K_d l kg^{-1})	Source	Best estimate	Comments
^{89,90}Sr	8–4 × 10^3	IAEA (1994)	10^3	Review.
	10^3	Coughtrey and Thorne (1983) Coughtrey et al. (1985)		Review.
	10^2–10^3	Mundschenk (1996)		Measurements at various suspended solids concentrations.
	250–500	Chittenden (1983)		'In situ' K_d from suspended sediment and river water.
	380–730	Joshi and McCrea (1992)		Two K_ds in the Ottawa River
	750–1,800	Konoplev et al. (1992a)		L. Lelev, 5 km from Chern. NPP
	1,600	Zeevaert et al. (1986)		'In situ' K_d from suspended sediment and river water.

Sorption is proportional to the cation exchange capacity (CEC) of the sorbent and inversely proportional to the strength of competing cations (usually Ca, Mg) in solution (Yasuda and Uchida, 1993). Relatively good agreement between different studies.

^{131}I	0–80	IAEA (1994)		Review.
	3 × 10^2	Coughtrey et al. (1983, 1985)		Review. Based on limited data.
	10.5 × 10^4 and 8.2 × 10^5	Milton et al. (1992)		Two experimental enclosures, Perch Lake, Canada.
	2.2 × 10^3 and 1.3 × 10^5	Estimated from data in Kryshev (1995) and assuming s values given in Sansone and Voitsekhovitch (1996)		Pripyat River, Kiev Reservoir, 1 May, 1986. May be influenced by hot particles, but most ^{131}I was in aerosol form.

Limited data and high variation between reported values.

^{134,137}Cs	5 × 10^1 to 8 × 10^4	IAEA (1994)		Review.
	2 × 10^4	Coughtrey and Thorne (1983), Coughtrey et al. (1985)		Review. K_d presented as an estimated value only.
	(3.7–9.4) × 10^3	Konoplev et al. (1992a)		L. Lelev, 5 km from Chern. NPP
	4.6 × 10^4 to 2.7 × 10^6	Konoplev et al. (2002)		L. Constance, Alpine Rhine, Rhine below L. Constance.
	8 × 10^3 to 4.2 × 10^5	J.T. Smith, unpubl. res. (see Figure 4.2)	8 × 10^4	Review of long term in situ measurements in 18 rivers and lakes.

Preferential sorption to illitic clay minerals. K_d believed to be inversely related to potassium and ammonium concentrations in water and to increase significantly over time. Best estimate is for long times after fallout.

^{238}Pu	10^2–10^7	IAEA (1994)	10^5	Review.
^{239,240}Pu	10^4–10^6	Coughtrey et al. (1984); Coughtrey et al. (1985)		Review.
	6 × 10^4	Allard et al. (1984)		Hudson river
	1.4 × 10^5	Zeevaert et al. (1986)		'In situ' K_d from suspended sediment and river water.
	2 × 10^2 to 6 × 10^5	Murdock et al. (1995)		'In situ' distribution from contaminated stream suspended sediment and water.

Most (~80%) Pu in Lake Michigan lake water was found to be in the dissolved phase (Sholkovitz, 1983) Possible remobilisation of Pu from anoxic sediments.

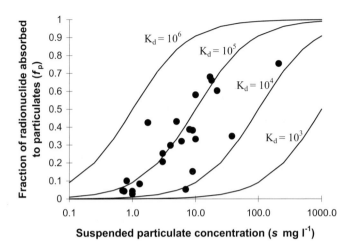

Figure 4.2. Fraction of a radionuclide absorbed to particulates as a function of suspended solids concentration in water for different values of K_d. The relationship is illustrated using measurements of the f_p for ^{137}Cs in 18 European rivers and lakes. In more than 75% of cases, most of the ^{137}Cs was found in the dissolved phase ($f_p < 0.5$).
From J.T. Smith, unpubl. res.

4.2 RADIONUCLIDES IN RIVERS AND STREAMS

The processes which control the radioactive contamination of rivers and streams have been identified by studies into the fate of ^{137}Cs and ^{90}Sr deposited as a result of atmospheric nuclear weapons testing (e.g., Carlsson, 1978; Helton et al., 1985). Following a radioactive fallout, deposition of radionuclides onto the water surface is combined with runoff of radioactivity from the catchment. These transfers of radioactivity from the catchment are due to washoff from plant surfaces and from the surface soils. After some weeks/months, infiltration of radionuclides to deeper soil layers, and binding to soil particles, considerably decreases the rate of radionuclide runoff.

In streams and small rivers, maximum radioactivity concentrations in water were observed during, and shortly after, the Chernobyl accident, with levels declining rapidly over the first few weeks. However, radionuclide deposition in large river catchments was often non-uniform, so in some rivers 'polluted' water took some time to travel downriver from contaminated to less contaminated areas. For example, radioactivity deposited on the upper Elbe River catchment on 29 April, 1986 took approximately 8 days to reach a sampling station near the river mouth at Hamburg (Schoer, 1988).

Over longer time periods after fallout, radionuclides held in catchment soils are slowly transferred to river water by erosion of soil particles and (in the dissolved phase) by desorption from soils. The rates of transfer are influenced by the extent of soil erosion, the strength of radionuclide binding to catchment soils and migration down the soil profile. The time changes in radionuclide activity concentrations in rivers can be modelled by a series of exponential functions, as shown in Box 4.1.

> **Box 4.1. Modelling time changes in radionuclide contamination of rivers**
>
> The time changes in radionuclide activity concentrations in rivers may be modelled by a series of exponential functions (Monte, 1997; Smith et al., 2004), as illustrated in Figure 4.3 for the Pripyat River.
>
> For radiocaesium (Figure 4.3(a)), the radionuclide concentration in runoff or river water C_R (Bq m^{-3}) is given by:
>
> $$C_R(t) = D_c(\alpha e^{-(\lambda+k_1)t} + \beta e^{-(\lambda+k_2)t} + \gamma e^{-(\lambda+k_3)t})$$
>
> where λ (y^{-1}) is the decay constant of the radionuclide and D_c is the radionuclide deposition to the catchment (Bq m^{-2}). α, β, γ (m^{-1}) and k_1, k_2, k_3 (y^{-1}) are empirically determined (radionuclide-specific) constants. The k values may be expressed as effective ecological half-lives T_{eff} where $T_{\mathit{eff}} \approx \ln 2/(k+\lambda)$. The three exponential terms represent, respectively: the fast 'flush' of activity as a result of rapid washoff processes; a slow decline as a result of soil fixation and redistribution processes; and the very long-term 'equilibrium' situation.
>
> For radiostrontium, the activity concentration in runoff water is also given by the above equation, though the parameter values are different to those for radiocaesium (Figure 4.3(b)).
>
> For ^{131}I (and other short-lived isotopes), the half-life is so short that there is no long-term component to the decline, so the model is a simple exponential decay:
>
> $$C_R(t) = D_c \alpha e^{-(\lambda+k_1)t}$$

4.2.1 Early phase

To our knowledge, there are few data of radionuclide concentrations in small streams in the Chernobyl area during the early phase of the accident. Most available data is for large rivers. Table 4.2 shows a summary of available measurements of radionuclide activity concentrations in a large river (the Pripyat) at Chernobyl at various times after the accident. Temporarily allowable levels of radionuclides in drinking water in the Ukraine at different times after Chernobyl are shown in Table 4.3.

Maximum radioactivity concentrations in rivers close to Chernobyl (the Pripyat, Teterev, Irpen and Dnieper) were approximately proportional to the amount of radioactivity released from the reactor (Figure 4.4). This relationship also approximately held for radionuclides in rivers in western Europe (the Glatt, Danube and Po) though it was less strong than in the the Ukrainian rivers.

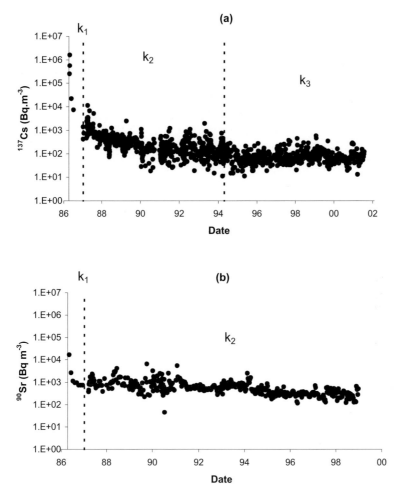

Figure 4.3. The change in activity concentration of ^{137}Cs and ^{90}Sr in the Pripyat River over time after the accident. The different phases in the exponential decline in activity concentrations (Box 4.1) are illustrated by the dotted lines and the 'k' values. ^{137}Cs shows a three-component exponential decline whilst, over this period, ^{90}Sr shows a two-component decline. Data from Ukrainian Hydrometeorological Institute, Kiev.

In the UK, maximum activity concentrations of ^{131}I in surface waters in most regions were of the order $10\,\mathrm{Bq\,l^{-1}}$, though one measurement of ^{131}I in surface water in the Strathclyde region of Scotland gave $1{,}315\,\mathrm{Bq\,l^{-1}}$ (c.f. $9{,}400\,\mathrm{Bq\,l^{-1}}$ observed in rainwater during 3–5 May in this region) (Jones and Castle, 1987).

As with initial maximum concentrations (Figure 4.4), a similar initial behaviour of different radionuclides was observed in the rate of their decline in concentrations in river water during the early phase after the accident. The rate of decline of radionuclide activity concentration in water is commonly measured as an effective

Table 4.2. Radionuclide levels (dissolved phase) in the Pripyat River at Chernobyl.[†] For some (radiologically important) radionuclides, doses are calculated assuming consumption at these concentrations over a one year period after the accident using ingestion rates and dose coefficients given in (NRPB, 1996).*

[†] From Vakulovsky et al. (1990), Voitsekhovitch et al. (1991), Vakulovsky et al. (1994), Kryshev (1995).

RN	Half-life	Radionuclide concentration in water (Bq l^{-1})						1987 (mean)	Committed effective dose during 1st year (mSv)
		01/05/86	02/05/86	06/05/86	03/06/86	16/07/86	09/08/86		
^{137}Cs	30.2 y	250	555	1591	22.2	7.4		1.8	0.57 (a)
^{134}Cs	2.1 y	130	289^1	827^1	11.5^1	3.8^1		0.94^1	0.43 (a)
^{131}I	8.1 d	2100	4440	814	33.3	<0.82^2			4.2 (i)
^{90}Sr	28 y	30			1.9			1.5	0.049 (c)
^{140}Ba	12.8 d	1400							
^{99}Mo	3 d	670							
^{103}Ru	40 d	550	814	170	26	15			0.053 (i)
^{106}Ru	365 d	183^3	271^3	57^3	8.7^3	5^3			0.29 (i)
^{144}Ce	284 d	380				37			
^{141}Ce	33 d	400		89		14.8			
^{95}Zr	65 d	400	1554	167	11	37			
^{95}Nb	35 d	420							
^{241}Pu	13 y	33^4					0.64		0.072 (a)
$^{239+240}$Pu	2.4 × 10^4 y, 6.6 × 10^3 y	0.4					0.0074		0.0046 (a)

* Doses from each radionuclide were calculated for infants (i), children (c) and adults (a), the result for the age group showing the highest dose is shown for each radionuclide. Note that radionuclide concentrations in water at the point of consumption, and consequently doses, are likely to be much lower than these measurements in rivers owing to dilution and water treatment.
^{1} From ^{137}Cs measurement and a ^{134}Cs : ^{137}Cs ratio ~0.52; ^{2} Assuming a decline from the 3 June, 1986 value by radioactive decay only; ^{3} From a ^{103}Ru measurement and assuming a ^{103}Ru : ^{106}Ru ratio (~3) for Chernobyl fallout; ^{4} From a ^{239,240}Pu measurement and a ^{241}Pu : ^{239,240}Pu ratio (~82) for Chernobyl fallout.

Table 4.3. Temporary allowable levels of radionuclides in drinking water in the Ukraine at different times after Chernobyl. The dates refer to the time at which each new regulation was implemented.
From Los et al. (1998).

Radionuclide (Bq l^{-1})	6 May, 1986	30 May, 1986	15 Dec, 1987	22 Jan, 1991
^{131}I	3,700			
Total beta activity		370		
^{137}Cs + ^{134}Cs			18.5	18.5
^{90}Sr				3.7

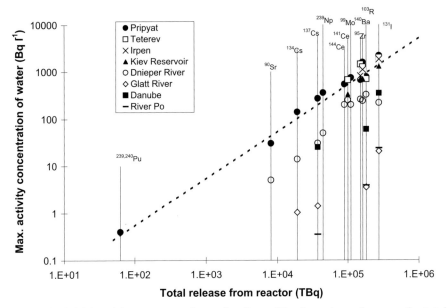

Figure 4.4. The initial activity concentrations of radionuclides in various rivers vs. the total amount released from the reactor. The measurements from the Pripyat, Irpen, Teterev and Dnieper Rivers show an approximately constant ratio between amount released and initial river water activity concentration.
From data in Voitsekhovitch et al. (1991), Kryshev (1995), Waber et al. (1987), Foulquier and Baudin-Jaulent (1990).

ecological half-life (Box 4.1), the time taken for the radioactivity concentration in water to decline by one-half. Measurements of the change in ^{137}Cs activity concentrations as a function of time after fallout for six European rivers were obtained from the literature. These gave values (Table 4.4) of effective ecological half-lives during the initial period after the accident of approximately 1–3 weeks. An additional study on measurements from the Rhine in Germany (Monte, 1995) gave a T_{eff} of 12.3 days for radiocaesium.

Table 4.4. Estimates of the initial rate of decline of radionuclides in river water (dissolved phase, except where indicated) after Chernobyl. Declines include a radioactive decay component: ecological half-lives (i.e., excluding radioactive decay) are given in brackets.

	Effective ecological half-life of decline in activity concentration (days)**					
RN	Pripyat, Ukraine[1]	Dnieper, Ukraine[1]	Po, Italy[2]	Glatt, Switzerland[3]	Danube, Hungary[4]	Elbe, Germany[5]
^{137}Cs	11.2 (11.2)	9.0 (9.0)	34.7* (34.7)	19.2 (19.2)	–	17.8† (17.8)
^{90}Sr	10.4 (10.4)	15.4 (15.4)	–	–	–	–
^{131}I	5.2 (14.3)	–	7.3* (72.3)	5.7 (19.5)	3.2 (5.2)	5.9† (21.4)
^{144}Ce	22.4 (24.3)	–	–	–	–	–
^{141}Ce	19.3 (46.0)	–	–	–	–	–
^{95}Zr	16.5 (22.2)	–	–	–	–	–
^{132}Te	–	–	–	2.3 (7.9)	1.9 (4.7)	–
^{103}Ru	13.8 (21.3)	–	17.1* (29.9)	28.7 (101)	5.3 (6.2)	7.5† (12.4)
^{241}Pu	17.3 (17.3)	–	–	–	–	–

* Measurements not begun until 16–20 May (Monte, 1995). † Dissolved and particulate phases.
** From data in [1] Voitsekhovitch et al. (1991); [2] Monte (1995); [3] Waber et al. (1987); [4] German (1986) quoted in Foulquier and Baudin Jaulent (1990); [5] Schoer (1988).

Though there is significant variation, there is little evidence of systematic differences in rates of decline between the different radionuclides. Generally faster rates of decline in ^{132}Te and ^{131}I are largely due to their rapid physical decay ($T_{1/2}$ = 3.2 and 8.05 d respectively) rather than to any obvious differences in their washoff and deposition behaviour. The observation of similar half-times of ^{137}Cs and ^{131}I, for example, is supported by measurements of the changes in their activity concentrations in grass in the UK (Cambray et al., 1987) which gave a removal half-time by washoff (excluding physical decay) of approximately 11 days for both elements (see also Chapter 3).

All of the radionuclides studied had effective ecological half-lives of less than one month (Table 4.4) in the initial period after the accident. The only estimate greater than one month is ^{137}Cs in the River Po (T_{eff} = 34.7 d), though this value is likely to be an overestimate of the initial effective ecological half-life since in this case measurements were not begun until around 3 weeks after the accident.

The consistent initial concentrations of different radionuclides in river water (per TBq released), and their similar rates of decline, would not be expected if different chemical interactions of the various radionuclides with the soil strongly controlled transport in the early period. Although there is significant variation, the measurements imply that in the early stages, physical transport processes such as rainfall onto the river surface, washoff from plants and from easily available fractions in the soil were more important than differing individual behaviours of the various radionuclides.

4.2.2 Intermediate phase

Following the initial rapid decline in radioactivity in rivers, longer term contamination was primarily due to ^{137}Cs and ^{90}Sr. Both radionuclides have relatively long half-lives (30.1 and 28.8 years, respectively) and were released from the reactor in significant quantities (Table 1.2). The temporal change in activity concentration of ^{137}Cs and ^{90}Sr in rivers is illustrated with measurements from the Pripyat River (Figure 4.3). Both radionuclides show a large decline in activity concentrations over time after fallout.

In the first few years after Chernobyl, the rate of decline in radiocaesium concentrations in river water (dissolved phase) was observed to be remarkably similar in the Dnieper, Pripyat, Rhine, Teterev and Uzh Rivers (Monte, 1995). This observation was confirmed by studies in a number of other surface waters in Europe (Smith et al., 1999a). Estimates of the rates of decline of radiocaesium activity concentrations during a five-year period after the accident are presented in Table 4.5. Almost all (95%) of the measurements show effective ecological half-lives within the range of 1–4 years in the intermediate period after the accident.

The decline in ^{90}Sr activity concentrations in rivers was much slower than for ^{137}Cs (Figure 4.3 and Table 4.5) with most estimates of T_{eff} being in the range 5.6–11.7 years. This rate of decline of ^{90}Sr did not change significantly over a 15 year period after the accident (1987–2001). Because the vast majority of ^{90}Sr fallout was close to the reactor and in the form of fuel particles, these measurements could have been affected by the change in the chemical availability of ^{90}Sr as the fuel particles broke down (Konoplev et al., 1992b; Kashparov et al., 1999). Surprisingly, however, measurements of T_{eff} for (chemically available) ^{90}Sr from nuclear weapons test (NWT) fallout showed similar rates of decline to the post-Chernobyl studies. In five catchments in Finland, mean T_{eff} was 7.7 y (Cross et al., 2002, from data in Salo et al., 1984) and in 11 Italian rivers the rate of decline showed relatively little variation, having mean value $T_{eff} = 5.3$ y (Monte, 1997). The low ^{90}Sr fallout at long distances from Chernobyl meant that measurements in areas unaffected by fuel particles are scarce, though data from two rivers in Finland give $T_{eff} = 4.9$ and 9.9 y (Cross et al., 2002) in general agreement with the pre-Chernobyl (NWT) studies.

4.2.3 Long-term ^{137}Cs contamination of water

The rate of decline in ^{137}Cs activity concentrations in the Pripyat river water (in contrast to ^{90}Sr) has slowed in recent years (Figure 4.3). The effective ecological half life of 1.2 years (dissolved phase) and 1.7 y (particulate phase) in the period 87–91, increased to 4.3 y (dissolved phase) and 11.2 y (particulate phase) between 1995 and 1998. This increase in T_{eff} has also been observed in rivers in Belarus (Kudelsky et al., 1998), Ukraine (Voitsekhovitch, 1998) and Finland (R. Saxén, pers. commun.) as illustrated in Table 4.5.

Table 4.5. Rates of change in ^{137}Cs and ^{90}Sr activity concentrations in different rivers in the medium to long term (1987–2001) after Chernobyl.

Study	Time period	Rate of decline k (y^{-1})	Effective ecological half-life (T_{eff}) (years)
^{137}Cs in dissolved phase (intermediate time period)			
9 Ukrainian rivers[1]	1987–1991	0.3–0.65	1.1–2.3
5 Finnish rivers[2]	1987–1991	0.14–0.39	1.8–5.0
5 Belarussian rivers[3]	1987–1991	0.49–0.65	1.1–1.4
Forest catchment in Sweden[4]	1987–1993	0.18	3.8
Inlet to Lake Sälgsjön, Sweden[5]	1987–1990	0.23	3.0
Upland catchment in the UK[6]	1987–1989	0.39	1.8
Dora Baltea River, Italy[7]	1987–1991	0.34	2.0
Rhine River, Germany[8]	1987–1991	0.52	1.3
^{137}Cs on particulates (intermediate time period)			
Rhine River, Germany[8]	1987–1991	0.35	2.0
Pripyat River, the Ukraine[9]	1987–1991	0.41	1.7
^{137}Cs in dissolved phase (long-term change)			
Pripyat River, the Ukraine[9]	1995–1998	0.16	4.3
Dnieper River, the Ukraine[9]	1995–1998	0.17	4.1
Desna River, the Ukraine[9]	1995–1998	0.047	14.7
5 Rivers in Finland[10]	1995–2002	0.07–0.11	6.3–9.9
5 Rivers in Belarus[3]	1994–1998	0.13–0.3	2.3–5.3
^{137}Cs on particulates (long-term change)			
Pripyat River, the Ukraine[9]	1995–1998	0.062	11.2
Dnieper River, the Ukraine[9]	1995–1998	0.07	9.9
Desna River, the Ukraine[9]	1995–1998	0.24	2.9
^{90}Sr in dissolved phase			
10 Rivers in the Ukraine[9]	1987–2001	0.025–0.124	5.6–28.8
Kymijoki (Finland)[11]	1987–1995		9.9
Kokemäenjoki (Finland)[11]	1987–1995		4.9

[1] From data in Vakulovsky *et al.* (1994); [2] Smith *et al.* (2000a); [3] from data in Kudelsky *et al.* (1998); [4] Nylén (1996); [5] Sundblad *et al.* (1991) quoted in Nylén (1996); [6] Hilton *et al.* (1993); [7] data from L. Monte, ENEA, Italy (pers. commun.); [8] Monte (1997); [9] Kanivets and Voitsekhovitch (2001) in Smith *et al.* (2001); [10] from data in Saxén and Ilus (2001) and Smith *et al.* (2004); [11] Cross *et al.* (2002).

4.2.4 Processes controlling declines in ^{90}Sr and ^{137}Cs in surface waters

There are three main mechanisms which may contribute to the decline in radionuclide transfers (in dissolved form) to runoff water. These are: (1) loss of radioactivity from the catchment; (2) vertical migration to deeper layers of soil; (3) slow chemical 'fixation' in the soil. The first studies of radionuclide washoff after Chernobyl were carried out by Borzilov *et al.* (1988), Bulgakov *et al.* (1990) and Konoplev *et al.* (1992b). These studies showed significant removal of radioactivity in both dissolved and particulate phases. Long-term estimates of rates of ^{137}Cs removal from catchments (e.g., Smith *et al.*, 1999a; Helton *et al.*, 1985; Kudelsky *et al.*, 1998) show that losses of ^{137}Cs from catchments are very slow – being at most around 0.5–2% of the total amount in the catchment per year. Removal rates of <2% (typically, 0.1–1%) of the radiocaesium in the catchment per year are not sufficient to cause a decline in activity concentration in river water by one-half every 1.5–2 years (as observed in Table 4.5). Thus the observed rates of change in activity concentrations in surface waters between 1987 and 1991 cannot be due simply to loss of the store of radiocaesium in the catchment.

Another possible mechanism for reduction in transfers of radionuclides to surface waters is transport into deeper layers of the soil, thus reducing the concentration in more erodible surface layers. If vertical migration were the controlling mechanism, however, we would expect to see more rapid declines in radiostrontium activity concentrations than radiocaesium since radiostrontium migrates in the soil more rapidly than radiocaesium (Chapter 2). From the measurements presented in Table 4.5, this is not the case, at least in the first five years after fallout. The observed slower declines in radiostrontium in river water compared to radiocaesium suggest that the change in radiocaesium concentration is primarily controlled by fixation to soil particles rather than vertical migration during the intermediate period after the accident (Smith *et al.*, 1999a).

The gradual slowing of the rate of decline (increase in T_{eff}) of radiocaesium in rivers (Table 4.5) is attributed to a long-term equilibration of sorption and desorption processes in the soil (Chapter 2). In other words, the rate of decrease in radiocaesium availability in the soil has slowed. In the coming decades, it is expected that ^{137}Cs activity concentrations in rivers will decline at a slow rate determined by physical decay and slow physical redistribution processes in the catchment (Smith *et al.*, 2000b).

Time changes in particulate phase radiocaesium appear to be similar to those observed for the dissolved phase (Table 4.5). We would expect radiocaesium absorbed to particles to change only by physical redistribution in the catchment, since the amount absorbed to particles (in the long term) is not significantly affected by sorption and desorption processes. The similar rates of decline of particulate and dissolved phase radiocaesium may reflect an equilibration of the two phases in river water (so that, on average, particulate phase concentrations are a constant multiple of dissolved, i.e., the two phases are 'coupled'). On the other hand, radiocaesium absorbed to particles may be controlled by entirely different mechanisms of erosional transport in the catchment, and the similarity in rates of decline may be coincidental. It is plausible that radiocaesium attached to the more 'erodible' soils in the catchment is lost in the early years after fallout, leading to declining radiocaesium erosion rates over time.

As discussed above (see Section 4.1), radiostrontium is found almost entirely in the dissolved phase in river water. The steady decline in radiostrontium concentrations in river waters is not expected to be due to long-term changes in the strength of its sorption to soils since it is generally assumed that strontium is rapidly and reversibly sorbed to soils. The relatively slow declines in ^{90}Sr concentrations in rivers ($T_{eff} \sim$ 5–10 years) may be partly explained by loss of the inventory of ^{90}Sr in the catchment. Studies of nuclear weapons test (NWT) and Chernobyl ^{90}Sr in rivers in Finland (Saxén and Ilus, 2001; Cross et al., 2002) suggest loss rates of the order of 1–2% per year, leading to a half-time of decline of approximately 35 years (including decay, $T_{eff} >$ 16 years). This is higher than typical observed T_{eff} values (Table 4.5), implying that vertical migration in soils and/or slow 'fixation' of ^{90}Sr may also play a role in its declining concentration in surface waters.

4.2.5 Influence of catchment characteristics on radionuclide runoff

The concentration of a radionuclide in surface water is usefully expressed as a concentration per unit of deposition ('normalised concentration', cf. aggregated transfer coefficient, Chapter 3):

$$R_c(t) = \frac{\text{Concentration of radionuclide in water (Bq m}^{-3})}{\text{Radionuclide deposition to catchment (Bq m}^{-2})} \, \text{m}^{-1} \quad (4.2)$$

Since the radionuclide concentration in water declines significantly over time, the normalised concentration also declines at the same rate. For simplicity, the fallout to the catchment is usually the value estimated for the time of the accident and is therefore constant.

Measurements of the normalised ^{137}Cs activity concentration in many different rivers (Figure 4.5(a)) show a range of approximately a factor of 30, even when temporal changes in concentrations have been accounted for. A number of studies (Hansen and Aarkrog, 1990; Hilton et al., 1993; Nylén, 1996; Kudelsky et al., 1996) have attributed this variation to the types of soil in the catchment, in particular, the proportion of highly organic peat bog soils. In mineral soils, ^{137}Cs is sorbed to highly selective 'Frayed Edge Sites' (FES) on the illitic clay fraction (Cremers et al., 1988) and becomes 'fixed' in the mineral lattice (Comans and Hockley, 1992). Work has shown, however, that in highly organic soils FES concentrations are low, leading to a reduced binding of ^{137}Cs to the solid phase (Valcke and Cremers, 1994). In field studies on small catchments (Hilton et al., 1993; Nylén, 1996; Kudelsky et al., 1996) it was found that highly organic soils (particularly saturated peats) released up to an order of magnitude more radiocaesium to surface waters than some mineral soils. The relationship between radiocaesium in surface water and the percentage of organic soils in the catchment is illustrated in Figure 4.5(b).

As discussed above, the assessment of concentrations of radiostrontium in surface waters is complicated by a lack of data for systems outside the fSU, and the large component of fallout in the form of fuel particles. In river catchments in Finland, it was found (Saxén and Ilus, 2001) that (per unit of deposition) the runoff of both NWT and Chernobyl radiostrontium was approximately one order of magnitude greater than for radiocaesium. In studies of NWT ^{90}Sr, runoff was highest in catchments with organic soils (Linsley et al., 1982; Hansen and Aarkrog, 1990) and those with a high proportion of surface waters (rivers, bogs and lakes) in the catchment (Salo et al., 1984; Smith et al., 2004).

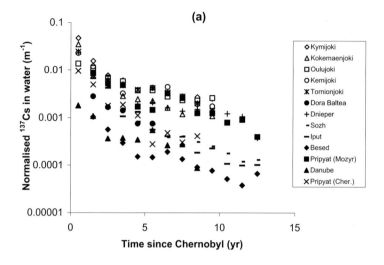

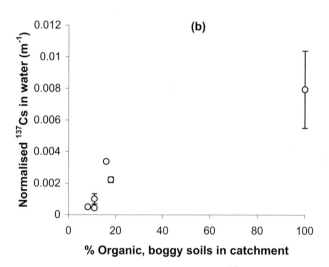

Figure 4.5. (a) Normalised activity concentration of ^{137}Cs in the dissolved phase of different rivers after Chernobyl. (b) Correlation between the normalised ^{137}Cs activity concentration and the percentage catchment coverage of organic, boggy soils in six different catchments.
(a) From data reviewed by Smith *et al.* (2004). (b) From data in Hilton *et al.* (1993); Kudelsky *et al.* (1996); Nylén (1996); Kudelsky *et al.* (1998).

Radionuclide activity concentrations in rivers can vary significantly throughout the year as a result of changing river and catchment conditions. For example, flooding of the Pripyat River, caused by blockages of the river by ice in late winter, led to temporary increases in ^{90}Sr activity concentrations in this system, but did not significantly affect ^{137}Cs concentrations (Vakulovsky *et al.*, 1994). The flooding caused increased washoff of ^{90}Sr from a highly contaminated flood plain area within the 30-km zone. For example, during winter 1991, concentrations of ^{90}Sr in the river water increased from around $1\,\text{Bq}\,\text{l}^{-1}$ to approximately $8\,\text{Bq}\,\text{l}^{-1}$ for a 5–10 day period (Vakulovsky et al., 1994).

4.3 RADIOACTIVITY IN LAKES AND RESERVOIRS

Maximum activity concentrations of radionuclides in lakes and reservoirs occured during and shortly after the accident as a result of direct deposition of activity to the water surface and (initially, to a much lesser extent) transport of radioactivity from the catchment (Figure 4.6). In most lakes, radionuclides were well mixed throughout the lake water during the first days/weeks after fallout. In deep lakes such as Lake Zurich (mean depth of 143 m), however, it took several months for full vertical mixing to take place (Santschi et al., 1990). In some areas of northern Europe, lakes were covered in ice at the time of the accident, so maximum concentrations in lake waters were only observed after the ice melted.

The initial concentration of radionuclides in lake and reservoir waters can be assessed by estimating the dilution of the surface deposited radioactivity in the body of water. The initial average activity concentration in the lake water C_T (Bq m^{-3}) is therefore estimated by:

$$C_T(t=0) = \frac{DA_L}{V_L} = \frac{D}{d} \qquad (4.3)$$

where D is the fallout (Bq m^{-2}), V_L is the lake volume, A_L the lake surface area and d ($= V_L/A_L$) the lake mean depth. It should be noted that C_T represents the total activity concentration in solid and aqueous phases and should not be confused with C_{aq} (the dissolved phase activity concentration).

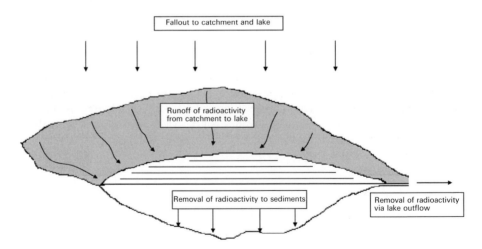

Figure 4.6. Radionuclide transfers in a catchment–lake system.

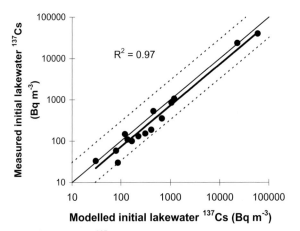

Figure 4.7. Comparison of initial ^{137}Cs activity concentration in 15 lakes determined from measurements with that estimated from a simple dilution model (Equation 4.3). Thick solid line shows best-fit regression line, thin solid line shows 1:1 relationship. Dotted lines show factor of 3 deviation from 1:1 relationship.

Estimates of initial ^{137}Cs in the water of a number of European lakes show that this simple dilution model (Equation 4.3) gives good estimates ($R^2 = 0.97$, $n = 15$, $p < 0.001$) of the initial average activity concentration, as shown in Figure 4.7 (Smith et al., 1999b). In general, however, estimates made on the basis of Equation 4.3 are slightly higher than the measurements (most points are below the $x = y$ line). The mean ratio of predicted C_T (at time = 0) to values extrapolated from measurements is 1.35. This difference is partly due to the fact that most of the measurements presented in Figure 4.7 are of radiocaesium in the dissolved phase only, whereas the model predicts the concentration in both phases. There may also be underestimation of initial measured values since most measurements were begun a few days after fallout and estimates by extrapolation to time zero may be slightly low. In spite of these differences, it is clear that initial radiocaesium activity concentrations in these lakes were determined primarily by dilution of radioactivity directly deposited on the lake surface.

It should be noted that Equation 4.3 predicts average activity concentrations in the whole lake. Activity concentrations in the surface waters of lakes are likely to be higher than the average concentration in the lake during the first few weeks after fallout. In addition, if the lake is stratified, radionuclides may initially be rapidly mixed in the upper layer of water, taking some time to mix throughout the whole lake (Santschi et al., 1990; Davison et al., 1993).

4.3.1 Initial removal of radionuclides from the lake water

Following deposition of radionuclides onto the lake surface, the concentration in lake water declines approximately exponentially (Figure 4.8):

$$C = C(0)\exp(-Kt) \quad (4.4)$$

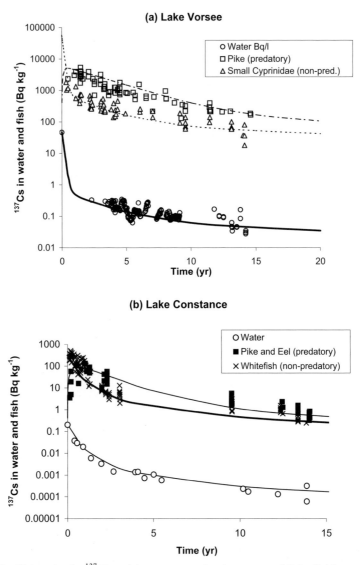

Figure 4.8. Change in the ^{137}Cs activity concentration in water and fish of: (a) a small shallow lake in Germany, Lake Vorsee and; (b) the large, deep Lake Constance (Bodensee).
Adapted from Klemt et al. (1998) and Zibold et al. (2002) using data kindly supplied by Gregor Zibold, Fachhochschule Weingarten.

where inputs of radionuclides to the lake from the catchment have been ignored since they are rarely significant in this early period. The rate of decline K (d^{-1}) in radionuclide concentration in water, termed the 'self-cleaning' capacity of a lake (Santschi et al., 1990) is determined by losses through the lake outflow, transfers of radioactivity to the bed sediments and physical decay with rate constant λ (d^{-1}). The removal rate K is given by (see, e.g., Smith et al., 1999b):

$$K = \frac{1}{T_w} + \frac{1}{T_s} + \lambda \qquad (4.5)$$

where T_w (d), the water residence or 'turnover' time of the lake, is defined as the ratio of the lake volume to the rate of water discharge through the outflow:

$$T_w = \frac{V_L}{q_o} \qquad (4.6)$$

and T_s is the transfer time of ^{137}Cs to bottom sediments. Rates of removal of radioactivity from the lake water typically vary in the range $4\text{--}25 \times 10^{-3}\,\text{d}^{-1}$ (i.e., T_{eff} (ln $2/K$) = 28–170 days (Santschi et al., 1990; Smith et al., 1999b)).

The process which predominantly determines pollutant transfers to sediments is still open to question. Most models assume that removal to bed sediments occurs primarily by absorption of the radionuclide to suspended particulate matter which subsequently falls to the lake bed. But some workers (Santschi et al., 1986; Hesslein, 1987) have shown that direct diffusion across the sediment–water interface may also be important for some radionuclides. Direct diffusion to bottom sediments is, however, ignored here since correct modelling of this process is complex (Smith and Comans, 1996). In addition, a study (Smith et al., 1999b) suggests that it is of minor importance for radiocaesium transfers to sediments.

The rate of transfer of radioactivity to bottom sediments by settling of suspended particles is given by:

$$\frac{1}{T_s} = \frac{f_p v_p}{d} \qquad (4.7)$$

where v_p (m d^{-1}) is the mean settling velocity of suspended particles, d is the lake mean depth, and f_p is the fraction of activity absorbed to suspended particles, defined in Equation 4.1.

The importance of different environmental processes to the transport of radionuclides in lakes was illustrated by a study of radiocaesium removal rates in 14 different European lakes following Chernobyl (Smith et al., 1999b). For each of these lakes, the rate of removal of radiocaesium from the lake water (K, units: d^{-1}) was estimated using time series measurements of ^{137}Cs in the lake water. This removal rate was then correlated with different lake characteristics to determine their influence on radiocaesium removal.

4.3.2 The influence of lake water residence time

The influence of water outflow on radiocaesium removal from the lake is demonstrated in a plot of the radiocaesium removal rate K (derived from field measurements) against the inverse of the lake water residence time ($1/T_w$), as shown in Figure 4.9(a) ($R^2 = 0.74$, $n = 14$, $p < 0.005$). The graph illustrates that, although there is a strong inverse correlation between removal rates and water residence times, the rate of ^{137}Cs removal is significantly higher than the rate of loss of

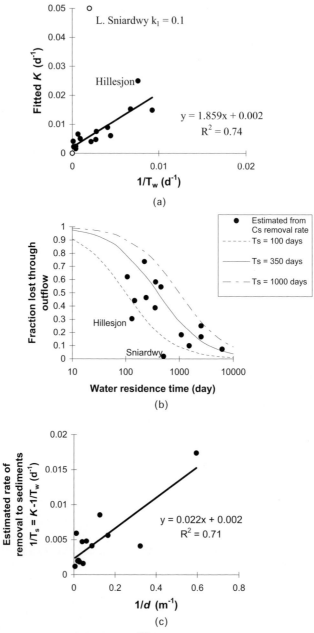

Figure 4.9. (a) The relationship between ^{137}Cs removal rate from 14 lakes and the removal rate of water through the outflow (the open circle shows an outlier, Lake Sniardwy, see text). (b) The relationship between the fraction of the total ^{137}Cs transferred to the outflow and the lake water residence time. (c) The relationship between ^{137}Cs removal rate (from lake water to sediments) and the lake mean depth.
Data is from a literature review carried out by Smith et al. (1999b).

water through the outflow (given by the inverse of the water residence time). Thus, as expected, lake water turnover ('flushing') alone is not sufficient to explain observed rates of ^{137}Cs removal, so transfers of activity to bottom sediments must be important.

It is useful to estimate the fraction f_o of the total activity deposited on the lake surface which was eventually lost via the outflow (and, therefore, the fraction $f_s = 1 - f_o$ of activity transported to the bottom sediments). This fraction can be estimated from the measured rate of removal to sediments ($1/T_s = K$(observed) $- 1/T_w$):

$$f_o = \frac{T_s}{T_s + T_w} \tag{4.8}$$

The fraction of activity lost through the outflow $f_o (= 1 - f_s)$ is shown as a function of T_w in Figure 4.9(b). The graph shows, unsurprisingly, that more radioactivity is lost through the outflow in lakes with shorter water residence time. Lakes which show unusually low losses through the outflow and hence unusually high transfers to sediments (Lakes Hillesjøn: Brittain et al., 1997; and Sniardwy: Robbins and Jasinski, 1995) are relatively shallow.

4.3.3 The influence of lake mean depth d

The influence of lake mean depth d on radiocaesium removal to sediments is demonstated by a plot of the rate of removal of activity to sediments ($1/T_s = K - 1/T_w - \lambda$, Equation 4.5) vs. $1/d$ (Figure 4.9(c)), showing removal rates in inverse proportion to the lake mean depth ($R^2 = 0.71$, $n = 14$, $p < 0.001$). Removal rates of radiocaesium to sediments are therefore greatest in shallow lakes, as implied by Equation 4.7.

4.3.4 The influence of sediment–water distribution coefficient K_d

Using measured values of T_w, s and v_p, Smith et al. (1999b) estimated the K_d value required to produce the observed radiocaesium removal rate K for different lakes. In Table 4.6, we compare these K_d values estimated from the ^{137}Cs removal rates with measured K_d values in six lakes. As shown in Table 4.6, the measured K_d values tend to be higher (but are of the same order as) than those required to give the observed rates of radiocaesium removal to sediments. Radiocaesium distribution coefficient (K_d) measurements in freshwaters are known to vary widely as a result of experimental error, environmental variability (water chemistry, clay content of particulate matter) and 'fixation' of ^{137}Cs over time (Comans and Hockley, 1992). The use of the K_d to represent the many processes that govern the partitioning of ^{137}Cs inevitably leads to discrepancies between model parameters (i.e., the K_d used in the model) and field or laboratory measurements (the measured K_d). However, the comparison in Table 4.6 indicates that the sorption of radiocaesium to solids is sufficiently strong (i.e., K_d is sufficiently high) to explain the rates of radiocaesium removal to sediments observed in these lakes.

Table 4.6. Comparison of radiocaesium K_d determined from removal rate measurements (assuming the particulate settling model) with K_d measured in the field or laboratory.

Lake	K_d required for observed removal (1/kg)	Measured K_d (1/kg)	Source and notes
Constance	9.0×10^4	2.0×10^4	(1) Laboratory measurement
Sniardwy	0.6×10^4	1.5×10^4	(2) Laboratory measurement
Zurich	4.3×10^4	4.5×10^4	(3) *In situ* measurement
Devoke	4.3×10^4	6.0×10^4	(4) *In situ* measurement
Esthwaite	1.5×10^4	1.4×10^5	(4) *In situ* measurement
Windermere	3.3×10^4	1.3×10^5	(4) *In situ* measurement

[1] Robbins *et al.* (1992); [2] Robbins and Jasinski (1995); [3] Santschi *et al.* (1990); [4] Smith *et al.* (1997).

4.3.5 Transport of ^{90}Sr in lakes

The water–sediment distribution coefficient (K_d) of radiostrontium is relatively low so the fraction absorbed to particulates f_p is expected to be less than 5% (see Section 4.1), assuming a suspended matter concentration $s < 50\,\text{mg}\,\text{l}^{-1}$. The low affinity of ^{90}Sr for particles compared to radiocaesium suggests that transfers to sediments will be relatively low in comparison with removal through the outflow in most lakes. This is illustrated by a comparison of removal of ^{137}Cs and ^{90}Sr in the Dnieper River–Reservoir system. The different affinities of these radionuclides for suspended matter influenced their transport through the 6 reservoirs of the system (Voitsekhovitch, 2001). Caesium-137 tends to become fixed onto clay sediments which are deposited in the deep sediments of the reservoirs, particularly in the Kiev Reservoir. Because of this process, very little ^{137}Cs flows through the cascade of reservoirs, the majority being trapped in the reservoir sediments. On the other hand, although ^{90}Sr concentration decreases with distance from the source (mainly due to dilution by water inflowing from less contaminated areas), about 40–60% passes through the cascade and reaches the Black Sea. Figure 4.10 shows the trend in average annual ^{90}Sr and ^{137}Cs concentrations in the Dnieper Reservoirs since the accident. As shown in Figure 4.10, ^{137}Cs is trapped by sediments in the reservoir system, so activity concentrations in the lower part of the system are orders of magnitude lower than in the Kiev Reservoir. ^{90}Sr is not strongly bound by sediments, so concentrations in the lower part of the river–reservoir system are much closer to those measured in the Kiev Reservoir.

The peaks in ^{90}Sr activity concentration in the reservoirs of the Dnieper cascade (Figure 4.10) were caused by flooding of the most contaminated floodplains in the Chernobyl exclusion zone. For example, concentrations of ^{90}Sr in the river water increased from around $1\,\text{Bq}\,\text{l}^{-1}$ to approximately $8\,\text{Bq}\,\text{l}^{-1}$ for a 5–10-day period during the winter of 1991 (Vakulovsky *et al.*, 1994) as a result of flooding due to blockages of the river by ice. Caesium-137 activity concentrations were unaffected.

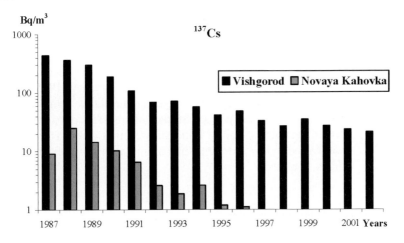
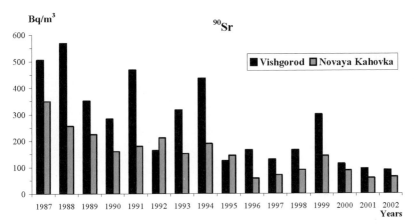

Figure 4.10. Changes in average annual content of ^{137}Cs and ^{90}Sr in the water of the first (Vishgorod, Kiev Reservoir) and last (Novaya Kahovka, Kahovka Reservoir) reservoirs of the Dnieper cascade.
From Voitsekhovitch (2001); Voitsekhovitch and Smith (2005).

Similar flood events took place during the winter flood of 1994, during summer rainfall in July 1993 and during the high spring flood in 1999.

4.3.6 Transport of ^{131}I in lakes

The rapid physical decay rate of ^{131}I (physical half-life, 8.02 days) controls its rate of removal from lake waters since, relative to physical decay, rates of transfers to sediments are low. In tracer studies in experimental lakes, it has been shown that, although ^{131}I can associate with large organic molecules in solution, the transfer to bed sediments was at a rate of only 0.018–0.033 d^{-1} (Milton et al., 1992). This is significantly lower than the ^{131}I physical decay rate of 0.086 d^{-1}. The importance of physical decay in determining ^{131}I removal from lakes is confirmed by measurements after Chernobyl. In the Kiev Reservoir, ^{131}I, with an initial concentration of 500 Bq l^{-1} declined to 20 Bq l^{-1} an estimated 37 days after the accident giving a rate of decline (K) of 0.087 d^{-1} (calculated from data in Kryshev, 1995). This rate of decline is close to the physical decay rate, suggesting negligible removal of radioactivity to the sediments during this period. Studies of transfers of radionuclides to bed sediments in the Chernobyl Cooling Pond showed that only 11% of the total inventory of ^{131}I was found in bed sediments one month after the accident (Kryshev, 1995; see also Table 4.9, p. 165).

4.3.7 Transport of ruthenium in lakes

There are relatively few data concerning ^{103,106}Ru in lakes following Chernobyl. In two lakes in the English Lake District (Windermere and Esthwaite Water), it was found that the ratio of ruthenium isotopes to ^{137}Cs in sediments during 1986 was not significantly different to the Ru : ^{137}Cs ratios in fallout (Hilton et al., 1994). These workers concluded that in these lakes the rate of Ru transfer to sediments was similar to that of ^{137}Cs. This conclusion was supported by similar reported distribution coefficients (K_ds) of the two radionuclides (Hilton et al., 1994). In studies on Lake Constance (Bodensee), however, ruthenium was removed from the water column two times faster than radiocaesium (Mangini et al., 1990). The K_d of ruthenium in this lake was of order 10^5 l kg^{-1}, significantly higher than the 2.0×10^4 l kg^{-1} measured for radiocaesium (Robbins et al., 1992).

4.3.8 Radionuclide balance in water of open lakes

Once the radioactivity which had directly deposited to the lake surface had been removed to the bed sediments or the outflow, the catchment (and potentially sediments) were a long-term source of radionuclides to the lake. In lakes with a relatively rapid inflow and outflow of water, it is postulated that average long-term radioactivity concentrations were primarily controlled by inflows in runoff water from the catchment. Under this assumption, in the long term (months/years) after fallout, the bed sediments would not (on average) act as a major source or sink of radionuclides. Therefore, average activity concentrations in the lake waters and outflow would be similar to average activity concentrations in inflowing streams.

Table 4.7 summarises measurements of the ^{137}Cs and ^{90}Sr balance in a number of European lakes. There are clearly variations in the ratio of activity concentrations in inflows compared to lake or outflow waters. A ratio greater than 1 implies a net loss of radioactivity to sediments, a ratio of less than 1 implies a net remobilisation of radioactivity from sediments. In Lake Hillesjøn in Sweden, for example, there appears to have been a significant remobilisation of ^{137}Cs from the sediments, at least during the spring period studied. In contrast, the Kiev Reservoir sediments act

Table 4.7. Mean ^{137}Cs and ^{90}Sr activity concentration (in dissolved and particulate phases) in inflow streams compared with concentrations in the lake water/outlet of different lakes.

Lake	Mean inlet conc. (Bq m^{-3}) ^{137}Cs	^{90}Sr	Mean lake or outlet conc. (Bq m^{-3}) ^{137}Cs	^{90}Sr	Ratio inlet/ (lake or outlet) ^{137}Cs	^{90}Sr	Sampling date
Brotherswater[1]	1.22	–	1.44	–	0.85	–	Spring/summer 1992
Devoke Water[1]	70.3	–	81.0	–	0.87	–	Spring/summer 1992
Loweswater[1]	12.0	–	12.2	–	0.98	–	Spring/summer 1992
Vorsee[2]	100*	–	151	–	0.66	–	Mean 1990–1994
Hillesjøn[3]	39.9	13.7	522	20.8	0.08	0.66	Spring 1991
Saarisjärvi[3]	372	4.4	215	3.8	1.73	1.15	Spring 1991
Øvre Heimdalsvatn[3]	251	13	113	9	2.22	1.44	Spring 1991
Örtrsket[4]	–	–	–	–	1.13	–	Total input 1987–1991
Kiev Reservoir[5]	369	476	225	384	1.64	1.24	Mean 1987–1993

*Approximate average of northern and southern inflows weighted by flow rate (G. Zibold, Fachhochschule Weingarten, pers. commun.).
[1] Smith et al. (1997); [2] G. Zibold, Fachhochschule Weingarten (pers. commun.); [3] Brittain et al. (1997); [4] Malmgren and Jansson (1995); [5] Sansone and Voitsekhovitch (1996).

as a sink for ^{137}Cs and, to a much lesser extent, ^{90}Sr (Voitsekhovitch, 2001), as discussed above.

Some of the variation in the ratios presented in Table 4.7 is probably due to sampling errors. For example, activity concentrations in inflow waters in particular may vary significantly, so averages of a few samples over a relatively short period of time (as is the case in some of the studies presented in Table 4.7) may be inaccurate. In spite of these problems, it can be concluded that there is little evidence of systematic differences in inlet compared to lake water/outflow activity concentrations across a number of lakes. This observation supports the hypothesis that, in the long term in most lakes, net transfers of radioactivity to and from the sediments do not have a major influence on activity concentrations in the lake water. Thus, activity concentrations in the lake and outlet are (on average) similar to those in inflowing water in the long term after fallout.

Because of the dominance of inputs of radioactivity from the catchment, long-term declines in radionuclide concentrations in open lakes are similar to those in rivers (Table 4.5). Thus, T_{eff} values for 9 open lakes were in the range 1–4 years during the period 1987–1992, as seen for rivers in Table 4.5 (Smith et al., 1999a).

4.3.9 Closed lake systems

In some lakes, where there is no (or only minor) surface inflow and outflow of water, the bed sediments play a major role in controlling radionuclide activity concentration in the water. Such lakes have been termed 'closed' lakes (Vakulovsky et al., 1994; Bulgakov et al., 2002) and, as shown in Table 4.8, have relatively much higher

Table 4.8. Normalised water concentrations (R_c, m^{-1}) of ^{137}Cs and ^{90}Sr in various water bodies 4–10 years after fallout.

Water body	Type	Date	R_c (m^{-1})
^{137}Cs after Chernobyl			
Iso Valkjärvi, Finland[1]	Closed lake	1990	26×10^{-3}
Lake Svyatoe, Bryansk, Russia[2]	Closed lake	1993–1994	19×10^{-3}
Lake Kozhanovskoe, Bryansk, Russia[2]	Closed lake	1993–1994	14×10^{-3}
Devoke Water, UK[3]	Open lake	1990–1995	2.2×10^{-3}
Ennerdale Water, UK[3]	Open lake	1990–1995	0.5×10^{-3}
Lake Constance, Switzerland[4]	Open lake	1990 1992	0.08×10^{-3}
Stream draining Opromokh peat bog, Belarus[5]	River	1993–1995	7.9×10^{-3}
River Sozh, Belarus (mineral catchment)[6]	River	1990–1995	0.4×10^{-3}
Range in 13 European rivers[7]	River	1990–1995	0.15–5.0×10^{-3}
Near-surface groundwater, Belarus (24 sites)[8]	Groundwater	1991–1997	0.03–0.7×10^{-3}
^{90}Sr after various fallout incidents			
Lake Svyatoe, Bryansk, Russia after Chernobyl[2]	Closed lake	1993–1994	14×10^{-3}
Lake Kozhanovskoe, Bryansk, after Chernobyl[2]	Closed lake	1993–1994	70×10^{-3}
Lake Uruskul, Siberia, Kyshtym accident 1957[9]	Closed lake	1962–1964	55×10^{-3}
Haweswater Reservoir, UK after NWT[10]	Open lake	1967–1971	13×10^{-3}
Range in 19 European rivers after NWT[7]	River	1968	2.1–31×10^{-3}
Two rivers in Finland after Chernobyl[11] *	River	1990–1994	30–40×10^{-3}

[1] IAEA (2000); [2] Sansone and Voitsekhovitch (1996); [3] Smith et al. (1997); [4] Zibold et al. (2002); [5] Kudelsky et al. (1996); [6] Kudelsky et al. (1998); [7] Smith et al. (2004); [8] Kudelsky et al. (2004); [9] Monte et al. (2002); [10] Linsley et al. (1982); [11] Cross et al. (2002). * These two rivers were observed to have high ^{90}Sr runoff compared to other European rivers after NWT, so these normalised concentrations for Chernobyl fallout are likely to be higher than in the majority of rivers.

^{137}Cs and ^{90}Sr activity concentrations (in the long term) than most rivers and open lake systems. There are a number of such lakes in the areas of the fSU most affected by Chernobyl, and fish from these lakes played an important part in radiocaesium intakes by some rural populations living nearby.

The most highly contaminated water bodies in the Chernobyl affected areas are the closed lakes of the Pripyat flood plain within the 30-km exclusion zone. There are many such lakes in the areas surrounding Chernobyl. During 1991, ^{137}Cs levels in these lakes were up to 74 Bq l^{-1} in Glubokoye Lake and ^{90}Sr activity concentrations were between 100 and 370 Bq l^{-1} in 6 of 17 studied water bodies (Vakulovsky et al., 1994).

As with other surface water systems, radionuclide activity concentrations in closed lakes tended to decline over time after fallout. In the early period, contamination declined only as a result of transfers to bed sediments since there are negligible losses through the lake outflow in these systems. In Lake Iso Valkjärvi in Finland, the rate of ^{137}Cs transfer to sediments K was 4.1×10^{-3} d^{-1} ($T_{eff} = 169$ days) after Chernobyl (Smith et al., 1999b) which is of the same order as rates observed in tracer experiments in Canadian Experimental Lakes of $K = 1/T_s = 14.5 \times 10^{-3}$ (Lake 226NE) and 8.4×10^{-3} (Lake 226SW) d^{-1} ($T_{eff} = 48$ and 83 days respectively, calculated from data in Hesslein, 1987). In the same study, the ^{89}Sr removal rate due to transfers to sediments was $K = 1/Ts = 10 \times 10^{-3}$ (L226NE) and 2.0×10^{-3} (L226SW) d^{-1} ($T_{eff} = 70$ and 347 days respectively). These removal rates are slower than those typically observed for open lakes of similar size because, by definition, there is no significant outflow of radioactivity in closed lake systems.

Over longer time periods after fallout, the activity concentration in closed lakes continues to decline at a slow rate as more radioactivity becomes incorporated in sediments and due to (slow) losses of water from the system. In Lake Svyatoe, Russia, it was observed that between 1993 and 1999 ^{137}Cs activity concentrations declined with an effective ecological half-life of 6.9 years (Bulgakov et al., 2002). In contrast, between 1993 and 1998, there was no significant decline in ^{137}Cs in the water of Lake Kozhanovskoe (A.V. Konoplev, A.A. Bulgakov, SPA Typhoon, unpubl. res.).

4.4 RADIONUCLIDES IN SEDIMENTS

Bed sediments are an important long term sink for radionuclides. In the Chernobyl Cooling Pond, approximately one month after the accident, most of the radioactivity was found in bed sediments (Table 4.9). In this area (i.e., within 10 km of the power plant), the majority of radionuclides were associated with hot particles (see Chapter 2), so the rapid transfer to bed sediments was largely due to sedimentation of these dense particles. This is illustrated by the contrasting behaviour of ^{90}Sr (89% in bed sediments) and ^{131}I (11% in bed sediments). Both isotopes have relatively low affinity for sediments when deposited in dissolved form. However, the majority of ^{90}Sr was deposited as fuel particles, whereas the volatile ^{131}I was primarily discharged as a vapour.

In the Cooling Pond, at present most radioactivity is found in the fine sediments in deeper areas (Table 4.10); sandy sediments along the shoreline have much lower radionuclide activity concentrations (Voitsekhovitch et al., 2002).

In the long term, approximately 99% of the radiocaesium in a lake is typically found in the bed sediment. From measurements in Lake Svyatoe (Kostiukovichy, Belarus), during 1997, it was estimated that there was 3×10^9 Bq in water and approximately 2.5×10^{11} Bq in sediments (A.V. Kudelsky, unpubl. res.). In Lake Kozhanovskoe, Russia, approximately 90% of the radiostrontium was found in the bed sediments during 1993–1994 (estimated from measurements of ^{90}Sr in water and sediment presented in Sansone and Voitsekhovitch, 1996).

Table 4.9. Radionuclides in Chernobyl Cooling Pond bed sediments approximately one month after the accident, expressed as a percentage of the total amount in both sediments and water.
From data in Kryshev (1995).

Date	^{90}Sr	^{95}Zr	^{95}Nb	^{103}Ru	^{106}Ru	^{131}I	^{134}Cs	^{137}Cs	^{140}Ba	^{140}La	^{141}Ce	^{144}Ce
30/5/86	89	96	94	95	92	11	67	65	77	78	93	97

Table 4.10. Typical radionuclide activity concentrations in the most contaminated silty sediments of the Cooling Pond (from data in Voitsekhovitch et al., 2002).

RN	Activity concentration in sediment (kBq kg^{-1})
Caesium-137	600
Strontium-90	110
Americium-241	5
Plutonium-239/240	2.4
Plutonium-238	1.2

Radiocaesium is relatively immobile in lake sediments, though some dispersion may occur by physical mixing of the sediments (by biota and water currents) or diffusion in the sediment pore waters. Studies into the sorption of Chernobyl radiocaesium to sediments have shown that the solid–aqueous distribution coefficient (K_d) is inversely proportional to the content of competing ions (specifically K^+ and NH_4^+) in the sediment pore waters (Comans et al., 1989). In stratified lakes, NH_4^+ in anoxic sediments and bottom waters can lead to remobilisation of radiocaesium from contaminated sediments (Evans et al., 1983), although it has been estimated that less than 3% of the sediment inventory could be remobilised per year by this mechanism (Smith and Comans, 1996).

The sediment composition also plays an important role in determining radiocaesium mobility. Figure 4.11 shows examples of ^{137}Cs activity–depth profiles in different sediments of marine and freshwater systems. In the Baltic Sea (Figure 4.11(a)), ^{137}Cs deposition was associated mainly with fine sediments, and dispersion in the sediment is relatively low in the muddy (fine) sediment, as indicated by a sharp decline in activity concentration with depth from the sediment surface. In Lake Constance, where bed sediments are relatively rich in illite clay minerals, clear peaks resulting from NWT and Chernobyl depositions are observed in the ^{137}Cs activity–depth profiles. The immobility of radiocaesium in many lake sediments means that the depth of the peaks representing the maxima of Chernobyl (1986) and NWT (1963) depositions (illustrated in Figure 4.11(b)) can be used to estimate the rate of sediment accumulation in lakes (e.g., Appleby, 1997).

The sediment profile from Lake Svyatoe (Kostiukovichy, Belarus, Figure 4.11(c)), in contrast to that from Lake Constance, shows a maximum ^{137}Cs concentration near the sediment surface, indicating that there has been little or no net sediment accumulation in this lake during the 14 years between fallout and the sampling date (16 February, 2000). The Lake Svyatoe activity–depth profile is also much more disperse (i.e., the change in activity concentration with depth is much less steep) than that from Lake Constance. This may be attributed to physical mixing of the sediments of this shallow lake, and/or greater molecular diffusion of radiocaesium since the sediments are high in organic matter and sand content and hence do not strongly 'fix' radiocaesium (Ovsiannikova et al., 2004). We would

Sec. 4.4]   Radionuclides in sediments   167

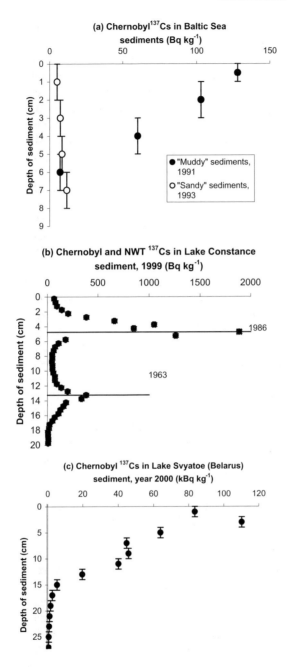

Figure 4.11. Graphs of ^{137}Cs activity–depth profiles in sediments in (a) Baltic Sea, muddy and sandy sediments; (b) Lake Constance; (c) Lake Svyatoe, Kostiukovichy, Belarus.
(a) Knapinska-Skiba et al. (2001). (b) G. Zibold and E. Klemt, Fachhochschule Weingarten (pers. commun.). (c) A.V Kudelsky, unpubl. res.

not expect to see any contribution from NWT in the Lake Svyatoe sediments as it is obscured by the much higher Chernobyl fallout.

Radionuclides which were deposited in the form of fuel particles are generally less mobile than those deposited in dissolved form since transport only occurs by physical mixing and accumulation of the sediment and not by molecular diffusion. In sediments of Glubokoye Lake (in the 30-km zone) in 1993, most fuel particles remained in the surface 5 cm of sediment (Sansone and Voitsekhovitch, 1996). It has also been observed (see Chapter 2) that fuel particle breakdown was at a much lower rate in lake sediments compared to soils.

4.5 UPTAKE OF RADIONUCLIDES TO AQUATIC BIOTA

4.5.1 ^{137}Cs in freshwater fish

There have been many studies of the levels of radiocaesium contamination of freshwater fish during the years after the Chernobyl accident. As a result of high radiocaesium bioaccumulation factors, fish have remained contaminated despite relatively low radiocaesium levels in water. In some cases, activity concentrations in fish have greatly exceeded the European Union intervention level for radiocaesium activity in fish of 1,250 Bq kg^{-1} wet weight (w.w.).

In the Chernobyl Cooling Pond, ^{137}Cs levels in carp (*Cyprinus carpio*), silver bream (*Blicca bjoerkna*), perch (*Perca fluviatilis*) and pike (*Esox lucius*) were of the order of 100 kBq kg^{-1} w.w. in 1986, declining to a few tens of kBq kg^{-1} in 1990 (Kryshev and Ryabov, 1990; Kryshev, 1995). In the Kiev Reservoir, activity concentrations in fish were in the range 0.6–1.6 kBq kg^{-1} w.w. (in 1987) and 0.2–0.8 kBq kg^{-1} w.w. (from 1990–1995) for adult non-predatory fish and 1–7 kBq kg^{-1} (in 1987) and 0.2–1.2 kBq kg^{-1} (from 1990–1995) for predatory fish species. In lakes in the Bryansk region of Russia, activity concentrations in a number of fish species varied within the range 0.215–18.9 kBq.kg^{-1} w.w. during the period 1990–1992 (Fleishman *et al.*, 1994; Sansone and Voitsekhovitch, 1996). It was estimated that about 14,000 lakes in Sweden had fish with ^{137}Cs concentrations above 1,500 Bq kg^{-1} (the Swedish guideline value) in 1987 (Håkanson *et al.*, 1992). In a small lake in Germany, levels in pike were up to 5 kBq kg^{-1} shortly after the Chernobyl accident (Klemt *et al.*, 1998). In Devoke Water in the English Lake District, perch and brown trout (*Salmo trutta*) contained around 1 kBq kg^{-1} in 1988 declining slowly to a few hundreds of Bequerels per kg in 1993 (Camplin *et al.*, 1989; Smith *et al.*, 2000b).

The contamination of fish following the Chernobyl accident was a cause for concern in the short term (months) for less contaminated areas (e.g., parts of the UK and Germany) and in the long term (years-decades) in the Chernobyl affected areas of the Ukraine, Belarus, Russia and parts of Scandinavia.

It is known that the bioaccumulation of radioactivity in fish is determined by numerous ecological and environmental factors such as the trophic level of the fish species, the length of the food chain, water temperature and the water chemistry.

Uptake may be via ingestion of contaminated food or direct transfers from the water via the gills. For many radionuclides, including radiocaesium, the food chain is the primary uptake pathway, so a food uptake model is usually used to estimate uptake rates. Since fish feeding rate is strongly influenced by temperature (Elliot, 1975), the uptake rate of radionuclides absorbed through food tends to be faster at higher water temperatures.

The level of radioactive contamination of aquatic biota is commonly defined in terms of a concentration factor (CF) where

$$\text{CF} = \frac{\text{Activity concentration per kg of fish (wet weight)}}{\text{Activity concentration per litre of water}} \, 1\,\text{kg}^{-1} \quad (4.9)$$

Previous studies on the accumulation of radiocaesium in fish have focused on the prediction of CF (sometimes termed the bioaccumulation factor (BAF) or aggregated concentration factor (ACF)). The equilibrium CF modelling approach is appropriate for cases in which the radionuclide activity concentration in fish can be assumed to be in equilibrium with that of water, for example at long times (years) after radionuclide fallout, or for continuous releases of radionuclides to a river. For short-term releases of radionuclides to an aquatic system, however, dynamic models may be required.

The activity concentration of a radionuclide in fish C_f (Bq kg^{-1}) is often modelled by a simple 'two-box' model describing uptake from the water C_w (Bq l^{-1}) and release from the fish (Figure 4.12):

$$\frac{dC_f}{dt} = k_f C_w - (k_b + \lambda) C_f \quad (4.10)$$

where k_f (1 kg^{-1} y^{-1}) is the rate constant describing transfers of ^{137}Cs to fish through its food and k_b (y^{-1}) is the backward rate constant describing excretion of radioactivity from the fish. The ratio of these rate constants gives the equilibrium CF (1 kg^{-1}) of the radionuclide in fish relative to water:

$$\frac{k_f}{k_b + \lambda} = \frac{C_f}{C_w} \text{ (at equilibrium)} = \text{CF} \quad (4.11)$$

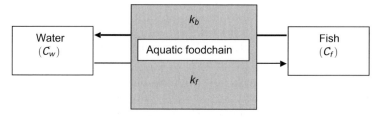

Figure 4.12. Illustration of a simple model for uptake in fish via the food chain.

4.5.2 Influence of trophic level on radiocaesium accumulation in fish

Maximum radiocaesium activity concentrations in non-predatory fish were observed within the first few months after the accident (Kryshev, 1995), indicating a rapid uptake. In predatory fish, however, there is a delay in the transfer of radiocaesium as it takes some time to accumulate up the food chain. In many systems, maximum ^{137}Cs activity concentrations in predatory fish were not observed until a period of between several months and one year after the accident (Brittain et al., 1991; Elliot et al., 1992; Kryshev, 1995). The dynamics of radiocaesium in water and fish are illustrated using data from Lake Vorsee and Lake Constance after Chernobyl (Figure 4.8; see Section 4.3). At long times (years) after fallout, the radionuclide uptake and excretion processes reached a steady state, so activity concentrations in fish changed at the same rate as those in water (Figure 4.8; Equation 4.11).

A comprehensive study of radiocaesium bioaccumulation in fish was carried out in Canadian aquatic systems contaminated by NWT fallout. In this study (Rowan et al., 1998) it was shown that there was a clear 'biomagnification' effect in radiocaesium activity concentrations as trophic level increased. In the Ottawa River, different organisms were assigned different trophic positions according to their diet. Zooplankton and molluscs were assigned position 2.0, planktivorous fish (one level higher in the trophic chain) occupied position 3.0 and wholly piscivorous fish occupied position 4.0. Omnivorous species were assigned positions between 3.0 and 4.0 according to the composition of their diet. It was found (Rowan et al., 1998) that there was an approximately four-fold difference in ^{137}Cs with each unit increase in trophic position (e.g., planktivorous fish had approximately four times as much ^{137}Cs (per kg) as zooplankton and molluscs). It was further observed in this study that, for a given trophic level, organisms with diets composed of benthic (bottom-dwelling) species had approximately 1.7 times greater ^{137}Cs activity concentration compared to organisms feeding from the open water food chain. This reflected the high accumulation of radiocaesium in bed sediments.

Similar observations have been made of radiocaesium accumulation in fish contaminated by the Chernobyl accident. In 1988, for example, radiocaesium concentration factors for predatory fish in the Chernobyl Cooling Pond and the Dnieper River were 4–5 times higher than for non-predatory fish (Kryshev and Sazykina, 1994).

4.5.3 Size and age effects on radiocaesium accumulation

The 'size effect' of radiocaesium accumulation in fish results in an increasing contamination (per unit weight of fish) with increasing fish size (Elliott et al., 1992; Hadderingh et al., 1997). In a study of ^{137}Cs in fish in the Pripyat River and the Kiev Reservoir during 1992, Hadderingh et al. (1997) found that CFs in non-predatory fish were generally independent of fish size. In contrast, predatory perch and pike showed an increase in ^{137}Cs concentration with increasing weight of fish (Figure 4.13). Increasing ^{137}Cs CFs with increasing fish size may be a result of environmental factors (e.g., different feeding habits, differing biological retention

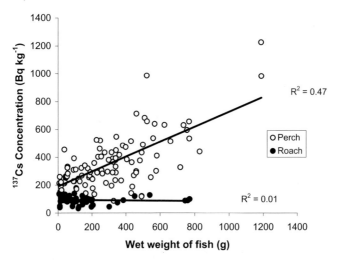

Figure 4.13. Radiocaesium in fish in the Kiev Reservoir after Chernobyl, illustrating the 'size effect' in predatory perch, but not in the non-predatory roach.
Adapted from Sansone and Voitsekhovitch (1996).

times in different sized fish), or the effect may be caused by non-equilibrium in the fish–water system. If radiocaesium concentrations in a water body are declining with time, then older, and hence larger, fish may retain relatively higher radiocaesium concentrations because they grew up in a scenario of higher water concentrations than younger fish.

The 'size effect' may also be caused by changes in fish diet as they grow. Ugedal et al. (1995) observed an increase in ^{137}Cs activity concentration in Arctic charr (*Salvelinus alpinus*) as the diet shifted from zooplankton in small fish to zoobenthos in larger fish. In perch, it was found that much of the 'size effect' could be explained by differences in feeding habits between small and large fish (Rowan et al., 1998; Smith et al., 2002). As perch mature, their diet changes from eating mainly invertebrates to being primarily piscivorous, though the size at which they do this varies depending on available food, competition and other environmental factors.

4.5.4 Influence of water chemistry on radiocaesium accumulation in fish

Because of its chemical similarity to potassium, radiocaesium can concentrate in organisms via the same accumulation mechanisms as potassium, an important nutrient. Potassium concentrations are regulated in organisms, so in waters with abundant potassium, bioaccumulation of potassium is weaker than in waters where it is scarce. Since caesium is bioaccumulated by organisms in the same way as potassium, caesium concentration factors tend also to be high in waters with low potassium. Studies on NWT derived ^{137}Cs found that the CF of radiocaesium in fish was inversely proportional to the potassium content of the surrounding water

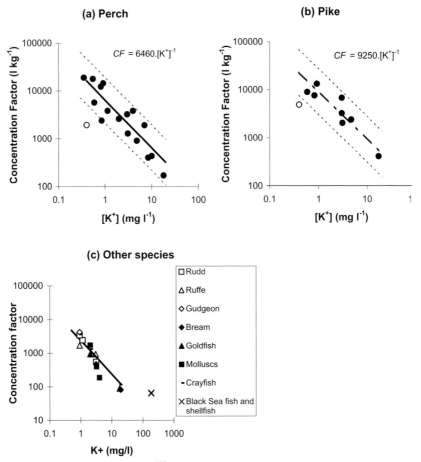

Figure 4.14. Relationship between ^{137}Cs concentration factor (w.w.) in fish and the potassium concentration in 17 lakes around Europe.* The open circles in (a) and (b) show data from the low pH Lake Iso Valkjärvi, see text. Notice that the measurement for the Black Sea (c)† does not follow the inverse trend observed for freshwaters (see Section 4.6.2).

* From Smith *et al.* (2000c, 2002). † From Baysal and Tunçer (1994; converted to w.w. basis).

(Kolehmainen, 1967; Fleishman, 1973; Blaylock, 1982; Rowan and Rasmussen, 1994). This relationship was also observed in fish contaminated after Chernobyl (Smith *et al.*, 2000c), as shown in Figure 4.14.

It is likely that other chemical conditions in water also affect radiocaesium uptake to fish, though the effect of potassium is dominant. In Lake Iso Valkjärvi in Finland, it was found that the radiocaesium CF was lower than expected for this low potassium concentration lake. This lake has low pH (high H^+) and low calcium concentration (IAEA, 1994). Interference in the potassium uptake mechanism (and therefore in caesium accumulation) has been hypothesised in conditions of low pH or low calcium concentration (Smith *et al.*, 2002).

4.5.5 ^{131}I in freshwater fish

Kryshev (1995) presented measurements of ^{131}I in fish in the Kiev Reservoir shortly after Chernobyl. Iodine-131 was rapidly absorbed, maximum concentrations in fish being observed at the beginning of May 1986. The fish–water CF was around 10 l kg^{-1}, and activity concentrations in fish muscle declined from around 6,000 Bq kg^{-1} on 1 May, 1986 to 50 Bq kg^{-1} on 20 June, 1986. This represents a rate of decline similar to the rate of physical decay of ^{131}I ($T_{1/2} = 8.1$ days). As with other vertebrate species, there is a tendency for fish to concentrate iodine in the thyroid gland (Vinogradov, 1953).

4.5.6 ^{90}Sr in freshwater fish

Strontium behaves chemically and biologically in a similar way to calcium. Strontium is most strongly bioaccumulated in low calcium ('soft') waters (this is analogous to the relationship between caesium and potassium discussed above). There is less quantitative information available for uptake and retention rates of ^{90}Sr in fish than for ^{137}Cs, however it has been shown that bioaccumulation is inversely proportional to the calcium concentration of water (Vanderploeg *et al.*, 1975, quoted in Blaylock, 1982). Like calcium, strontium is primarily absorbed in the bony parts of the fish (skeleton, fins, scales). Using CFs estimated by Vanderploeg *et al.* (1975, quoted in Blaylock, 1982), it is estimated that approximately 95% of the strontium in a fish is found in the bony parts and only 5% in the soft tissues. It is known (Chowdhury and Blust, 2001) that strontium can be absorbed through the gills of fish, though it is believed that intake of food plays an important role in strontium uptake (Kryshev, 2003).

Estimates of ^{90}Sr CF values for fish in some lakes in the Ukraine, Russia and Belarus are given in Table 4.11. The model of Vanderploeg *et al.* (1975, quoted in

Table 4.11. ^{90}Sr concentration factors in freshwater fish after Chernobyl (whole fish, w.w.).

Water body	Sampling date	^{90}Sr CF (l kg^{-1})	Calcium conc. in water (μM l^{-1})
Cooling Pond	July–December 1986	100[1]	1,175[2]
Kiev Reservoir	1987–1988	99[1] (Pike-perch)	1,300[3]
		46[1] (Bream)	
Lake Kozhanovskoe	1993–1994	140[2] (Crucian carp)	1,010
		173[2] (Goldfish)	
		365[2] (Perch)	
Lake Perstok	2003	452 (Roach, Rudd)	660
		239 (Perch)	
Glubokoye Lake	2003	190 (Roach)[4]	738[2]

[1] Kryshev (1995); [2] I.N. Ryabov, Severtsov Institute, Moscow (pers. commun.); [3] Sansone and Voitsekhovitch (1996); [4] N.V. Belova, Severtsov Institute, Moscow (pers. commun)

Blaylock, 1982) using the observed Ca concentration in these lakes predicts CF values in the range 30–70 $l\,kg^{-1}$ which are significantly lower than the measurements in Table 4.11, particularly for measurements made many years after the accident. This suggests that accumulation in bones may continue to increase ^{90}Sr activity concentrations in fish in the long term.

In general, whole fish–water CFs of ^{90}Sr (of order $10^2 l\,kg^{-1}$; Table 4.11) were significantly lower than for radiocaesium. In addition to its generally lower fallout, this meant that ^{90}Sr levels in fish were typically much lower than those of ^{137}Cs. In the Chernobyl Cooling Pond, ^{90}Sr activity concentrations were around $2\,kBq\,kg^{-1}$ in fish during 1986, compared with around $100\,kBq\,kg^{-1}$ for ^{137}Cs (Kryshev, 1995). Freshwater molluscs showed significantly stronger bioaccumulation of ^{90}Sr than fish. In the Dnieper River, molluscs had approximately 10 times more ^{90}Sr in their soft tissues than was found in fish muscle (Kryshev and Sazykina, 1994).

4.5.7 Radiocaesium and radiostrontium in aquatic plants

The accumulation of radiocaesium in aquatic plants in various freshwater and marine systems is summarised in Table 4.12. CFs in aquatic plants tend to be lower than in fish (a result of the 'biomagnification' effect with increasing trophic level), but, as with fish, a strong inverse relationship with potassium concentration is observed.

CFs of ^{137}Cs and ^{90}Sr in various aquatic plants in lakes in the 30-km zone were studied by Gudkov et al. (2001). Accumulation of both radionuclides varied between species. ^{137}Cs accumulation was higher in summer than in spring or autumn, though there was little obvious pattern in seasonal changes of ^{90}Sr. Mean CF values (w.w. basis) in different species of aquatic plants varied from around 130–580 $l\,kg^{-1}$ for ^{137}Cs and from 8–55 $l\,kg^{-1}$ for ^{90}Sr (Gudkov et al., 2001).

4.5.8 Bioaccumulation of various other radionuclides

There are relatively few data available on the accumulation of radionuclides other than Cs, Sr and I in freshwater biota. For short-lived radionuclides, bioaccumulation factors must be interpreted with care: they are unlikely to represent an equilibrium condition since activity concentrations in water and biota are changing rapidly. It is also believed that accumulation of Cs, Sr and I is by biological uptake whereas other radionuclides are adsorbed to the surfaces of organisms (Kryshev and Sazykina, 1994).

Accumulation factors of different radionuclides in aquatic organisms of the Dnieper River illustrate the difficulties of predicting bioaccumulation of radionuclides other than the more intensively studied caesium, strontium and iodine (Table 4.13). Measurements in June 1986 of accumulation factors in the Dnieper (Kryshev and Sazykina, 1994) showed that isotopes of Ce, Zr, Nb, Ba, La and Ru had much higher accumulation factors in molluscs and aquatic plants than isotopes of Cs, Sr and I (Table 4.13). Kryshev and Sazykina (1994) noted that 'owing to the processes of radioactive decay and deposition of radionuclides to the bottom with

Table 4.12. Mean CF of radiocaesium in aquatic plants (w.w. basis).

Site	Date	Species sampled	CF ($l\,kg^{-1}$)	[K$^+$] ($\mu M\,l^{-1}$)	Reference
Cooling Pond	1986–1987	*Potamogeton pectinatus* *Potamogeton perfoliatus*	294*	103	(1)
Dnieper River	1986–1987	*Myriophyllum spicatum* *Ceratophyllum spicatum* *Cladophora glomerata Kuetz*	294*	82	(1)
Crummock water	1986–1987	Aquatic moss (unspecified) *Juncus Bulbous*	7,640	9.0	(2)
Devoke water	1986–1987	Aquatic moss (unspecified) *Fontinalis* *Juncus Bulbous* *Spirogyra* *Subularia Aquatica*	3,770	14	(2)
Loweswater	1986–1987	*Elodea canadensis* *Fontinalis* *Potamogeton gramineus* *Subularia aquatica*	4,750	21	(2)
Black Sea, Turkish coast	1992	*Zostera marina*	50*	4,750†	(3, 4)
Baltic Sea at Kämpinge, Sweden	1987–1988	*Fucus vesiculosus*	40*	2,370**	(5)

* Estimated from CF measured on a dry weight basis and converted to wet weight by assuming dry weight/wet weight = 0.07 for aquatic plants. † Estimated assuming a salinity of 17.0 in the Black Sea, Turkish coast. ** Estimated assuming a salinity of 8.5 in the Southern Baltic.
(1) Kryshev and Sazykina (1994); (2) Camplin *et al.* (1989); (3) Baysal and Tunçer (1994); (4) Vakulovsky *et al.* (1994); (5) Carlson and Holm (1992).

suspended matter, these estimates are highly arbitrary'. They also note, importantly, that '[the estimates] may differ ... from the so called equilibrium accumulation coefficients which are measured in laboratory experimental conditions, but practically never occur in nature'. With these caveats, however, it appears that most of the measurements for fish are within or relatively close to the ranges of CF values for fish given by IAEA (1994).

4.6 RADIOACTIVITY IN MARINE SYSTEMS

Marine systems were not seriously affected by fallout from Chernobyl, the nearest seas to the reactor being the Black Sea (approximate distance 520 km) and the Baltic Sea (approximate distance 750 km). The primary pathway of contamination of these seas was atmospheric fallout, with smaller inputs from riverine transport occurring over the years following the accident. Surface deposition of ^{137}Cs was approximately

Table 4.13. Radionuclide CFs[†] in biota of the Dnieper River in June 1986. For fish, measurements are compared with estimated CF's from IAEA (1994).
From Kryshev and Sazykina (1994).

RN	Molluscs (w.w.)	Aquatic plants (w.w)*	Fish (w.w.)	Fish CF from IAEA (1994)
^{90}Sr	440	16.8	50	10^0–10^3
^{103,106}Ru	750, 1,000	770, 1190	120, 130	10^1–2×10^2
^{131}I	120	4.2	30	2×10^1–6×10^2
^{134,137}Cs	270–300	189, 210	300, 300	3×10^1–3×10^3
^{95}Zr	2,900	1,400	190	3–3×10^2
^{95}Nb	3,700	1,540	220	1×10^2–3×10^4
^{140}Ba	2,800	252	420	4–2×10^2
^{140}La	2,400	210	400	30
^{141,144}Ce	4,600	1,680	900	30–5×10^2

[†] Estimates are means of more than 20 samples. Standard deviations of measured values were typically in the range 20–50%. *Estimated from CF measured on a dry weight basis and converted to wet weight by assuming dry weight/wet weight = 0.07 for aquatic plants.

2.8 PBq over the Black Sea (Eremeev et al., 1995) and 3.0 PBq over the Baltic Sea (Vakulovsky et al., 1994). Using estimates of the seas' surface areas of 423,000 km^2 and 415,000 km^2, this gives average surface fallout of ^{137}Cs of 6,630 Bq m^{-2} and 7,130 Bq m^{-2} for the Black and Baltic Seas respectively. The total fallout of ^{137}Cs from Chernobyl to the world's seas and oceans was estimated to be 15–20 PBq (Aarkrog, 1998).

Fallout onto the seas' surface was not uniform: in the Black Sea surface water, concentrations of ^{137}Cs ranged from 14.8–503.2 mBq l^{-1} in June–July of 1986. By 1989, horizontal mixing of surface waters had resulted in relatively uniform concentrations in the range 40.7–77.7 mBq l^{-1} (Vakulovsky et al., 1994).

The maximum radioactive fallout onto the Baltic Sea occurred in the northern Gulf of Bothnia and Gulf of Finland where ^{137}Cs activity concentrations of up to 930 mBq l^{-1} were observed (Vakulovsky et al., 1994). By 1988–1989, ^{137}Cs activity concentrations were much more uniform being in the range 100–200 mBq l^{-1} (Vakulovsky et al., 1994; Carlson and Holm, 1992), as shown in Figure 4.15. Vakulovsky et al. (1994) report only a minor increase of 20% in pre-accident levels of ^{90}Sr in this area. Holm (1995) observed that Chernobyl fallout was a less significant source of Pu in the Baltic Sea than NWT.

Vertical mixing of surface deposited radioactivity also reduced maximum radioactivity concentrations observed in water over the months to years after fallout. Removal of radioactivity to deeper waters steadily reduced ^{137}Cs activity concentrations in the surface (0–50 m) layer of the Black Sea (Figure 4.15). Profiles of ^{137}Cs with depth (Vakulovsky et al., 1994) showed that by October of 1986 the fallout had mixed vertically within the surface 30–40-m layer of water and by June of 1987, significant amounts of radioactivity had penetrated below a depth of 50 m. The reduction in radiocaesium in the northern Baltic Sea, and the slight increase in

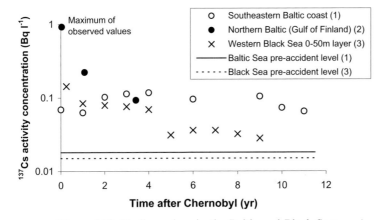

Figure 4.15. Radiocaesium in the Baltic and Black Seas.
From data in: (1) Styro et al. (2001); (2) Vakulovsky et al. (1994); (3) Kanivets et al. (1999).

concentrations in the southern Baltic (Figure 4.15) may be due to horizontal mixing of these differently contaminated waters.

The influence of transfers of Chernobyl radionuclides to seabed sediments has not to our knowledge been quantified. Low K_d's of ^{137}Cs and ^{90}Sr in saline waters suggest that the influence will be low for these isotopes. For example, it is estimated that <1% of radiocaesium in the Baltic is in the dissolved phase of seawater (Knapinska-Skiba et al., 2001). These authors observed accumulation of radiocaesium in fine, muddy sediments but little accumulation in sandy sediments of the southern Baltic. In the Gulf of Finland, up to $80\,\text{kBq}\,\text{m}^{-2}$ of ^{137}Cs was observed in bed sediments of coastal waters. In the Black Sea, anoxic bed sediments and bottom waters are likely to have resulted in low transfer rates of radiocaesium to bed sediments. Pu isotopes are more strongly bound to sediments, and studies in the Baltic Sea indicate a residence time of Pu in the water column of 8–10 years (Holm, 1995).

Outflow of water from the Baltic to the North Sea and from the Black Sea through the Bosphorus Strait had little effect on activity concentrations in the seas since water removal is slow. The residence time of water in the Baltic is 25–35 years and in the Black Sea it is of order 1,000 years (the latter estimated from data in Kanivets et al., 1999).

4.6.1 Riverine inputs to marine systems

Riverine input of radioactivity (primarily from the Danube and Dnieper Rivers) to the Black Sea was much less significant than direct atmospheric fallout to the sea surface. Over the period 1986–1995, riverine input for ^{137}Cs was only 4% of the atmospheric deposition, though ^{90}Sr inputs were more significant, being approximately 25% of the total inputs from atmospheric deposition and rivers together (Kanivets et al., 1999). For the Baltic Sea, riverine inputs were similar to those observed in the Black Sea, being approximately 4% and 35% of atmospheric fallout for ^{137}Cs and ^{90}Sr, respectively (Nielsen et al., 1999). The greater relative riverine input of ^{90}Sr compared to ^{137}Cs is due to its weaker adsorption to catchment soils and to lake and river sediments. In addition, a greater proportion of the ^{90}Sr fallout was deposited near to the reactor compared with ^{137}Cs (Figure 2.1) so increasing the influence of transport from the catchment compared to atmospheric fallout to the (distant) Black and Baltic Seas.

4.6.2 Transfers of radionuclides to marine biota

Bioaccumulation of radiocaesium and radiostrontium in marine systems is generally lower than in freshwater because of the much higher concentration of competing ions in saline waters. This is illustrated by comparison of the CFs of ^{137}Cs in freshwater and marine systems for aquatic plants (Table 4.12) and fish (Figure 4.14). Notice, however, that the CF of marine plants and animals is only slightly lower than in freshwaters of high potassium concentration, even though the potassium concentration of marine systems is more than one order of magnitude higher than the maximum typically observed in freshwater. It appears that, relative to potassium, radiocaesium is more strongly accumulated in marine biota than in freshwaters. This confirms similar observations made prior to Chernobyl (Fleishman, 1973).

The relatively low fallout of ^{90}Sr on marine systems and the relatively low bioaccumulation of ^{90}Sr in high calcium content saline waters (Fleishman, 1973) imply that ^{90}Sr contamination of marine species was not significant after Chernobyl. Activity concentrations of other radionuclides in marine systems were also relatively low, though ^{110m}Ag, ^{95}Zr and ^{103,106}Ru were observed in marine macroalgae (*Fucus vesiculosus*) in the Baltic Sea during July 1986 (Carlson and Holm, 1992). Mean ratios of ^{134}Cs, ^{103,106}Ru and ^{95}Zr (corrected for decay) were similar to those observed in atmospheric fallout in this area (Table 4.14), though there was considerable variability in these ratios between different samples.

Table 4.14. Radionuclides in marine macroalgae and fallout compared to ^{137}Cs in July 1986 and August–September 1987.

Calculated from data in Carlson and Holm (1992).*

RN	Ratio RN: ^{137}Cs in macroalgae	Ratio RN: ^{137}Cs in fallout†
^{137}Cs	1.0	1.0
^{134}Cs (July 1986)	0.54	0.55
^{134}Cs (August–September 1987)	0.56	0.55
^{103}Ru (July 1986)	1.19	1.27
^{106}Ru (July 1986)	0.32	0.28
^{106}Ru (August–September 1987)	0.17	0.28
^{95}Zr (July 1986)	0.084	0.051

* We estimated ratios in marine macroalgae from averages of the ratio at 7 sites which showed the highest activity concentrations in macroalgae and were decay corrected to the date of fallout. † Calculated from relationships in Mück et al. (2002) for a distance 1,200 km from Chernobyl.

No evidence was found in the study of Carlson and Holm (1992) for increases in ^{99}Tc, ^{238}Pu, $^{239+240}$Pu and ^{241}Am in macroalgae (compared to background levels from other sources) following Chernobyl. It was, however, suggested that ^{241}Pu from Chernobyl could potentially be observable since the ^{241}Pu:$^{239+240}$Pu activity ratio was about 86 in Chernobyl fallout compared to 6 in the Baltic Sea before the accident (Carlson and Holm, 1992). Though ^{241}Pu was not directly measured, it was suggested that this may lead to an observable increase in ^{241}Am as a result of ingrowth from ^{241}Pu in the coming decades.

4.7 RADIONUCLIDES IN GROUNDWATER AND IRRIGATION WATER

4.7.1 Radionuclides in groundwater

Transfers of radionuclides to groundwaters has occurred from waste disposal sites in the exclusion zone. After the accident, radioactive debris as well as trees from the 'Red Forest' were buried in shallow unlined trenches. At these waste disposal sites, ^{90}Sr activity concentrations in groundwaters were in some cases of the order of 1,000 Bq l^{-1} (Voitsekhovitch et al., 1996). Health risks from groundwaters to hypothetical residents of these areas, however, were shown to be low in comparison to external radiation and internal doses from foodstuffs (Bugai et al., 1996). Although there is a potential for off-site (i.e., outside the 30-km zone) transfer of radionuclides from the disposal sites, these workers concluded that this will not be significant in comparison to washout of surface deposited radioactivity. Off-site transport of contaminated groundwater around the Sarcophagus is also expected to be insignificant since radioactivity in the Sarcophagus is separated from ground waters by an unsaturated zone of thickness 5–6 m, and groundwater velocities are low (Bugai et al., 1996).

Radionuclides could potentially contaminate groundwater by migration of radioactivity deposited on the surface soils. It is known (see Chapter 2), however, that long-lived radionuclides such as ^{137}Cs and ^{90}Sr are relatively immobile in surface soils and transfers from surface fallout to deep groundwaters are expected to be very low in comparison to transfers from surface runoff to rivers and lakes. After fallout from NWT, it was observed that ^{90}Sr in Danish groundwater was approximately 10 times lower than in surface streams (Hansen and Aarkrog, 1990). These authors also observed that after Chernobyl, despite measureable quantities of ^{137}Cs in surface streams, activity concentrations were below detection limits in groundwater. Short-lived radionuclides are not expected to affect groundwater supplies since ground-water residence times are much longer than their physical decay time.

In Belarus, the ^{137}Cs activity concentration of water in the saturated zone of soils had no observable correlation with the water level (which was between 35 and 130 cm below the soil surface) but had a significant, but weak ($R^2 = 0.44$, $p < 0.05$) correlation with the ^{137}Cs contamination density of the soils (Smith et al., 2001; Kudelsky et al., 2004). At 10 different sites, the activity concentration in groundwater (per unit of radiocaesium deposition) was significantly lower than in most river and lake systems (Table 4.8).

4.7.2 Irrigation water

There is little information available on the use of contaminated water for irrigation purposes. If the water originates from contaminated areas and is used to irrigate similarly contaminated soils, it will add little to the radioactivity in crops since activity levels in the irrigation water will be comparable to those in the *in situ* soil water. In areas where water originating in a contaminated area is used to irrigate much less contaminated soils, there is a potential problem. This latter scenario is not unlikely, since upland reservoirs often supply irrigation water for lowland soils.

A large amount (1.8 million hectares) of agricultural land in the Dnieper basin is irrigated. Almost 72% of this area is irrigated with water from the Kakhovka Reservoir in the Dnieper River–Reservoir system. Accumulation of radionuclides in plants on irrigated fields can take place because of root uptake of radionuclides introduced with irrigation water and due to direct incorporation of radionuclides through leaves after sprinkling. However, recent studies (O.V. Voitsekhovitch, unpubl. res.) have shown that, in the case of irrigated lands of the southern Ukraine, radioactivity in irrigation water did not add significant radioactivity to crops in comparison with that which had been initially deposited in atmospheric fallout and subsequently taken up *in situ* from the soil.

4.8 RADIATION EXPOSURES VIA THE AQUATIC PATHWAY

Doses from ^{137}Cs and ^{90}Sr contamination of waterbodies in the countries of the fSU are difficult to quantify. Doses from the freshwater pathway (including fish and irrigation water) to the people of Kiev were relatively low, being around 5–10% of doses via terrestrial foodstuffs (Konoplev *et al.*, 1996). As radioactivity (mainly ^{90}Sr) was transferred to areas not significantly contaminated by fallout to terrestrial systems in the lower reaches of the Dnieper cascade of reservoirs, the relative contribution to dose via freshwaters increased to 10–20%.

Radionuclides in the Pripyat River could potentially have led to significant doses in the first months after the accident through consumption of drinking water (Table 4.2). The most significant potential dose calculated (assuming the drinking water activity concentration was equal to that in the river) was 4.2 mSv from ^{131}I (Table 4.2). It is likely, though, that there was significant reduction in activity concentrations within the water supply system, so these doses are likely to be overestimates. Studies in the UK after Chernobyl indicated that in one water supply area, 80% of the radioactivity in raw water was removed by the water treatment process (Jones and Castle, 1987). In an experiment carried out in a water treatment plant in Belgium, it was found that removal efficiencies by conventional water treatment were 73%, 61%, 17% and 56% for Ru, Co, I and Cs respectively (Goosens *et al.*, 1989).

Contamination of fish led to significant doses in some areas. The critical group amongst the 30 million users of Dnieper water were commercial fishermen in the Kiev Reservoir who, in 1986, received a dose of 5 mSv from ^{137}Cs contamination of fish (Berkovski et al., 1996). These commercial fishermen were estimated to consume up to 360 kg of fish per year, fish being the main component of their diet. For the general population living around the Dnieper river-reservoir system, consumption (and hence ^{137}Cs ingestion from fish) was much lower, being around 5–7 kg per year. These consumption estimates can be compared with critical group freshwater fish consumption estimates in the UK of 20 kg per year (NRPB, 1996).

In rural parts of the Chernobyl contaminated areas of the fSU during 1994–1995, it was found that so-called 'wild foods' (mushrooms, berries, freshwater fish, game animals) had radiocaesium contents which were around one order of magnitude higher than agricultural products (e.g., milk or meat). Whole body monitoring of people living close to Lake Kozhanovskoe, Bryansk, showed (Travnikova et al., 2004) that ^{137}Cs intake by the population was strongly correlated with levels of consumption of freshwater fish. In rare situations like this, where people consume fish from the (few) highly contaminated 'closed' lakes, the ingestion dose can be dominated by ^{137}Cs from fish.

In western Europe, consumption of freshwater fish does not form an important part of the diet, but sports and commercial fisheries may be of economic importance in some areas. Though not necessarily dependent on the consumption of fish, angling is one of Europe's most popular leisure activities. In the UK, fallout from Chernobyl had little effect on fisheries, though anecdotal evidence suggests that it was a cause of (unfounded) concern amongst anglers in the more heavily contaminated areas such as the western part of Cumbria. In Norway, where fallout levels were up to one order of magnitude higher, consumption of freshwater fish declined by up to 50% in the more contaminated areas, and the sale of freshwater fish to the general public was prohibited in these areas (Brittain et al., 1991). These authors also reported that the sale of fishing licences in parts of Norway declined by 25% after Chernobyl.

Doses from various sources of radioactivity in the Baltic Sea have been estimated by Nielsen et al. (1999). The maximum dose from Chernobyl derived radionuclides to the critical group of seafood consumers was estimated to be 0.2 mSv in the Bothnian Sea and the Gulf of Finland during the first year after the accident. This was a factor of 3–4 times lower than estimated annual doses from naturally occuring ^{210}Po in the Baltic Sea (Nielsen et al., 1999).

4.9 REFERENCES

Aarkrog, A. (1998) A retrospect of anthropogenic radioactivity in the global marine environment. *Radiation Protection Dosimetry*, **75**, 23–31.

Allard, B., Olofsson, U. and Torstenfelt, B. (1984) Environmental Actinide Chemistry. *Inorganica Chimica Acta*, **94**, 205–221.

Appleby, P.G. (1997) Sediment records of fallout radionuclides and their application to studies of sediment–water interactions. *Water, Air and Soil Pollution*, **99**, 573–585.

Balonov, M.I. and Travnikova, I.G. (1990) The role of agriculture and natural ecosystems in forming of the internal irradiation of citizens from the contaminated zone. *Proceedings of the First International Working Group on Severe Accidents and their Consequences, 30 October–3 November 1989*, pp. 153–160. Nauka, Moscow.

Baysal, A. and Tunçer, S. (1994) Radioactivity levels in fish, shellfish, algae and seagrass collected from the Eastern Black Sea coast of Turkey, 1992. *Toxicological and Environmental Chemistry*, **42**, 149–153.

Berkovski, V., Voitsekhovitch, O.A., Nasvit, O., Zheleznyak, M. and Sansone, U. (1996) Exposures from aquatic pathways. In: Karaoglou, A., Desmet, G., Kelly, G.N. and Menzel, H.G. (eds), *The Radiological Consequences of the Chernobyl Accident: EUR 16544 EN*, pp. 283–294. European Commission, Luxembourg.

Blaylock, B.G. (1982) Radionuclide data bases available for bioaccumulation factors for freshwater biota. *Nuclear Safety*, **23**, 427–438.

Borzilov, V.A., Konoplev, A.V., Revina, S.K., Bobovnikova, Ts.I., Lyutik, P.M., Shveikin, Yu.I. and Scherbak, A.V. (1988). Experimental research on the washout of radionuclides deposited on the soil following the Chernobyl accident. *Soviet Meteorology & Hydrology*, **11**, 43–53 (in Russian).

Brittain, J.E., Storruste, A. and Larsen, E. (1991) Radiocaesium in Brown Trout (*Salmo trutta*) from a subalpine lake ecosystem after the Chernobyl reactor accident. *Journal of Environmental Radioactivity*, **14**, 181–191.

Brittain, J.E., Bjørnstad, H.E., Sundblad, B. and Saxén R. (1997) The characterisation and retention of different transport phases of ^{137}Cs and ^{90}Sr in three contrasting Nordic lakes. In: Desmet, G., Blust, R., Comans, R.N.J., Fernandez, J., Hilton, J. and de Bettencourt, A. (eds), *Freshwater and Estuarine Radioecology*, pp. 87–96. Elsevier, Amsterdam.

Bugai, D.A., Waters, R.D., Dzhepo, S.P. and Skal'skij, A.S. (1996) Risks from radionuclide migration to groundwater in the Chernobyl 30-km zone. *Health Physics*, **71**, 9–18.

Bulgakov, A.A., Konoplev, A.V., Popov, V.E. and Scherbak, A.V. (1990) Removal of long-lived radionuclides from the soil by surface runoff near the Chernobyl Nuclear Power Station. *Soviet Soil Science*, **4**, 47–54.

Bulgakov, A.A., Konoplev, A.V., Smith, J.T., Hilton, J., Comans, R.N.J., Laptev, G.V. and Christyuk, B.F. (2002) Modelling the long-term dynamics of radiocaesium in a closed lake system. *Journal of Environmental Radioactivity*, **61**, 41–53.

Cambray, R.S., Cawse, P.A., Garland, J.A., Gibson, J.A.B., Johnson, P., Lewis, G.N.J., Newton, D., Salmon, L. and Wade, B.O. (1987) Observations on radioactivity from the Chernobyl accident. *Nuclear Energy*, **26**, 77–101.

Camplin, W.C., Leonard, D.R.P., Tipple, J.R. and Duckett, L. (1989) *Radioactivity in Freshwater Systems in Cumbria (UK) Following the Chernobyl Accident*. MAFF Fisheries Research Data Report No. 18. 90 pp.

Carlson, L. and Holm, E. (1992) Radioactivity in *Fucus vesiculosus* L. from the Baltic Sea following the Chernobyl accident. *Journal of Environmental Radioactivity*, **15**, 231–248.

Carlsson, S. (1978). A model for the movement and loss of ^{137}Cs in a small watershed. *Health Physics*, **34**, 33–37.

Chittenden, D.M. (1983) Factors affecting the soluble–suspended distribution of Strontium-90 and Cesium-137 in Dardanelle Reservoir, Arkansas. *Environmental Science and Technology*, **17**, 26–31.

Chowdhury, M.J. and Blust, R. (2001) A mechanistic model for the uptake of waterborne strontium in the common carp. *Environmental Science and Technology*, **35**, 669–675.

Comans, R.N.J., Middelburg, J.J., Zonderhuis, J., Woittiez, J.R.W., De Lange, G.J., Das, H.A. and Van Der Weijden, C.H. (1989) Mobilization of radiocaesium in pore water of lake sediments. *Nature*, **339**, 367–369.

Comans, R.N.J. and Hockley, D.E. (1992) Kinetics of caesium sorption on illite. *Geochimica et Cosmochimica Acta*, **56**, 1157–1164.

Coughtrey, P.J. and Thorne, M.C. (1983) *Radionuclide Distribution and Transport in Terrestrial and Aquatic Ecosystems. A Critical Review of Data* (Volume 1). A.A. Balkema, Rotterdam.

Coughtrey, P.J., Jackson, D. and Thorne, M.C. (1983) *Radionuclide Distribution and Transport in Terrestrial and Aquatic Ecosystems. A Critical Review of Data* (Volume 3). A.A. Balkema, Rotterdam.

Coughtrey, P.J., Jackson, D., Jones, C.H., Kane, P. and Thorne, M.C. (1984) *Radionuclide Distribution and Transport in Terrestrial and Aquatic Ecosystems. A Critical Review of Data* (Volume 4). A.A. Balkema, Rotterdam.

Coughtrey, P.J., Jackson, D., Jones, C.H., Kane, P. and Thorne, M.C. (1985) *Radionuclide Distribution and Transport in Terrestrial and Aquatic Ecosystems. A critical review of data* (Volume 6). A.A. Balkema, Rotterdam.

Cremers, A., Elsen, A., De Preter, P. and Maes, A. (1988) Quantitative analysis of radiocaesium retention in soils. *Nature*, **335**, 247–249.

Cross, M.A., Smith, J.T., Saxén, R. and Timms, D.N. (2002) An analysis of the time dependent environmental mobility of radiostrontium in Finnish river catchments. *Journal of Environmental Radioactivity*, **60**, 149–163.

Davison, W., Hilton, J., Hamilton-Taylor, J., Kelly, M., Livens, F., Rigg, E., Carrick, T.R. and Singleton, D.L. (1993) The transport of Chernobyl-derived radiocaesium through two freshwater lakes in Cumbria, UK. *Journal of Environmental Radioactivity*, **19**, 125–133.

Elliott, J.M. (1975) Number of meals per day, maximum weight of food consumed per day and maximum rate of feeding of brown trout, Salmo Trutta. *Freshwater Biology*, **5**, 287–303.

Elliott, J.M., Hilton, J., Rigg, E., Tullett, P.A., Swift, D.J. and Leonard D.R.P. (1992) Sources of variation in post-Chernobyl radiocaesium in fish from two Cumbrian lakes (northwest England). *Journal of Applied Ecology*, **29**, 108–119.

Evans, D.W., Alberts, J.J. and Clark, R.A. (1983) Reversible ion-exchange fixation of cesium-137 leading to mobilization from reservoir sediments. *Geochimica et Cosmochimica Acta*, **47**, 1041–1049.

Eremeev, V.N., Ivanov, L.M., Kirwan, A.D. and Margolina, T.M. (1995) Amount of ^{137}Cs and ^{134}Cs radionuclides in the Black Sea produced by the Chernobyl accident. *Journal of Environmental Radioactivity*, **27**, 49–63.

Fleishman, D.G. (1973) Radioecology of marine plants and animals. In: Klechkovskii, V.M., Polikarpov, G.G. and Aleksakhin, R.M. (eds), *Radioecology*, pp. 347–370. Wiley, Chichester, UK.

Fleishman, D.G., Nikiforov, V.A., Saulus, A.A. and Komov, V.T. (1994) ^{137}Cs in fish of some lakes and rivers of the Bryansk region and North-West Russia in 1990–1992. *Journal of Environmental Radioactivity*, **24**, 145–158.

Foulquier, L. and Baudin-Jaulent, Y. (1990) *Impact Radioécologique de l'Accident de Tchernobyl sur les Écosystèmes Aquatiques Continentaux* (European Commission Radioprotection series. Volume 50). European Commisssion, Luxembourg.

Goossens, R., Delville, A., Genot, J., Halleux, R. and Masschelein, W.J. (1989). Removal of the typical isotopes of the Chernobyl fall-out by conventional water treatment. *Water Research*, **23**, 693–697.

Gudkov, D.I., Derevets, V.V., Kuz'menko, M.I. and Nazarov, A.B. (2001) ^{90}Sr and ^{137}Cs in higher aquatic plants of the Chernobyl nuclear plant exclusion zone. *Radiation Biology and Radioecology*, **41**, 265–262 (in Russian).

Hadderingh, R.H., van Aerssen, G.H.F.M, Ryabov, I.N., Koulikov, O.A., Belova, N. (1997) Contamination of fish with ^{137}Cs in Kiev reservoir and old river bed of Pripyat near Chernobyl. In: Desmet, G., Blust, R., Comans, R.N.J., Fernandez, J., Hilton, J. and de Bettencourt, A. (eds), *Freshwater and Estuarine Radioecology*, pp. 339–351. Elsevier Studies in Environmental Science 68, Amsterdam.

Håkanson, L., Andersson, T. and Nilsson, Å. (1992) Radioactive caesium in fish from Swedish lakes 1986–1988 – general pattern related to fallout and lake characteristics. *Journal of Environmental Radioactivity*, **15**, 207–229.

Håkanson, L. (1997) Testing different sub-models for the partition coefficient and the retention rate for radiocesium in lake ecosystem modelling. *Ecological Modelling*, **101**, 229–250.

Hansen, H.J.M. and Aarkrog, A. (1990) A different surface geology in Denmark, The Faroe Islands, and Greenland influences the radiological contamination of drinking water. *Water Research*, **24**, 1137–1141.

Helton, J.C., Muller, A.B. and Bayer, A. (1985) Contamination of surface-water bodies after reactor accidents by the erosion of atmospherically deposited radionuclides. *Health Physics*, **48**, 757–771.

Hesslein, R.H. (1987) Whole-lake radiotracer movement in fertilized lake basins. *Canadian Journal of Fisheries and Aquatic Science*, **44**, 74–82.

Hilton, J., Livens, F.R., Spezzano, P. and Leonard, D.R.P. (1993) Retention of radioactive caesium by different soils in the catchment of a small lake. *Science of the Total Environment*, **129**, 253–266.

Hilton, J., Davison, W., Hamilton-Taylor, J., Kelly, M., Livens, F.R., Rigg, E. and Singleton, D.L. (1994) Similarities in the behaviour of Chernobyl derived Ru-103, Ru-106 and Cs-137 in two freshwater lakes. *Aquatic Sciences*, **56**, 133–144.

Holm, E. (1995) Plutonium in the Baltic Sea. *Applied Radiation and Isotopes*, **46**, 1225–1229.

IAEA (1994) *Handbook of Parameter Values for the Prediction of Radionuclide Transfer in Temperate Environments*. IAEA Technical Reports Series No. 364. IAEA, Vienna.

IAEA (2000) *Modelling of the Transfer of Radiocaesium from Deposition to Lake Ecosystems*. Technical document of the VAMP aquatic working group. IAEA TecDoc 1143. IAEA, Vienna.

Jones, F. and Castle, R.G. (1987) Radioactivity monitoring of the water cycle following the Chernobyl accident. *Journal of the Institute of Water and Environmental Management*, **1**, 205–207.

Joshi, S.R. and McCrea, R.C. (1992) Sources and behaviour of anthropogenic radionuclides in the Ottawa river waters. *Water, Air and Soil Pollution*, **62**, 167–184.

Kanivets, V.V., Voitsekhovitch, O.V., Simov, V.G. and Golubeva, Z.A. (1999) The post-Chernobyl budget of ^{137}Cs and ^{90}Sr in the Black Sea. *Journal of Environmental Radioactivity*, **43**, 121–135.

Kanivets, V.V and Voitsekhovitch, O.V. (2001) Effective half lives in Ukranian rivers. In: Smith, J.J. (ed.), *Aquascope Final Report*. Centre for Ecology and Hydrology, Dorset, U.K., 116 pp.

Kashparov, V.A., Oughton, D.H., Zvarich, S.I., Protsak, V.P. and Levchuk, S.E. (1999) Kinetics of fuel particle weathering and ^{90}Sr mobility in the Chernobyl 30 km exclusion zone. *Health Physics*, **76**, 251–299.

Klemt, E., Drissner, J., Kaminski, S., Miller, R., and Zibold, G. (1998) Time dependency of the bioavailability of radiocaesium in lakes and forests. In: Linkov, I. and Schell, W.R. (eds), *Contaminated Forests*, pp. 95–101. Kluwer, Dordrecht.

Kolehmainen, S., Häsänen, E. and Miettinen, J.K. (1967) ^{137}Cs in the plants, plankton and fish of the Finnish lakes and factors affecting its accumulation. In: Snyder, W.S. (ed), *Proceedings of the 1st International Congress on Radiation Protection*, pp. 407–415. Pergamon Press, Oxford.

Konoplev, A.V., Kopylova, L.P., Bobovnikova, Ts.I., Bulgakov, A.A., Siverina, A.A. (1992a) Distribution of ^{90}Sr and ^{137}Cs within the system of bottom sediments-water of the reservoirs in the areas adjacent to the Chernobyl NPP. *Soviet Meteorology and Hydrology*, **1**, 35–42 (in Russian).

Konoplev, A.V., Bulgakov, A.A., Popov, V.E. and Bobovnikova, Ts.I. (1992b) Behaviour of long-lived radionuclides in a soil-water system. *Analyst*, **117**, 1041–1047.

Konoplev, A.V., Comans, R.N.J., Hilton, J., Madruga, M.J., Bulgakov, A.A., Voitsekhovitch, O.V., Sansone, U., Smith, J.T. and Kudelsky, A.V. (1996) Physicochemical and hydraulic mechanisms of radionuclide mobilization in aquatic systems. In: Karaoglou, A., Desmet, G., Kelly, G.N. and Menzel, H.G. (eds), *The Radiological Consequences of the Chernobyl Accident: EUR 16544 EN*, pp. 121–135. European Commission, Luxembourg.

Konoplev, A.V., Kaminski, S., Klemt, E., Konopleva, I., Miller, R. and Zibold, G. (2002). Comparative study of ^{137}Cs partitioning between solid and liquid phases in Lake Constance, Lugano and Vorsee. *Journal of Environmental Radioactivity*, **58**, 1–11.

Knapinska-Skiba, D., Bojanowski, R., Radecki, Z. and Millward, G.E. (2001) Activity concentrations and fluxes of radiocaesium in the Southern Baltic Sea. *Estuarine, Coastal and Shelf Science*, **53**, 779–786.

Kryshev, A.I. (2003) Model reconstruction of ^{90}Sr concentrations in fish from 16 Ural lakes contaminated by the Kyshtym accident of 1957. *Journal of Environmental Radioactivity*, **64**, 67–84.

Kryshev, I.I. (1995) Radioactive contamination of aquatic ecosystems following the Chernobyl accident. *Journal of Environmental Radioactivity*, **27**, 207–219.

Kryshev, I.I. and Ryabov, I.N. (1990) About the efficiency of trophic levels in the accumulation of Cs-137 in fish of the Chernobyl NPP cooling pond. In: Ryabov, I.N. and Ryabtsev, I.A. (eds), *Biological and Radioecological Aspects of the Consequences of the Chernobyl Accident*, pp. 116–121. USSR Academy of Sciences, Moscow.

Kryshev, I.I. and Sazykina, T.G. (1994) Accumulation factors and biogeochemical aspects of migration of radionuclides in aquatic ecosystems in the areas impacted by the Chernobyl accident. *Radiochimica Acta*, **66/67**, 381–384.

Kudelsky, A.V., Smith, J.T., Ovsiannikova, S.V. and Hilton, J. (1996) The mobility of Chernobyl-derived ^{137}Cs in a peatbog system within the catchment of the Pripyat River, Belarus. *Science of the Total Environment*, **188**, 101–113.

Kudelsky, A.V., Smith, J.T., Zhukova, O.M., Matveyenko, I.I. and Pinchuk, T.M. (1998) Contribution of river runoff to the natural remediation of contaminated territories (Belarus). *Proceedings of the Academy of Sciences of Belarus*, **42**, 90–94 (in Russian).

Kudelsky, A.V., Smith, J.T., Ovsiannikova, S.V. and Pashkevitch, V.I. (2004) ^{137}Cs migration in soils of the aeration zone and level of ^{137}Cs contamination of groundwaters in Belarus. *Geoecology*, **3**, 223–236.

Linsley, G.S., Haywood, S.M., and Dionan, J. (1982) *Use of Fallout Data in the Development of Models for the Transfer of Nuclides in Terrestrial and Freshwater Systems*. IAEA-SM-257/8, pp. 615-633. IAEA, Vienna.

Los, I.P., Segeda, I.I., Shepelevitch, K.I and Voitsekhovitch, O.V. (1998) Assessment of radiation risks and reasonability for justification of the water remedial countermeasure (Radiological and social-economical criteria for decision making). In: Voitsekhovitch, O.V. (ed.), *Radioecology of Water Bodies of the Chernobyl Affected Areas* (Volume 2), pp. 136–168. Chernobyl Inter-inform, Kiev (in Russian).

Malmgren, L. and Jansson, M. (1995) The fate of Chernobyl radiocesium in the River Öre catchment, Northern Sweden. *Aquatic Sciences*, **57**, 144–160.

Mangini, A., Christian, U., Barth, M., Schmitz, W. and Stabel, H.H. (1990) Pathways and residence times of radiotracers in Lake Constance. In: Tilzer, M.M. and Serruya, C. (eds), *Large Lakes*, pp. 245–264. Springer, Berlin.

Milton, G.M., Cornett, R.J., Kramer, S.J. and Vezina, A. (1992) The transfer of iodine and technetium from surface waters to sediments. *Radiochimica Acta*, **58/59**, 291–296.

Monte, L. (1995) Evaluation of radionuclide transfer functions from drainage basins of freshwater systems. *Journal of Environmental Radioactivity*, **26**, 71–82.

Monte, L. (1997) A collective model for predicting the long-term behaviour of radionuclides in rivers. *Science of the Total Environment*, **201**, 17–29.

Monte, L., Kryshev, I. and Sazykina, T. (2002) Quantitative assessment of the long term behaviour of Sr-90 in Lake Uruskul, Southern Urals, Russia. *Journal of Environmental Radioactivity*, **62**, 61–74.

Mück, K., Pröhl, G., Likhtarev, I., Kovgan, L., Meckbach, R. and Golikov, V. (2002) A consistent radionuclide vector after the Chernobyl accident. *Health Physics*, **82**, 141–156.

Mundschenk, H. (1996) Occurrence and behaviour of radionuclides in the Moselle River. Part II: Distribution of radionuclides between aqueous phase and suspended matter. *Journal of Environmental Radioactivity*, **30**, 215–232.

Murdock, R.N., Johnson, M.S., Hemingway, J.D. and Jones, S.R. (1995) The distribution of radionuclides between the dissolved and particulate phases of a contaminated freshwater stream. *Environmental Technology*, **16**, 1–12.

Nielsen, S.P., Bengtson, P., Bojanowsky, R., Hagel, P., Herrmann, J., Ilus, E., Jakobson, E., Motiejunas, S., Panteleev, Y., Skujina, A. and Suplinska, M. (1999) The radiological exposure of man from radioactivity in the Baltic Sea. *The Science of the Total Environment*, **237/238**, 133–141.

NRPB (1996) *Generalised Derived Limits for Radioisotopes of Sr, Ru, I, Cs, Pu, Am, Cm.* Documents of the NRPB Vol. 7 No. 1, HMSO, London.

Nylén, T. (1996) *Uptake, Turnover and Transport of Radiocaesium in Boreal and Forest Ecosystems.* FOA-R-96-00242-4.3-SE. Report of the National Defence Research Establishment, Umeå, Sweden.

Ovsiannikova, S.V., Smith, J.T., Kudelsky, A.V., Madruga, M.J., Pashkevitch, V.I. and Yankov, A.I. (2004) Behaviour of radioactive caesium in the bottom sediments of an undrained lake. *Proceedings of the Academy of Sciences of Belarus*, **48**, 95–101 (in Russian).

Robbins, J.A., Lindner, G., Pfeiffer, W., Kleiner, J., Stabel, H.H. and Frenzel, P. (1992) Epilimnetic scavenging of Chernobyl radionuclides in Lake Constance. *Geochimica et Cosmochimica Acta*, **56**, 2339–2361.

Robbins, J.A. and Jasinski, A.W. (1995) Chernobyl fallout radionuclides in Lake Sniardwy, Poland. *Journal of Environmental Radioactivity*, **26**, 157–184.

Rowan, D.J. and Rasmussen, J.B. (1994) Bioaccumulation of radiocesium by fish: The influence of physicochemical factors and trophic structure. *Canadian Journal of Fisheries and Aquatic Science*, **51**, 2388–2410.

Rowan, D.J., Chant, L.A. and Rasmussen, J.B. (1998) The fate of radiocaesium in freshwater communities – why is biomagnification variable both within and between species? *Journal of Environmental Radioactivity*, **40**, 15–36.

Salo, A., Saxén, R. and Puhakainen, M. (1984) Transport of airborne ^{90}Sr and ^{137}Cs deposited in the basins of the five largest rivers in Finland. *Aqua Fennica*, **14**, 21–31.

Sansone, U. and Voitsekhovitch., O.V. (1996) *Modelling and Study of the Mechanisms of the Transfer of Radionuclides from the Terrestrial Ecosystem To and In Water Bodies Around Chernobyl*. EUR 16529 EN. European Commission, Luxembourg.

Santschi, P.H., Nyffeler, U.P., Anderson, R.F., Schiff, S.L., O'Hara, P. and Hesslein, R.H. (1986) Response of radioactive trace metals to acid-base titrations in controlled experimental ecosystems: Evaluation of transport parameters for application to whole-lake radiotracer experiments. *Canadian Journal of Fisheries and Aquatic Science*, **43**, 60–77.

Santschi, P.H., Bollhalder, S., Zingy, S., Luck, A. and Farrenkother, K. (1990) The self cleaning capacity of surface waters after radionuclide fallout. Evidence from European lakes after Chernobyl 1986–1988. *Environmental Science and Technology*, **24**, 519–527.

Saxén, R. and Ilus, E. (2001) Discharge of ^{137}Cs and ^{90}Sr by Finnish rivers to the Baltic Sea in 1986–1996. *Journal of Environmental Radioactivity*, **54**, 275–291.

Schoer, J. (1988) Investigation of transport processes along the Elbe River using Chernobyl radionuclides as tracers. *Environmental Technology Letters*, **9**, 317–324.

Sholkovitz, E.R. (1983) The Geochemistry of Plutonium in Fresh and Marine Water Environments. *Earth-Science Reviews*, **19**, 95–161.

Smith, J.T. and Comans, R.N.J. (1996) Modelling the diffusive transport and remobilization of Cs-137 in sediments: The effects of sorption kinetics and reversibility. *Geochimica et Cosmochimica Acta*, **60**, 995–1004.

Smith, J.T., Leonard, D.R.P., Hilton, J. and Appleby, P.G. (1997) Towards a generalised model for the primary and secondary contamination of lakes by Chernobyl – derived radiocaesium. *Health Physics*, **72**, 880–892.

Smith, J.T., Fesenko, S.V., Howard, B.J., Horrill, A.D., Sanzharova, N.I., Alexakhin, R.M., Elder, D.G. and Naylor, C. (1999a) Temporal change in fallout ^{137}Cs in terrestrial and aquatic systems: A whole ecosystem approach. *Environmental Science and Technology*, **33**, 49–54.

Smith, J.T., Comans, R.N.J. and Elder, D.G. (1999b) Radiocaesium removal from European lakes and reservoirs: Key processes determined from 16 Chernobyl contaminated lakes. *Water Research*, **33**, 3762–3774.

Smith, J.T., Clarke, R.T. and Saxén, R. (2000a) Time dependent behaviour of radiocaesium: A new method to compare the mobility of weapons test and Chernobyl derived fallout. *Journal of Environmental Radioactivity*, **49**, 65–83.

Smith, J.T., Comans, R.N.J., Beresford, N.A., Wright, S.M., Howard, B.J. and Camplin, W.C. (2000b) Chernobyl's legacy in food and water. *Nature*, **405**, 141.

Smith, J.T., Kudelsky, A.V., Ryabov, I.N. and Hadderingh, R.H. (2000c) Radiocaesium concentration factors of Chernobyl-contaminated fish: A study of the influence of potassium, and 'blind' testing of a previously developed model. *Journal of Environmental Radioactivity*, **48**, 359–369.

Smith J.T., Konoplev, A.V., Bulgakov, A.A., Kudelsky, A.V., Voitsekhovitch, O.V., Zibold, G., Madruga, M-J., Comans, R.N.J. and de Koning, A. (2001) *Aquifers and Surface Waters in the Chernobyl Area: Observations and Predictive Evaluation*, 116 pp. Final report to the EU. Centre for Ecology and Hydrology, Dorset.

Smith, J.T., Kudelsky, A.V., Ryabov, I.N., Daire, S.E., Boyer, L., Blust, R.J., Fernandez, J.A., Hadderingh, R.H. and Voitsekhovitch, O.V. (2002) Uptake and elimination of radiocaesium in fish and the 'size effect'. *Journal of Environmental Radioactivity*, **62**, 145–164.

Smith, J.T., Wright, S.M., Cross, M.A., Monte, L., Kudelsky, A.V., Saxén, R., Vakulovsky, S.M. and Timms, D.N. (2004) Global analysis of the riverine transport of ^{90}Sr and ^{137}Cs. *Environmental Science and Technology*, **38**, 850–857.

Styro, D., Bumeliene, Zh., Lukinskiene, M. and Morkuniene, R. (2001) ^{137}Cs and ^{90}Sr behavioural regularities in the southeastern part of the Baltic Sea. *Journal of Environmental Radioactivity*, **53**, 27–39.

Sundblad, B., Bergström, U. and Evans, S. (1991) Long-term transfer of fallout nuclides from the terrestrial to the aquatic environment. In: Moberg, L. (ed.), *The Chernobyl Fallout in Sweden*. Arprint, Stockholm, ISBN 91-630-0721-5.

Travnikova, I.G., Bazjukin, A.N., Bruk, G.Ja., Shutov, V.N., Balonov, M.I., Skuterud, L., Mehli, H. and Strand, P. (2004) Lake fish as the main contributor of internal dose to lakeshore residents in the Chernobyl contaminated area. *Journal of Environmental Radioactivity*, **77**, 63–75.

Ugedal, O., Forseth, T., Jonsson, B. and Njåstad, O. (1995) Sources of variation in radiocaesium levels between individual fish from a Chernobyl contaminated Norwegian lake. *Journal of Applied Ecology*, **32**, 352–361.

Vakulovsky, S.M., Voitsekhovitch, O.V., Katrich, I.Yu., Medinets, V.I., Nikitin, A.I. and Chumichev, V.B. (1990) Radioactive contamination of water systems in the area affected by releases from the Chernobyl nuclear power plant accident. In: *Proceedings of an International Symposium on Environmental Contamination Following a Major Nuclear Accident*, pp. 231–246. IAEA, Vienna.

Vakulovsky, S.M., Nikitin, A.I., Chumichev, V.B., Katrich, I.Yu., Voitsekhovitch, O.A., Medinets, V.I., Pisarev, V.V., Bovkum, L.A. and Khersonsky, E.S. (1994) Cs-137 and Sr-90 contamination of water bodies in the areas affected by releases from the Chernobyl Nuclear Power Plant accident: An overview. *Journal of Environmental Radioactivity*, **23**, 103–122.

Valcke, E. and Cremers, A. (1994) Sorption-desorption dynamics of radiocesium in organic-matter soils. *Science of the Total Environment*, **157**, 275–283.

Vinogradov, A.P. (1953) *The Elementary Composition of Marine Organisms*, 647 pp. Sears Foundation, New Haven.

Voitsekhovitch, O.V., Borzilov, V.A. and Konoplev, A.V. (1991) Hydrological aspects of radionuclide migration in water bodies following the Chernobyl accident. In: *Proceedings of Seminar on Comparative Assessment of the Environmental Impact of Radionuclides Released during Three Major Nuclear Accidents: Kyshtym, Windscale, Chernobyl*, pp. 528–548. Radiation Protection-53, EUR 13574, European Commission, Luxembourg.

Voitsekhovitch, O.V., Sansone, U., Zhelesnyak, M. and Bugai, D. (1996) Water quality management of contaminated areas and its effects on doses from aquatic pathways. In: Karaoglou, A., Desmet, G., Kelly, G.N. and Menzel, H.G. (eds), *The Radiological Consequences of the Chernobyl Accident: EUR 16544 EN*, pp. 401–410. European Commission, Luxembourg.

Voitsekhovitch, O.V. (1998) Water quality management of the radioactive contaminated waters (Strategy and Technology). In: Voitsekhovitch, O.V. (ed.), *Radioecology of The Water Bodies at the Areas Affected as a Result of Chernobyl Accident*, pp. 169–217. Chernobyl Inter-inform, Kiev (in Russian).

Voitsekhovitch, O.V. (2001) *Management of Surface Water Quality in the Areas Affected by the Chernobyl Accident*, 135 pp. Ukrainian Hydrometeorological Institute, Kiev (in Russian).

Voitsekhovitch, O.V., Dutton, M. and Gerchikov, M. (2002) *Experimental Studies and Assessment of the Present State of Chernobyl Cooling Pond Bottom Topography and Bottom Sediment Radioactive Contamination*. Final report, Contract C647/D0426, Center for Monitoring Studies and Environmental Technologies/Ukrainian Hydrometeorological Institute.

Voitsekhovitch, O.V. and Smith, J.T. (2005) *Radioactivity in aquatic systems*. IAEA Chernobyl Forum Report. IAEA, Vienna (in press).

Waber, U., von Gunten, H.R. and Krähenbühl, U. (1987) The impact of the Chernobyl accident on a river/groundwater aquifer. *Radiochimica Acta*, **41**, 191–198.

Yasuda, H. and Uchida, S. (1993) Statistical approach for the estimation of Strontium distribution coefficient. *Environmental Science and Technology*, **27**, 2462–2465.

Zeevaert, Th., Vandecasteele, C.M., Konings, J., Kirchmann, R., Colard, J., Koch, G. and Hurtgen, C. (1986) Radiological studies of a river receiving radioactive wastes: results of recent field experiments. In: Sibley, T.H. and Myttenaere, C. (eds), *Application of Distribution Coefficients to Radiological Assessment Models*. Elsevier Applied Science Publishers, London.

Zibold, G., Kaminski, S., Klemt, E. and Smith, J.T. (2002). Time-dependency of the ^{137}Cs activity concentration in freshwater lakes, measurement and prediction. *Radioprotection – Colloques*, **37**, 75–80.

5

Application of countermeasures

Nick A. Beresford and Jim T. Smith

The clean-up operation around the Chernobyl Nuclear Power Plant (NPP) and evacuation of people from the most contaminated areas of Belarus, Russia and the Ukraine has been discussed in Chapter 1. The social, economic and psychological consequences of these actions are discussed in Chapter 7. In this chapter we will concentrate on the longer term measures used to reduce the radionuclide content in the diet of human populations, although methods of reducing external exposure and doses to the thyroid are also discussed. Following an introduction to countermeasure techniques, an overview of measures used to reduce ingestion doses in the former Soviet Union (fSU) and their effectiveness is given.

Whilst many countermeasures had previously been suggested, and indeed some used following the 1957 Windscale and Kyshtym (Mayak) accidents, the Chernobyl accident provided probably the greatest impetus to develop countermeasure techniques. Much of this work is described or summarised in Howard and Desmet (1993), Voigt (2001) and Cécille (2000).

5.1 COUNTERMEASURE TECHNIQUES

In the radiological context the term 'countermeasure' is applied to approaches which can be used to reduce the exposure of populations to internal or external radiation exposure. These encompass a wide range of techniques including removing contamination, manipulation of the uptake of radionuclides by plants or animals, altering farming or other practices and providing advice to affected populations. The main countermeasure techniques are described below. Recent reviews of countermeasures for application in food production (Howard *et al.*, 2004; Nisbet *et al.*, 2004), urban (Andersson *et al.*, 2003; Howard *et al.*, 2002; 2004) and freshwater systems (Voitsekhovitch *et al.*, 1997; Smith *et al.*, 2001) are used as the main sources of information for this section.

The justification for application of a countermeasure is generally accepted to be that the intervention should result in more good than harm (ICRP, 1999; see also Segal, 1993). Examples of 'harm' in this context may include social disruption, risk to countermeasure implementers, monetary cost and environmental damage. In addition to reduced doses, examples of 'good' may include reassurance and maintenance of lifestyles and trade.

5.1.1 Methods of reducing uptake of radioiodine to the thyroid

Potassium iodide tablets can significantly reduce the transfer of radioiodine to people's thyroid. As discussed in Chapter 1, however, there was 'no systematic distribution' (IAEA, 1991) of tablets to the population of Pripyat or to the other affected populations, although tablets were distributed to power plant workers within half an hour of the accident (UNSCEAR, 2000).

The government of Poland administered stable iodine to 18.5 million people in three days from 29 April, 1986 'the greatest prophylactic action in the history of medicine performed in such a short period of time' (Jaworowski, 2004). However, because of the relatively low thyroid doses to the population of Poland, Jaworowski (the instigator of this action) now believes it to have been 'nonsensical' (Jaworowski, 2004). Administration of stable iodine to populations in such circumstances has been advised against as the risks of stable iodine prophylaxis may be greater than the benefits of reduced radioiodine (Rubery and Smales, 1990).

Widespread restrictions on the consumption of milk and leafy vegetables would have significantly reduced radiation doses to the thyroid in the three most affected countries. After the 1957 Windscale accident in the UK, restrictions were placed on consumption of milk contaminated by ^{131}I, and this action led to a major reduction in the potential dose to the population (Jackson and Jones, 1991).

5.1.2 Methods of reducing the soil-to-plant transfer of radionuclides

Ploughing of agricultural soils mixes contaminated surface soils with uncontaminated sub-soils, thereby reducing the radionuclide activity concentration in the rooting zone of plants. Use of a standard single-furrow mouldboard plough at a typical ploughing depth of 20–30 cm reduces the radiocaesium and radiostrontium activity concentration in plants by approximately 50% (although this figure can vary between 0 and 75%). Ploughing to deeper soil depths may be more effective than this. Ploughing also has the additional benefit that it reduces external doses (see Section 2.2.3). Ploughing was extensively used, often in combination with other measures, in the fSU in response to the Chernobyl accident (see below). Whilst information on the effectiveness of ploughing is limited to radiocaesium and radiostrontium it is probable that it would also effectively reduce plant activity concentrations of most other radionuclides. A disadvantage of ploughing is that it may increase the rate of transfer of radionuclides to groundwaters, though this is unlikely to significantly impact on drinking water supplies (see Chapter 4).

A specialist plough (the 'skim and burial plough') has been developed which has two ploughshares (Roed et al., 1996). A top, thin layer of contaminated soil is skimmed-off and buried at a depth of approximately 45 cm. The deeper soil layer is then lifted by the second plough share without inversion. Small scale use of the skim and burial plough has demonstrated reductions in plant activity concentrations in the range 80–90% with the external dose reduced by approximately 95%.

Lime (CaO or $CaCO_3$) or potassium fertilisers may be applied to arable land or pastures to reduce the uptakes of radiostrontium or radiocaesium respectively. Calcium and potassium are chemically analogous to strontium and caesium (respectively) and their application to soils reduces plant root uptakes by competition. Reductions of up to 80% in the plant uptake of radiocaesium have been observed as a consequence of potassium fertilisation. The countermeasure is most effective when used on soils with comparatively low exchangeable potassium contents. Liming to increase soil pH from 5 to 7 is most effective on soils with a low pH or calcium status and may reduce ^{90}Sr transfer to plants by approximately 50 to 70%, depending on soil type. The effectiveness of both measures is increased by ploughing or harrowing after application; treatment may require repeating in subsequent years. Both of these measures were used in the fSU following the Chernobyl accident (see below).

5.1.3 Methods of reducing the radionuclide content of animal-derived foodstuffs

Administration of binders and competing elements

The administration of binding agents or analogous elements have both been suggested for use to reduce radionuclide transfer across the gastrointestinal tract and/or to animal products following absorption. A number of clay minerals (e.g., bentonite, vermiculite, mordenite and clinoptilite) have been orally administered to animals to bind radiocaesium in the gastrointestinal tract making it unavailable for absorption (see reviews by Voigt (1993) and Hove (1993)). At administration rates of 0.5 g bentonite per kg of animal body weight per day, a reduction of approximately 50% can be achieved. Some minerals may also be effective in reducing radiostrontium absorption. Hansen et al. (1995) obtained a 40% reduction in the radiostrontium activity concentration in goats' milk by administering sodium-aluminiumsilicate at an administration rate of $0.5 \, \text{g kg}^{-1}$ body weight d^{-1}. However, the influence of this compound on the absorption of essential elements has, to date, not been adequately considered.

The most effective binders to reduce radiocaesium absorption in the gut are the hexacyanoferrate compounds commonly referred to as 'Prussian Blue'. Of these, ammonium-ferric(III)-cyano-ferrate (II) (AFCF) (often referred to as Giese salt: Giese, 1988) is the most commonly tested (and used) compound in western Europe. A reduction in radiocaesium absorption of approximately 90% is obtained at administration rates of $6-40 \, \text{mg kg}^{-1}$ body weight d^{-1} with a 60% reduction having been observed for administration rates as low as $1 \, \text{mg kg}^{-1}$ body weight d^{-1} (Hove, 1993).

In western Europe, the production of foodstuffs in excess of national intervention limits and the consequent need to implement countermeasures was predominantly associated with animal production in semi-natural ecosystems (Howard et al., 1991). Post-Chernobyl measures to reduce the entry of ^{137}Cs into the human foodchain are (in 2005) still in place in the UK and Scandinavia. Production systems affected included sheep grazing upland and mountain pastures and herding of semi-domesticated reindeer. The management of such animals is extensive, with free grazing over large areas and infrequent handling. Procedures for the administration of radiocaesium binders to livestock devised prior to the Chernobyl accident relied on daily administration in feed.

In the UK experimental trials in 1987 investigated the possibility of spreading bentonite on upland pastures (Beresford et al., 1989). Although radiocaesium activity concentrations in sheep grazing the pastures were reduced, there was a simultaneous reduction in herbage intake by the sheep. However, in Norway effective methods of administering AFCF to free grazing animals were developed. NaCl saltlicks, incorporating 2.5% AFCF were placed on mountain pastures (grazed by sheep or reindeer) in areas where standard NaCl licks are routinely used (Hove, 1993). Whilst the use of these AFCF–saltlicks reduced the mean radiocaesium activity concentration in animals in areas where they were used, there was considerable variability in effectiveness dependent upon the degree of utilisation of the licks by individual animals.

The other method developed to deliver AFCF to infrequently handled animals was incorporation into a sustained release bolus for oral administration (Hove and Hansen, 1993; Solheim Hansen et al., 1996). Because of their high density, the boli (containing 15% AFCF) remain in the reticulo-rumen where abrasion against each other and the gut wall causes slow degradation of the bolus and a gradual release of AFCF. The boli (usually 2–3 per animal) can be given to animals when they are being handled as part of routine farming practices. During extensive field trials, administration of AFCF boli reduced radiocaesium activity concentrations in Norwegian mountain sheep by 48–65% over periods of 9–11 weeks (Solheim Hansen et al., 1996). Boli containing AFCF have been routinely used to reduce radiocaesium levels in the meat of sheep and reindeer in Norway (Brynildsen et al., 1996). For application to reindeer, a special delivery device had to be developed to allow administration of boli (Hove et al., 1991).

The original boli developed in Norway were typically 16–20 mm in diameter and 50–65 mm in length, comprising a compressed mixture of 15% AFCF, 10% beeswax and 75% barite (Hove and Hansen, 1993). These were later coated with wax to delay the initial degradation of the bolus and lengthen the effective period. When tested in the UK, Norwegian AFCF boli were found to be too large for the lambs of native breeds of hill sheep. A smaller bolus (14 mm by 50 mm) was therefore developed containing 20% AFCF (Beresford et al., 1999) and successfully used on an upland farm (Howard et al., 2001). In 2001, permanent authorisation was given by the European Communities for AFCF to be used as a feed additive for the purposes of binding radiocaesium (Regulation 2013/2001). The use of hexacyanoferrate feed additives in the fSU is discussed below.

As discussed in Chapter 3, the behaviour of radiostrontium in animals is governed by that of its homeostatically controlled analogue, calcium. The transfer coefficient (F_m) of radiostrontium from feed to milk can be expressed as (Beresford et al., 1998; see Figure 3.4):

$$F_m(Sr) = \frac{0.11 \times [Ca]_{milk}}{I_{Ca}} \quad (5.1)$$

where $[Ca]_{milk}$ (g kg^{-1}) is the concentration of calcium in milk and I_{Ca} the daily intake of Ca (g d^{-1}).

This relationship allows predictions to be made of the reduction in transfer of radiostrontium in milk which would be achieved by different administrations of additional calcium. In general terms, if the dietary intake of calcium is doubled, then the radiostrontium activity concentration in milk would be halved. Typically, in countries with farming systems similar to those of the UK, the calcium intake for dairy goats is in the range 51–30 g d^{-1}, whilst that of dairy cows is 70–150 g d^{-1} (Beresford et al., 2000). Application of Equation (5.1) shows that a reduction in the feed-to-milk transfer coefficient (F_m) of 40–60% would be expected if dairy cattle were supplemented with 100–200 g d^{-1} of calcium.

The upper limit on effectiveness of this measure is likely to be the maximum prolonged supplement rate that can be used to avoid exceeding the advised upper limit for calcium intake by ruminants (1–2% of their daily dry matter intake). Higher calcium intake rates than this may reduce the absorption of other essential nutrients (NRC, 1980, 1989). Similarly, it had been suggested that stable iodine administration was a possible method of reducing the transfer of ^{131}I to milk (Howard et al., 1996). However, on the basis of recent studies, this recommendation has been withdrawn (Howard et al., 2001) since misjudged administration rates may result in *increased* ^{131}I levels in milk. In addition, high stable iodine administration rates may lead to stable iodine concentrations in milk in excess of internationally advised values (Beresford et al., 1997; Crout et al., 2000; Vandecasteele et al., 2000).

Animal management

Radiostrontium and radiocaesium activity concentrations in milk decline rapidly (with biological half-lives in the order of days) when contaminated animals are placed on an uncontaminated diet. Radiocaesium biological half-lives in meat are typically in the order of tens of days for ruminant animals and pigs; larger animals have longer biological half-lives than smaller ones (Coughtrey and Thorne, 1983). Radiocaesium concentrations in meat therefore also decline comparatively quickly when a contaminated diet is substituted with uncontaminated feed. Feeding of uncontaminated feed as a countermeasure is generally referred to as 'clean feeding'.

For meat-producing animals, clean feeding is most effective if targeted at the period prior to slaughter. Increased feeding rates and live-weight gains during clean feeding may result in more rapid biological half-lives (Jones, 1993; Beresford et al., 2001). Note the effectiveness of clean feeding for some radionuclides will be limited as a consequence of their long biological half-lives (e.g., biological half-lives of Pu may be similar to the lifespan of the animal). Movement of animals to

Figure 5.1. Nick Beresford live-monitoring a sheep in upland west Cumbria in 1993 to determine ^{137}Cs activity concentration in muscle.
Photo: Cath Barnett.

uncontaminated, or less contaminated, pastures is an effective alternative to clean feeding with stored or harvested foodstuffs.

Where the supply of clean feed or availability of uncontaminated pasture is limited, the best use of limited resources can be made by designing feeding strategies such that uncontaminated feed is preferentially supplied to milk-producing animals or meat animals prior to slaughter (e.g., see Lepicard and Hériard Dubreuil, 2001). Indeed, the feeding of contaminated crops or milk in excess of agreed intervention limits to unproductive animals has been suggested as a method of reducing the need to dispose of contaminated produce without significantly increasing doses to human populations (Nisbet *et al.*, 2004).

The use of animal management measures in the fSU is discussed below. In Norway and Sweden clean feeding was used as a post-Chernobyl countermeasure to reduce radiocaesium activity concentrations in sheep, reindeer and cattle grazing unimproved pastures (Tveten *et al.*, 1998; Åhman, 1999). In the UK, animals above the national intervention limit (1,000 Bq kg^{-1} f.w.) have to be grazed on less contaminated pastures for a period prior to slaughter (Nisbet and Woodman, 2000; Wright *et al.*, 2003). These measures are usually combined with live-monitoring to determine radiocaesium activity concentrations in animals. Rapid live-monitoring techniques to determine the radiocaesium activity concentration in the meat of animals (see Figure 5.1) have been substantially improved since the Chernobyl accident (Meredith *et al.*, 1988; Brynildsen and Strand, 1994).

The manipulation of slaughtering times of free-ranging animals to times when they are consuming less contaminated diets was used in reindeer production systems in Scandinavia as a post-Chernobyl measure (Åhman, 1999). Similarly, in Sweden hunters are allowed to shoot roe deer (*Capreolus capreolus*) bucks during an additional season (May to mid-June) when ^{137}Cs activity concentrations are 5-times lower than those of bucks shot during the normal hunting season (August–September) (Johanson, 1994). (See Chapter 3 for further information on seasonal variation of radiocaesium in game animals.)

5.1.4 Countermeasures for freshwater systems

Countermeasures for freshwater systems may be grouped into two categories: those aimed at reducing doses from radioactivity in drinking water and those aimed at reducing doses from consumption of aquatic foodstuffs, principally freshwater fish.

Drinking water

Switching to less contaminated supplies is an accepted and effective countermeasure to radioactive contamination of drinking water (Smith *et al.*, 2001). If the water supply system is sufficiently integrated, supplies may rapidly be switched to groundwater sources, or less contaminated surface water sources. The effectiveness of this countermeasure is also dependent on a rapid assessment of the extent of contamination of different water supplies. A summary of the measures taken by the Ukrainian authorities to switch to alternative supplies from less contaminated rivers and from groundwater can be found in Voitsekhovitch (1998).

Conventional water treatment has been found to remove 73% of Ru, 61% of Co, 17% of I and 56% Cs from water (Goosens *et al.*, 1989). The mechanism for removal was by attachment to and settling of solid particles during flocculation, as well as, for Cs, removal during sand filtration. After the Chernobyl accident, at the Dnieper River Waterworks Station (Ukraine) activated charcoal and zeolite were added to water filtration systems. Charcoal effectively removed ^{131}I and ^{106}Ru, and zeolite removed ^{137,134}Cs and ^{90}Sr (Tsarik, 1993). After three months, however, the sorbents became saturated and their efficiency declined (Tsarik, 1993; Voitsekhovitch, 1998).

After the Chernobyl accident, the surface level gates on the Kiev Reservoir dam were opened. It was believed at the time that the surface water was relatively low in radionuclide content and that the release of water would allow room in the reservoir to contain runoff water which was believed to be highly contaminated. However, because of direct atmospheric deposition to the reservoir, surface waters were much more contaminated than deep waters (Voitsekhovitch *et al.*, 1997). It was later suggested that a better approach to lowering the reservoir water level would have been to open the bottom dam gates and close the surface gates. This would also have reduced the levels of radioactivity in downstream drinking water in the first weeks after the accident (Voitsekhovitch *et al.*, 1997).

A number of methods to reduce runoff and prevent flooding of contaminated land have been suggested to prevent secondary contamination of surface waters (Smith et al., 2001). Standard anti-soil erosion measures could be used to reduce runoff of radionuclides attached to eroded soil particles (see, e.g., Schwab et al., 1966; Morgan, 1986). Typically, however, suspended solids concentrations in river and lake waters are $<50\,\mathrm{mg\,l^{-1}}$. In this case, less than 50% of radiocaesium and less than 10% of radiostrontium and radioiodine are expected to be in the particulate phase (see Chapter 4), limiting the potential effectiveness of this countermeasure. Furthermore the dissolved, and not the particulate, form of these radionuclides determines activity concentrations in drinking water (and freshwater biota). Dredging of canal-bed traps to intercept suspended particles in contaminated rivers was carried out after the Chernobyl accident in the fSU (Voitsekhovitch et al., 1988). However, canal-bed traps were found to be highly inefficient because flow rates were too high to trap the most contaminated small suspended particles and, as already noted, the dissolved forms contribute most to the radioactivity in drinking water and fish.

Buffer strips adjacent to rivers can be used to trap eroded particles and encourage infiltration of overland flow to increase sorption of radionuclides to the soil. Buffer strips have been used to reduce runoff of fertilisers and pesticides from agricultural land, but are untested for radionuclides. Zeolite-containing dykes were constructed on smaller rivers and streams around the Chernobyl 30-km zone to intercept dissolved radionuclides. These were found to be very ineffective: only 5–10% of the ^{90}Sr and ^{137}Cs having been adsorbed (Voitsekhovitch et al., 1997; Voitsekhovitch, 1998). In addition, the rivers and streams on which they were placed were later found to contribute only a few percent to the total activity load in the Pripyat–Dnieper River system.

Spring flooding of the highly contaminated Pripyat flood plain resulted in increases in ^{90}Sr activity concentrations in the Pripyat River from annual average activity concentrations of around $1\,\mathrm{Bq\,l^{-1}}$ to a maximum of around $8\,\mathrm{Bq\,l^{-1}}$; the flood event covering an approximately 2-week period (Vakulovsky et al., 1994). The construction of a dyke around the most contaminated areas of the flood plain has proved effective in reducing ^{90}Sr loads during subsequent flood events (Voitsekhovitch et al., 1997). Annual average ^{90}Sr activity concentrations in Kiev Reservoir water, however, have been below $1\,\mathrm{Bq\,l^{-1}}$ from 1987. The radiological significance of the ^{90}Sr activity concentrations in Kiev Reservoir water, even during the short flood events, is therefore low, though it has been argued that the averted *collective* dose to the large number of users of the river–reservoir system is significant. Flooding of meadows has also been reported to result in increases in ^{137}Cs activity concentrations in the milk of grazing cattle (Burrough et al., 1999) although mechanisms for this increase are unclear.

Measures to reduce uptake by fish and aquatic foodstuffs

Liming of 18 Swedish lakes with different types of lime and by various application methods was tested as a method of reducing the ^{137}Cs activity concentration in fish

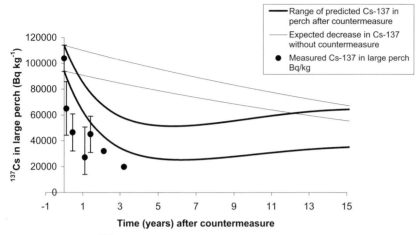

Figure 5.2. Decrease in ^{137}Cs activity concentrations in perch in Lake Svyatoe over a 15-year period after a potassium countermeasure was applied.* Expected decrease in the absence of countermeasures is also shown assuming an effective ecological half-life of 20 years, as observed in other lakes in the area.†

* From Kudelsky et al. (2002) and Smith et al. (2003). † From Smith et al. (2003).

(Håkanson and Andersson, 1992). The results of the experiments showed that liming had no significant effect on uptake of ^{137}Cs in fish in comparison with control lakes. Although uptake of ^{90}Sr was not studied in these experiments, the increased Ca concentration in lakes may be expected to reduce ^{90}Sr concentrations in fish (see Chapter 4). Experience of lake liming, in conjunction with artificial feeding of fish in the Ukraine has been summarised by Voitsekhovitch (1998).

The concentration factor of radiocaesium in fish is inversely related to the potassium content of the surrounding water (as discussed in Chapter 4). After the Chernobyl accident, potassium was added to 13 lakes in Sweden either as potash or as an additive in 'mixed lime' (Håkanson and Andersson, 1992). In most of the treated lakes, mean potassium concentrations were increased from $<0.4\,\mathrm{mg\,l^{-1}}$ to $>0.8\,\mathrm{mg\,l^{-1}}$. The results of the potash treatment were somewhat inconclusive, with a small reduction in activity concentrations in perch fry being observed during the two-year experiment. It was found that in lakes with short water retention times it was difficult to maintain high levels of potassium in the lake.

In an experiment on Lake Svyatoe in Belarus, Kudelsky et al. (2002) and Smith et al. (2003) added KCl-fertiliser onto the frozen lake surface. This resulted in an increase of the lake water potassium concentration from $1\,\mathrm{mg\,l^{-1}}$ to $10\,\mathrm{mg\,l^{-1}}$. Because of the long water residence time of this lake (it has no outflow), potassium concentrations were still $7\text{--}8\,\mathrm{mg\,l^{-1}}$ two years after application. Results showed a 50% reduction in ^{137}Cs concentration in fish during the first year after the experiment as shown in Figure 5.2. The efficiency of the countermeasure, was reduced by the release of ^{137}Cs from sediments by competition with the increased water K^+ concentrations: the ^{137}Cs activity concentration in water increased by a factor of 2–3 after the countermeasure application. The increased ^{137}Cs in water is unlikely to be acceptable in lakes where water is abstracted for drinking, irrigation or livestock use. In addition, it is likely that potassium treatment is only feasible in lakes with very long water residence times, allowing increased potassium concentrations to be maintained.

Fertilisation using 'Osmocoat' (5% P and 15% N) was also tested in two Swedish lakes (Håkanson and Andersson, 1992); no effect on fish ^{137}Cs activity concentrations or long-term water P concentrations were observed.

In some areas of the 30-km zone, radionuclide activity concentrations in groundwaters are significant. Modelling studies of health risks from groundwater contamination (Bugai et al., 1996), however, concluded that 'there is little current or future health risk basis for ... complex and costly groundwater remediation measures in the 30-km zone'.

5.1.5 Reduction of the external dose in residential areas

A compendium of countermeasures for application in contaminated residential areas to (principally) reduce the external dose has recently been published by Andersson et al. (2003). Those measures used or tested in the fSU following the Chernobyl accident are considered here.

If applied within the first few weeks of deposition, the vacuuming of roads using vehicular road cleaners can reduce contamination on hard surfaces by 50–70%. Andersson et al. (2003) report that this measure was applied in the fSU following the Chernobyl accident.

Removal of the top few centimetres of soil either manually or by mechanical digger from gardens, communal parks etc., can remove more than 90% of deposited contamination with minimal impact on soil fertility. The effectiveness of this measure depends upon the time between deposition and implementation. In addition, the requirement to dispose of the removed soil needs to be considered (see below for a discussion of the application in the fSU). Triple digging, which involves manually (by spade) redistributing the top soil layers such that the top most contaminated layer (approximately 5 cm) is placed at the bottom of the profile, the bottom layer (approximately 15–20 cm) is placed in the middle and the middle layer at the top. The uncontaminated soil layers provide shielding from the contaminated layer, reducing doses by up to 90%. An additional benefit of these methods is that they will also reduce radionuclide activity concentration in garden fruits and vegetables.

Addition of a 10-cm layer of uncontaminated soil in communal areas or around dwelling can reduce dose rates by 75–85%. Similarly, surfacing of such areas with 5–6 cm of asphalt can reduce doses by 50–75%. Andersson et al. (2003) report that this measure was widely used in the fSU.

Roed et al. (1998) present a summary of decontamination efforts conducted by the Soviet army (although other references state that activities were conducted by civil defence troops see, e.g., Balonov et al. (1999)). The decontamination measures studied by Roed et al. (1998) were carried out during the summer of 1989 in 93 settlements of the Bryansk region of Russia. They report that the army consistently carried out only two procedures: (i) removal of the topsoil layer; (ii) subsequent application of a layer of sand or gravel to shield against residual contamination. Decontamination generally only took place in the yards of private houses, around public buildings and along roads (a total road length of about 190 km being treated). The dose reduction as a consequence of these efforts is reported as 'disappointingly low', decreases generally being by 10–30%. Similarly low efficiencies (an 8% reduction in dose) for these same procedures in a Belarussian settlement are also referred to. Roed *et al.* (1998) speculate that the low reductions may have been the result of a number of factors:

- Individual areas treated may not have been large enough.
- Insufficient planning to take into account local radiological conditions.
- Insufficiently thick layer of top soil may have been removed.
- Lack of identification and treatment of 'hotspots' (e.g., soil adjacent to buildings, under gutters, etc.).

An international study (Roed *et al.*, 1998; see also Fogh *et al.*, 1999) is reported to have achieved an overall reduction in dose of up to 64% in a village in Bryansk using the following approach:

- Removal of the top 5–10-cm soil layer from a 20 m × 20 m area around three wooden houses using hand tools.
- Cleaning of (asbestos) roofs of the three houses, firstly removing loose litter and secondly using a specially constructed scrubbing device.
- Application of clean gravel to decontaminated land.

The total number of settlements decontaminated by Soviet civil defence troops between 1988 and 1990 is quoted by Antsipov *et al.* (2000) as 472 in Russia, 560 in the Ukraine and 500 in Belarus. Work after 1990 was more targeted, as follows:

- Russia – construction and improvement of waste storage; targeting highly contaminated plots.
- Ukraine – measures (removal of roofs, paving land, removing top soil and covering ground with uncontaminated soil) employed in 44 settlements where annual dose was approximately 1 mSv yr^{-1} or higher. Measures were especially targeted at buildings used by children and medical facilities.
- Belarus – largely restricted to replacement of contaminated soil especially around schools and kindergartens.

5.1.6 Social countermeasures

Social countermeasures are here defined as those which do not target mainstream agricultural production or require large-scale physical intervention (such as those described above). The category is broad-ranging and may include:

- Education programmes, for example, the inclusion of radiological protection/ radioecology into the school curriculum of affected areas (Hériard Dubreuil *et al.*, 1999).

- Food labelling, for example, with source or radioactivity content.
- Raising intervention limits to 'protect' a given way of life or allow marketing of a foodstuff. For example, after Chernobyl intervention limits were raised for reindeer meat compared to other meats in Scandinavia (Mehli et al., 2000).
- Banning of foodstuffs because they are in excess of intervention limits.

Governments within the fSU placed bans on the collection of highly contaminated wild foodstuffs such as fungi and freshwater fish following the Chernobyl accident (Ryabov, 1992; Beresford et al., 2001; Smith et al., 2001). It is reported that such bans were often ignored by the population for whom these practices are traditional.

There are some other countermeasures falling in the 'social countermeasures' category which can be implemented by people living in affected areas themselves if they want to reduce their exposure to radioactivity. These are often referred to as 'self-help' countermeasures and include many measures such as:

- Dietary advice, for instance, which species of fungi or berries to avoid or how much of a given foodstuff to consume (e.g., Strand et al., 1992; Beresford et al., 2001).
- Provision of monitoring equipment so that people can measure levels in home grown or collected foodstuffs (e.g., Lepicard and Hériard Dubreuil, 2001).
- Advice on animal management (e.g., Lepicard and Hériard Dubreuil, 2001).
- Advice on how to prepare and cook foods to best reduce the radionuclide content (see Long et al., 1995a,b).
- Some of the methods to reduce external doses in residential areas discussed above (e.g., triple digging).

Obviously these measures need effective dialogue with affected people (Hunt and Wynne, 2002): the local population also have to have the available resources to implement the countermeasure. Some self-help countermeasures such as those involving the use of monitoring equipment will require training. Hunt and Wynne (2002) also advise that:

> Communicative feasibility depends not only on the material requirements of communication (such as production and distribution in appropriate forms and through appropriate networks), but also on the extent to which information is authoritative, and the extent to which it is meaningful to relevant groups, and takes account of existing and interrelated practices. This implies the need for two-way communication strategies, where information is both provided and sought.

Provision of clearly explained dietary advice has been effective in some countries. For instance, following the Chernobyl accident the Norwegian Directorate of Health published a brochure aimed at groups, such as hunters, fishermen and the Sami reindeer breeders, who were most likely to consume large quantities of comparatively highly contaminated foods such as reindeer meat and freshwater fish. The brochure gave examples of the best ways to prepare food to limit radiocaesium intake and presented advised intake rates in easily understandable units (e.g., 'meals per week'). Follow-up surveys estimated that up to 80% of individuals within the target groups changed their diet as a consequence of the advice (Strand et al., 1992).

Table 5.1. The reduction achieved in the radiocaesium content of fungi following commonly used cooking procedures.
Adapted from Rantavara (1987) and Jacob et al. (1995).

Cooking procedure	Radiocaesium removed compared with freshly collected fungi
Wash	20%
Boil	50%
Salt	50%
Dry then soak	75%
Parboil, salt then soak	90%

It was estimated that the radiocaesium intakes of the Sami and hunters would have been, respectively, 400–700% and up to 50% higher if these dietary changes had not been made (Strand et al., 1992). Advice on restricting intakes of various foodstuffs from some areas was also produced in Sweden together with information on how to have game, berries and fungi monitored for radioactivity (Bengtsson, 1986). This document, which was published in 12 languages for immigrant populations, also included some basic information on the Chernobyl accident, radioactivity, radioecology and radioprotection.

The effect of commercial and cooking techniques on the radionuclide content of consumed foods has already been introduced in Chapter 3 (see Table 3.13 for dairy produce). Reviews of such techniques (some of which could be employed as countermeasures in commercial processing plants (Nisbet et al., 2004)) are presented by Long et al. (1995a,b). Some of those which have been suggested as self-help countermeasures are discussed here.

Generally, cooking and preparation practices such as frying, boiling and salting all remove radiocaesium from many foodstuffs. For instance, Table 5.1 presents the percentage of radiocaesium lost from fungi as a consequence of commonly used cooking procedures for wild collected species in the fSU (Beresford et al., 2001). However, doses to the consumer will only be reduced if the cooking liquors are disposed of (as these contain the radiocaesium 'lost' from the food itself). Other preparation methods increase the concentrations of radionuclides (per unit of weight consumed), for instance smoking and drying of fish (Ryabov, 1992) and advice may therefore be given to avoid these. Although radiocaesium is relatively evenly distributed throughout meat and fish, radiostrontium is largely concentrated in bone and there is some evidence to suggest that cooking meat on the bone can increase the strontium content of the meal (Long et al., 1995b).

Some developing approaches to the application of social countermeasures within the fSU (Hériard Dubreuil et al., 1999; Lepicard and Hériard Dubreuil, 2001) are discussed further below.

5.2 COUNTERMEASURES TO REDUCE INTERNAL DOSES APPLIED WITHIN THE AGRICULTURAL SYSTEMS OF THE fSU

At the time of the Chernobyl accident, the main agricultural producers within the areas of the USSR affected by the Chernobyl accident were state owned collective farms producing meat, milk, potatoes, cereals, fodder crops and forage, and sugar beet.

Soon after the accident, the principle countermeasure employed in animal production was abandonment of unimproved/low yielding pastures. Milk consumption was banned in areas with $>185\,\mathrm{kBq\,m^{-2}}$ $^{137}\mathrm{Cs}$ (IAEA, 1994). Contaminated milk was processed into butter, cheese and other products which have low residual radionuclide levels (see Table 3.12). All milk produced in contaminated areas was bought by the state. In total, approximately 120,000 livestock were evacuated from the 30-km exclusion zone. Because of the lack of deposition data, evacuated cattle were allowed to grazed contaminated areas outside of the exclusion zone. There was insufficient uncontaminated feed available to enable activity concentrations to decline to acceptable levels via biological turnover and excretion. Consequently these livestock were slaughtered in May–June 1986 generating 30,000 tonnes of meat contaminated to levels in excess of national intervention limits (IAEA, 1994). Subsequently, 90% of this meat was used in processed products although approximately 3,000 t was disposed of.

The main countermeasures used to reduce the radionuclide (predominantly targeting radiocaesium) activity concentration in agricultural products in the areas of the fSU affected by the Chernobyl accident were (Alexakhin, 1993): ploughing; liming of acid soils; application of organic and mineral fertilisers (especially potassium and phosphorous); 'radical improvement' of meadows (ploughing, draining, fertilisation and reseeding of low productivity 'natural' pastures); processing of products (especially milk) to end products with activity concentration below intervention limits; development of feeding plans to limit contamination at the time of slaughter; administration of binding agents to reduce radiocaesium transfer from the diet to tissues/milk of animals; and increasing the sorption capacity of soils. The effectiveness of some of these measures for radiocaesium, as summarised by Alexakhin (1993), is presented in Table 5.2. The effectiveness for many countermeasures was dependent on soil and crop type (Prister et al., 1993; Beresford and Wright 1999). Animal products were reported to be the main contributor to the dose resulting from agricultural production. Milk and meat contributed 90% of the internal dose (Prister et al., 1993). The most effective countermeasure for reducing doses via milk was reported to be the radical improvement of pasture (Alexakhin, 1993).

It has been stated (Prister et al., 1993) that the effectiveness of the application of zeolites and sodium humates to land, without prior technological preparation, was low in Belarus and the Ukraine. Because of the high cost of these materials they concluded that 'these methods cannot be regarded as suitable for practical application'. These workers (Prister et al., 1993), did, however, comment on the effectiveness of a material called 'sapropell'. Sapropell is obtained from lake bottom deposits and

Table 5.2. Summary of the effectiveness of different agricultural countermeasures to reduce ^{137}Cs activity concentrations employed within the fSU.
Adapted from Alexakhin (1993).

Countermeasure	Reduction factor*
Radical improvement of pasture	2.5–8.0
Liming of acidic soils	1.5–2.0
Application of mineral fertilisers	1.5–2.5
Application of zeolites	1.5–2.0
Processing of milk into butter	up to 10–15
Selection of crops with low ^{137}Cs uptake	up to 4.5–10

* Ratio of activity concentration prior to countermeasure implementation to activity concentration after implementation.

comprises of plant residues decomposed under anaerobic conditions together with mineral materials. Sapropell was found to decrease the radiocaesium activity concentration in plants by a factor of 1.8 to 4.5 when applied at a rate of $100\,\text{t}\,\text{ha}^{-1}$ (Prister et al., 1993).

As part of the regime to manage land in contaminated areas, intervention limits were derived for contamination levels in soil in addition to those derived for activity concentrations in (human and animal) foodstuffs (Alexakhin, 1993; Prister et al., 1993, 2000). These limits were dependent upon soil type and usage and, as with intervention limits for foodstuffs, were reduced over time after the Chernobyl accident (Prister et al., 2000).

Selective feeding regimes were also initiated such that contaminated feed could be used early in the life of animals destined for slaughter with less contaminated feed being used during the final fattening stages (Firsakova, 1993; Prister et al., 1993, 2000). Table 5.3 presents recommendations on the maximum daily intake of

Table 5.3. Suggested feeding regime for beef cattle at various times prior to slaughter and the effect on the activity concentration in meat. Values are mean ± standard deviation.
Adapted from Prister et al. (1993).

Stage of fattening	Dietary radiocaesium intake (kBq d^{-1})	Radiocaesium activity concentration in muscle (Bq kg^{-1} fw)		Length of feeding period (days) for animals at different slaughter ages		
		At start of each stage	At end of each stage	1.5 years	2.5 years	2.5–9 years
Initial	74 ± 3	n/a	2,960 ± 73	–	–	–
Intermediate	33 ± 2	2,960 ± 73	1,300 ± 65	15	15	30
Final	15 ± 1	1,300 ± 59	600 ± 48	50	60	60–120

radiocaesium by beef cattle at different stages prior to slaughter with the aim of producing meat below the EC intervention level of 600 Bq kg^{-1} (Prister et al., 1993). Prister et al. (2000) report that in the Ukrainian Zhitomir and Kiev regions in 1996, 1,600 animals were fed using a regime such as this. No restrictions were placed on the dietary ^{137}Cs intake over the first 12–16 months of the animal's life. Clean feeding subsequently reduced the ^{137}Cs activity concentration in muscle from 3,000 Bq kg^{-1} to 130 Bq kg^{-1} over 2–3 months. Clean feeding was combined with live-monitoring to determine that the ^{137}Cs activity concentration in meat was acceptable for slaughter. Firsakova (1993) gives brief details of the development of live-monitoring techniques in Belarus.

The cost of importing AFCF into the fSU was prohibitive. Instead, a locally manufactured product called 'ferrocyn' (a mixture of 95% Fe4[Fe(CN)$_6$] and 5% KFe[Fe(CN)$_6$]) was developed and commercially produced. It has been administered to animals as 15% ferrocyn boli, 10% ferrocyn salt licks, 98% pure ferrocyn powder and as a preparation called 'bifege' (sawdust with 10% adsorbed ferrocyn) (Ratnikov et al., 1998). Ferrocyn powder and bifege were mixed into the daily feed. The administration of 30–60 g d^{-1} bifege was the most effective delivery system reducing the transfer to cows' milk by 90–95% (probably because the ferrocyn was more uniformly mixed into feed than ferrocyn powder).

The application of countermeasures effectively reduced the volume of food products in excess of intervention limits produced by collective farms (Alexahkin, 1993; Firsakova et al., 1996; Prister et al., 2000) as shown in Table 5.4. The apparent increases in the amount of milk exceeding the limit in Belarus in 1991 and again in 1993 (Table 5.4) coincided with lower intervention limits being adopted. However, a reduced spending on countermeasure implementation as a consequence of the break-

Table 5.4. Changes in the amount (in kilotonnes) of meat and milk produced by collective farms with ^{137}Cs activity concentrations in excess of intervention limits.
From Firsakova et al. (1996).

Year	Belarus		Russia		Ukraine	
	Milk (kt)	Meat (kt)	Milk (kt)	Meat (kt)	Milk (kt)	Meat (kt)
1986	524.6*	21.1‡	110.9*	5.4†		6.41†
1987	308.9*	6.9‡	96.8*	5.7†		1.28†
1988	193.3*	1.45‡	95.9*	1.5†	78*	0.168†
1989	69.6*	0.6‡	74.8*	0.3†	61*	0.064†
1990	7.2*	0.08‡	48.8*	0.04†	62*	0.017†
1991	22.1**		10.97*		1*	
1992	9.4**		6.2*		0*	
1993	14.9***		1.6*		0*	
1994	12.4***	0.0026*	0.59*	0.012†	0*	0†

† Intervention limit = 740 Bq ^{137}Cs kg^{-1}; ‡ Intervention limit = 600 Bq ^{137}Cs kg^{-1}; * Intervention limit = 370 Bq ^{137}Cs kg^{-1}; ** Intervention limit = 185 Bq ^{137}Cs kg^{-1}; *** Intervention limit = 111 Bq ^{137}Cs kg^{-1}.

up of the USSR and a deteriorating financial situation of the affected countries was noted in the mid-late 1990s (Alexhahkin et al., 1996; Prister et al., 2000; Fesenko et al., 2001). Fertilisation rates within the contaminated areas may have reduced to below those used in normal agriculture (Beresford and Wright, 1999). Alexahkin et al. (1996) report that K_2O fertilisation rates in the Bransk oblast of Russia decreased from $81\,kg\,ha^{-1}$ in 1991 to $18\,kg\,ha^{-1}$ in 1993, with a consequent three-fold increase in the transfer of ^{137}Cs to cereals and potatoes.

5.2.1 Key foodstuffs contributing to ingestion doses

The rural populations within the fSU produce, or gather, much of their own diet. Many families own a dairy cow, and keep pigs and poultry. Vegetables and fruit are produced in the garden or on nearby land allocated to villagers by the collective farm. There is a common tradition of collecting edible fungi and berries from forests (often termed *forest gifts*) and, in some areas, fish from lakes or rivers. Therefore, products from collective farms are usually of little importance to the rural diet. With respect to radiocaesium intake, the significance of wild foodstuffs has increased over time because the effective ecological half-lives of these foods are often much greater than those of other foodstuffs and can be similar to the physical decay of radiocaesium (Shutov et al., 1996; Gillett and Crout, 2000; see also Chapter 3).

Additionally, privately owned cows often graze poor quality pastures, forest clearings or meadows along watercourses (Strand et al., 1996). The soils in such pastures allow a comparatively high root uptake of radiocaesium. Consequently, a greater proportion of privately produced milk has ^{137}Cs activity concentrations in excess of national intervention limits than milk produced on collective farms. Fesenko et al. (2001) report an increase in internal doses within rural settlements following the reintroduction of privately owned cows (private production having been banned in the early years after the accident).

Frank et al. (1998) recommended that restrictions on the consumption of forest fungi would significantly reduce the internal dose in 18 of 36 study settlements within the fSU where they estimated a total annual dose of >1 mSv. However, many people have ignored bans placed by national governments on the collection of fungi from contaminated regions. Beresford et al. (2001) and Lepicard and Hériard Dubreuil (2001) report that although ferrocyn products are generally available to private milk producers, they are sometimes being inefficiently used, if used at all.

Self-help approaches to reducing radiocaesium ingestion via collected and self-produced foods have been suggested for rural populations in the fSU (Beresford et al., 2001). Hériard Dubreuil et al. (1999) and Lepicard and Hériard Dubreuil (2001) describe a programme initiated in a Belarussian village to develop sustainable living conditions in contaminated areas. In trying to improve the quality of life, they attempted to not only address radiological issues but to also encompass the social, psychological, economic, political and ethical aspects of living in contaminated areas. Their aim was a transfer of practical skills in radiation protection to the local population enabling them to regain control and contribute to an improvement of their living conditions.

208 Application of countermeasures [Ch. 5

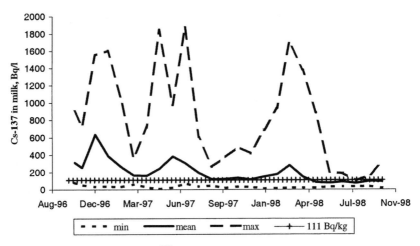

Figure 5.3. Variability within the ^{137}Cs activity concentration of private milk within a Belarussian village. The intervention limit within Belarus at that time (111 Bq kg^{-1} of ^{137}Cs) is indicated.

Data from the programme of Hériard Dubreuil et al. (1999); figure reproduced from Beresford et al. (2001).

As an example of the potential success of such approaches, Lepicard and Hériard Dubreuil (2001) report a solution to the problem of contaminated milk. The population had expressed concern over the radiological quality of the milk they produced and a desire to produce uncontaminated milk for their children. A question which first had to be asked is 'did the population have the possibility to reduce contamination levels in milk?' Figure 5.3 shows that whilst some milk samples greatly exceeded the intervention limit, there were samples below this throughout the measurement period and on occasions the median value was also less than 111 Bq kg^{-1}. Therefore, the village could produce milk with ^{137}Cs activity concentrations below the intervention limit. In this case, by working with the private milk producers, it was established that during the summer grazing period the most contaminated milk samples were coming from cows within two of the villages' seven herds (during the summer the villagers collectively graze their cows) (Hériard Dubreuil et al., 1999; Lepicard and Hériard Dubreuil, 2001). As a consequence, the villagers negotiated with the local collective farm and in August 1997 the two herds were allocated grazing on pastures which had been deep ploughed and reseeded as a countermeasure following the Chernobyl accident. A general decline in the ^{137}Cs activity concentrations of milk was subsequently observed, which also offered the possibility for the private producers to sell their milk (Lepicard and Hériard Dubreuil, 2001).

5.3 DISCUSSION

There has been considerable development of countermeasures in response to the Chernobyl accident. From necessity these have predominantly considered radiocaesium, although development of countermeasures for other radionuclides (namely radioiodine and radiostrontium) has also progressed (see, e.g., review by Howard et al., 2001).

There has also been an increasing appreciation that countermeasure application is not only a 'radiological problem'. For the implementation of robust and effective strategies in contaminated areas, wider social and ethical aspects, environmental considerations and quality of life must be taken into account (Howard et al., 2004). The potential effectiveness of holistic and inclusive approaches to coping with contaminated environments has been demonstrated in the fSU in the work of Lepicard and Hériard Dubreuil (2001) discussed above. In some western European countries, stakeholder groups have now been established to consider acceptable countermeasures and their implementation should they be needed in the future (Nisbet et al. in press).

5.4 REFERENCES

Åhman, B. (1999) Transfer of radiocaesium via reindeer meat to man – effect of countermeasures applied in Sweden after the Chernobyl accident. *Journal of Environmental Radioactivity*, **46**, 113–120.

Alexakhin, R.M. (1993) Countermeasures in agricultural production as an effective means of mitigating the radiological consequences of the Chernobyl accident. *Science of the Total Environment*, **137**, 9–20.

Alexakhin, R., Firsakova, A., Rauret, G., Arkhipov, N., Vandecasteele, C.M., Ivanov, Yu., Fesenko, S. and Sanzharova, N. (1996) Fluxes of radionuclides in agricultural environments: Main results and still unsolved problems. In: Karaoglou, A., Desmet, G., Kelly, G.N. and Menzel, H.G. (eds), *The Radiological Consequences of the Chernobyl Accident*, pp. 39–47. European Commission, Brussels.

Andersson, K.G., Roed, J., Eged, K., Kis, Z., Voigt, G., Meckbach, R., Oughton, D.H., Hunt, J., Lee, R., Beresford, N.A. and Sandalls, F. J. (2003) *Physical Countermeasures to Sustain Acceptable Living and Working Conditions in Radioactively Contaminated Residential Areas*. Risø-R-1396(EN); ISBN 87-550-3190-0. Risø National Laboratory, Roskilde.

Antsipov, G., Tabachny, L., Balonov, M. and Roed, J. (2000) Evaluation of the effectiveness of decontamination work carried out in the CIS countries on buildings and land contaminated as a result of the Chernobyl accident. In: Cécille, L. (ed.), *Procceding of the Workshop on Restoration Strategies for Contaminated Territories Resulting from the Chernobyl Accident*. EUR 18193EN, European Commission, Brussels.

Balonov, M.I., Anisimova, L.I. and Perminova, G.S. (1999) Strategy for population protection and area rehabilitation in Russia in the remote period after the Chernobyl accident. *Journal of Radiological Protection*, **19**, 261–269.

Bengtsson, G. (1986) After Chernobyl? Implications of the Chernobyl accident for Sweden. *Invandrartidningen på engelska News & views*. ISSN 0349-5515. Stiftelsen Invandrartidningen, Stockholm.

Beresford, N.A., Lamb, C.S., Mayes, R.W., Howard, B.J. and Colgrove, P.M. (1989) The effect of treating pastures with bentonite on the transfer of Cs-137 from grazed herbage to sheep. *Journal of Environmental Radioactivity*, **9**, 251–264.

Beresford, N.A., Mayes, R.W., Barnett, C.L., Lamb, C.S., Wilson, P., Howard, B.J. and Voigt, G. (1997). The effectiveness of oral administration of potassium iodide to lactating goats in reducing the transfer of radioiodine to milk. *Journal of Environmental Radioactivity*, **35**, 115–128.

Beresford, N.A., Mayes, R.W., Hansen, H.S., Crout, N.M.J., Hove, K. and Howard, B.J. (1998) Generic relationship between calcium intake and radiostrontium transfer to milk of dairy ruminants. *Radiation and Environmental Biophysics*, **37**, 129–131.

Beresford, N.A. and Wright, S.M. (eds) (1999). *Self-help Countermeasure Strategies for Populations Living Within Contaminated Areas of the Former Soviet Union and an Assessment of Land Currently Removed from Agricultural Usage.* Joint deliverable of EC Projects RESTORE and RECLAIM. Institute of Terrestrial Ecology, Grange-over-Sands, U.K.

Beresford, N.A., Hove, K., Barnett, C.L., Dodd, B.A, Fawcett, R.H. and Mayes, R.W. (1999) The development and testing of an intraruminal slow-release bolus designed to limit radiocaesium absorption by small lambs grazing contaminated pastures. *Small Ruminant Research*, **33**, 109–115.

Beresford, N.A., Mayes, R.W., Colgrove, P.M., Barnett, C.L., Bryce, L., Dodd B.A. and Lamb, C.S. (2000) A comparative assessment of the potential use of alginates and dietary calcium manipulation as countermeasures to reduce the transfer of radiostrontium to the milk of dairy animals. *Journal of Environmental Radioactivity*, **51**, 69–82.

Beresford, N.A., Voigt, G., Wright, S.M., Howard, B.J., Barnett, C.L., Prister, B., Balonov, M., Ratnikov, A., Travnikova, I., Gillett, *et al.* (2001) Self-help countermeasure strategies for populations living within contaminated areas of Belarus, Russia and the Ukraine. *Journal of Environmental Radioactivity*, **56**, 215–239.

Brynildsen, L. and Strand, P. (1994) A rapid method for the determination of radioactive caesium in live animals and carcasses and its practical application in Norway after the Chernobyl accident. *Acta Veterinaria Scandinavica*, **35**, 401–408.

Brynildsen, L.I., Selnaes, T.D., Strand, P. and Hove, K. (1996) Countermeasures for radiocesium in animal products in Norway after the Chernobyl accident – techniques, effectiveness, and costs. *Health Physics*, **70**, 665–672.

Bugai, D.A., Waters, R.D., Dzhepo, S.P. and Skal'skij, A.S. (1996) Risks from radionuclide migration to groundwater in the Chernobyl 30-km zone. *Health Physics*, **71**, 9–18.

Burrough, P.A., Van Der Perk, M., Howard, B.J., Prister, B.S., Sansone, U. and Voitsekhovitch, O.V. (1999) Environmental mobility of radiocaesium in the Pripyat catchment, Ukraine, Belarus. *Water Air and Soil Pollution*, **110**, 35–55.

Cécille, L. (ed.) (2000) *Proceedings of the Workshop on Restoration Strategies for Contaminated Territories Resulting from the Chernobyl Accident.* EUR18193EN, European Commission, Brussels.

Coughtrey, P.J. and Thorne, M.C. (1983) *Radionuclide Distribution and Transport in Terrestrial and Aquatic Ecosystems* (Volume 1). A.A. Balkema, Rotterdam.

Crout, N.M.J., Beresford, N.A., Mayes, R.W., MacEachern, P.J., Barnett, C.L., Lamb, C.S. and Howard, B.J. (2000) A model of radioiodine transfer to goat milk incorporating the influence of stable iodine. *Radiation and Environmental Biophysics*, **39**, 59–65.

Fesenko, S., Jacob, P., Alexakhin, R., Sanzharova, N.I., Panov, A., Fesenko, G. and Cecille, L. (2001) Important factors governing exposure of the population and countermeasure application in rural settlements of the Russian Federation in the long term after the Chernobyl accident. *Journal of Environmental Radioactivity*, **56**, 77–98.

Firsakova, S.K. (1993) Effectiveness of countermeasures applied in Belarus to produce milk and meat with acceptable levels of radiocaesium after the Chernobyl accident. *Science of the Total Environment*, **137**, 199–203.

Firsakova, S., Hove, K., Alexakhin, R., Prister, B., Arkhipov, N. and Bogdanov, G. (1996) Countermeasures implemented in intensive agriculture. In: Karaoglou, A., Desmet, G., Kelly, G.N. and Menzel, H.G. (eds), *The Radiological Consequences of the Chernobyl Accident*, pp. 379–387. European Commission, Brussels.

Frank, G., Jacob, P., Pröhl, G., Smith-Briggs, J.L., Sandalls, F.J., Holden, P.L., Firsakova, S.K., Zhuchenko, Y.M., Jouve, A., Tikhomirov, F.A., Mamikhin, S., et al. (1998) *Optimal Management Routes for the Restoration of Territories Contaminated During and After the Chernobyl Accident*. Final report COSU-CT94-0101, COSU-CT94-0102, B7-6340/001064/MAR/C3 and B7-5340/96/000178/MAR/C3. EUR 17627 EN. ISBN 92-828-2237-0. European Commission, Brussels.

Fogh, C.L., Andersson, K.G., Barkovsky, A.N., Mishine, A.S., Ponamarjov, A.V., Ramzaev, V.P. and Roed, J. (1999) Decontamination in a Russian settlement. *Health Physics*, **76**, 421–430.

Giese, W.W. (1988) Ammonium-ferric-cyano-ferrate(II) (AFCF) as an effective antidote against radiocaesium burdens in domestic animals and animal derived foods. *British Veterinary Journal*, **144**, 363.

Gillett, A.G. and Crout, N.M.J. (2000) A review of ^{137}Cs transfer to fungi and consequences for modelling environmental transfer. *Journal of Environmental Radioactivity*, **48**, 95–121.

Goossens, R., Delville, A., Genot, J., Halleux, R. and Masschelein, W.J. (1989) Removal of the typical isotopes of the Chernobyl fall-out by conventional water treatment. *Water Research*, **23**, 693–697.

Håkanson, L. and Andersson, T. (1992) Remedial measures against radioactive Caesium in Swedish lake fish after Chernobyl. *Aquatic Sciences*, **54**, 141–164.

Hansen, H.S., Saether, M., Asper, N.P. and Hove, K. (1995) In vivo testing of compounds with possible strontium binding effects in ruminants. In: *Proceedings of a Symposium on Environmental Impact of Radioactive Releases*, pp. 719–721. IAEA-SM-339/198P. IAEA, Vienna.

Hériard Dubreuil, G., Lochard, J., Girard, P., Guyonnet, J.F., Le Cardinal, G., Lepicard, S., Livolsi, P., Monroy, M., Ollagon, H., Pena-Vega, A., et al. (1999). Chernobyl post-accident management: The Ethos project. *Health Physics*, **77**, 61–272.

Hove, K., Staaland, H., Pedersen, Ø., Ensby, T. and Sæthre, O. (1991) Equipment for placing a sustained release bolus in the rumen of reindeer. *Rangifer*, **11**, 49–52.

Hove, K. (1993) Chemical methods for reduction of the transfer of radionuclides to farm animals in semi-natural environments. *Science of the Total Environment*, **137**, 235–248.

Hove, K. and Hansen, H.S. (1993) Reduction of radiocaesium transfer to animal products using sustained release boli with ammoniumiron(III)-hexacyanoferrate(II). *Acta vetinaria scandinavia*, **34**, 287–297.

Howard, B.J., Beresford, N.A. and Hove, K. (1991) Transfer of radiocesium to ruminants in unimproved natural and semi-natural ecosystems and appropriate countermeasures. *Health Physics*, **61**, 715–725.

Howard, B.J. and Desmet, G.M. (eds) (1993) Special issue: Relative effectiveness of agricultural countermeasure techniques – REACT. *Science of the Total Environment*, **137**.

Howard, B.J., Voigt, G., Segal, M. and Ward, G. (1996) A review of countermeasures to reduce radioiodine in milk of dairy animals. *Health Physics*, **71**, 661–673.

Howard, B.J., Beresford, N.A. and Voigt, G. (2001) Countermeasures for animal products: A review of effectiveness and potential usefulness after an accident. *Journal of Environmental Radioactivity*, **56**, 115–137.

Howard, B.J., Andersson., K.G., Beresford, N.A., Crout, N.M.J., Gil., J.M., Hunt, J., Liland, A., Nisbet, A., Oughton, D. and Voight, G. (2002) Sustainable restoration and long-term management of contaminated rural, urban and industrial ecosystems. *Radioprotection – colloques*, **37**, 1067–1072.

Howard, B.J., Liland, A., Beresford, N.A., Andersson, K.G., Cox, G., Gil, J. M., Hunt, J., Nisbet, A., Oughton, D. and Voigt, G. (2004) A Critical Evaluation of the STRATEGY Project. *Radiation Protection Dosimetry*, **109**, 63–67.

Hunt, J. and Wynne, B. (2002) *Social assumptions in remediation strategies*. Deliverable 5 of the STRATEGY project FIKR-CT-2000-00018. University of Lancaster, Lancaster (available from www.strategy-ec.org.uk).

IAEA (1991) *The International Chernobyl Project Technical Report*, 640 pp. IAEA, Vienna.

IAEA (1994) *Guidelines for Agricultural Countermeasures Following an Accidental Release of Radionuclides*. Technical report series 363. IAEA, Vienna.

ICRP (1999) Protection of the public in situations of prolonged radiation exposure. Publication 82. *Annals of the ICRP*, **29**.

Jackson, D. and Jones, S.R. (1991) Reappraisal of environmental countermeasures to protect members of the public following the Windscale nuclear reactor accident 1957. In: *Comparative Assessment of the Environmental Impact of Radionuclides Released During Three Major Nuclear Accidents: Kyshtym, Windscale and Chernobyl*. EUR 13574, 1015–1055. CEC, Luxembourg.

Jacob, P., Pröhl, G., Likhtarev, L., Kovgan, L., Gluvchinsky, R., Perevoznikov, O., Balonov, M.I., Golikov, A., Ponomarev, A., Erkin, V., et al. (1995) *Pathway Analysis and Dose Distributions*. Joint study project No 5. Final report EUR 16541 EN. European Commission, Brussels.

Jaworowski, Z. (2004) Radiation folly. In: Okonski, K. and Morris, J. (eds), *Environment & Health: Myths & Realities*, pp. 69–91. International Policy Press, London.

Johanson, K.J. (1994) Radiocaesium in game animals in the Nordic countries. In: Dahlgaard, H. (ed.), *Nordic radioecology – The transfer of radionuclides through Nordic Ecosystems to Man*, pp. 287–301. Studies in Environmental Science 62. Elsevier, Oxford.

Jones, B.E. (1993) Management methods of reducing the radionuclide contamination of animal food product. *Science of the Total Environment*, **137**, 227–233.

Kudelsky, A.V., Smith, J.T. and Petrovich, A.A. (2002). An experiment to test the addition of potassium to a non-draining lake as a countermeasure to ^{137}Cs accumulation in fish. *Radioprotection Colloques*, **37**, 621–626.

Lepicard, S. and Hériard Dubreuil, G. (2001) Practical improvement of the radiological quality of milk produced by peasant farmers in the territories of Belarus contaminated by the Chernobyl accident. The ETHOS project. *Journal of Environmental Radioactivity*, **56**, 241–251.

Long, S., Pollard, D., Cunningham, J.D., Astasheva, N.P., Donskaya, G.A. and Labetsky, E.V. (1995a) The effects of food processing and direct decontamination techniques on the radionuclide content of foodstuffs: A literature review. Part 1: Milk and milk products. *Journal of Radioecology*, **3**, 15–30.

Long, S., Pollard, D., Cunningham, J.D., Astasheva, N.P., Donskaya, G.A. and Labetsky, E.V. (1995b) The effects of food processing and direct decontamination techniques on the radionuclide content of foodstuffs: A literature review. Part 2: Meat, fruit, vegetables, cereals and drinks. *Journal of Radioecology*, **3**, 15–38.

Mehli, H., Skuterud, L., Mosdøl, A. and Tønnessen, A. (2000) The impact of Chernobyl fallout on the Southern Saami reindeer herders in Norway in 1996. *Health Physics*, **79**, 682–690.

Meredith, R.C.K., Mondon, K.J. and Sherlock, J.C. (1988) A rapid method for the in vivo monitoring of radiocaesium activity in sheep. *Journal of Environmental Radioactivity*, **7**, 209–214.

Morgan, R.P.C. (1986) *Soil Erosion and Conservation*, 298 pp. Longman, Harlow, UK.

NRC (National Research Council) (1980) *Mineral Tolerance of Domestic Animals*. National Academy Press, Washington.

NRC (National Research Council) (1989) *Nutrient Requirements of Dairy Cattle* (6th revised edition). National Academy Press, Washington.

Nisbet, A.F. and Woodman, R. (2000) Options for the management of Chernobyl-restricted areas in England and Wales. *Journal of Environmental Radioactivity*, **51**, 239–254.

Nisbet, A.F., Mercer, J.A., Hesketh, N., Liland, A., Thorring, H., Bergan, T., Beresford, N.A., Howard, B.J., Hunt, J. and Oughton, D.H. (2004) *Datasheets on Countermeasures and Waste Disposal Options for the Management of Food Production Systems Contaminated following a Nuclear Accident*. NRPB Report NRPB-W58. NRPB, Didcot.

Nisbet, A.F., Howard, B., Beresford, N. and Voigt, G. (eds) (2005) Workshop to extend the involvement of stakeholders in decisions on restoration management (WISDOM). *Journal of Environmental Radioactivity*. (Available online at time of writing.)

Prister, B., Perepelyatnikov, G.P. and Perepelyatnikova, L.V. (1993) Countermeasures used in the Ukraine to produce forage and animal food products with radionuclide levels below intervention limits after the Chernobyl accident. *Science of the Total Environment*, **137**, 183–198.

Prister, B., Alexakhin, R., Firsakova, S. and Howard, B.J. (2000) Short and long term environmental assessment. In: Cécille, L. (ed.), *Proceedings of the Workshop on Restoration Strategies for Contaminated Territories Resulting from the Chernobyl Accident*, pp. 103–114. EUR18193EN, DG. Environment of the European Commission, Brussels.

Rantavara, A. (1987) *Radioactivity of Vegetables and Mushrooms in Finland After the Chernobyl Accident in 1986*. STUK-A5 Finnish Centre for Radiation and Nuclear Safety, Helsinki.

Ratnikov, AN., Vasiliev, A.V., Krasnova, E.G., Pasternak, A.D., Howard, B.J., Hove, K. and Strand, P. (1998) The use of hexacyanoferrates in different forms to reduce radiocaesium contamination of animal products in Russia. *Science of the Total Environment*, **223**, 167–176.

Roed, J., Andersson, K.G. and Prip, H. (1996) The Skim and Burial Plough: A new implement for reclamation of radioactively contaminated land. *Journal of Environmental Radioactivity*, **33**, 117–128.

Roed, J., Andersson, K.G., Barkovsky, A.N., Fogh, C.L., Mishine, A.S., Olsen, S.K., Ponamarjov, A.V., Prip, H., Ramzaev, V.P. and Vorobiev, B.F. (1998) *Mechanised Decontamination Tests in Areas Affected by the Chernobyl Accident*. Risø-R-1029(EN); ISBN 87-550-1361-4. Risø National Laboratory, Roskilde.

Rubery, E. and Smales, E. (eds) (1990) *Iodine Prophylaxis Following Nuclear Accidents*. Proceedings of joint WHO/CEC workshop July 1988. Pergamon Press, Oxford.

Ryabov, I.N. (1992) Analysis of countermeasures to prevent intake of radionuclides via consumption of fish from the region affected by the Chernobyl accident. *Proceedings of the International Seminar on Intervention Levels and Countermeasures for Nuclear Accidents*, pp. 379–390. EUR 14469. European Commission, Luxembourg.

Schwab, G.O., Frevert, R.K., Edminster, T.W. and Barnes, K.K. (1966) *Soil and Water Conservation Engineering*. Wiley, New York.

Segal, M.G. (1993) Agricultural countermeasures following deposition of radioactivity after a nuclear accident. *The Science of the Total Environment*, **137**, 31–48.

Shutov, V.N., Bruk, G.Ya., Basalaeva, L.N., Vasilevitskiy, V.A., Ivanova, N.P. and Kaplun, I.S. (1996) The role of fungi and berries in the formulation of internal exposure doses to the population of Russia after the Chernobyl accident. *Radiation Protection Dosimetry*, **67**, 55–64.

Smith, J.T., Voitsekhovitch, O.V., Håkanson, L. and Hilton, J. (2001) A critical review of measures to reduce radioactive doses from drinking water and consumption of freshwater foodstuffs. *Journal of Environmental Radioactivity*, **56**, 11–32.

Smith, J.T., Kudelsky, A.V., Ryabov, I.N., Hadderingh, R.H. and Bulgakov, A.A. (2003) Application of potassium chloride to a Chernobyl – contaminated lake: Modelling the dynamics of radiocaesium in an aquatic ecosystem and decontamination of fish. *The Science of the Total Environment*, **305**, 217–227.

Solheim Hansen, H., Hove, K. and Barvik, K. (1996) The effect of sustained release boli with ammoniumiron(III)-hexacyanoferrate)II) on radiocesium accumulation in sheep grazing contaminated pasture. *Health Physics*, **71**, 705–712.

Strand, P., Selnæs, T.D., Bøe, E., Harbitz, O. and Andersson-Sørlie, A. (1992) Chernobyl fallout: Internal doses to the Norwegian population and the effect of dietary advice. *Health Physics*, **63**, 385–392.

Strand, P., Howard, B.J. and Averin, V. (eds) (1996) *Transfer of Radionuclides to Animals, their Comparative Importance Under Different Agricultural Ecosystems and Appropriate Counteremeasures.* Experimental collaboration project No. 9. Final Report EUR 16539EN. European Commission, Luxembourg.

Tsarik, N. (1993) Supplying water and treating sewage in Kiev after the Chernobyl accident. *Journal of the American Water Works Association*, **85**, 42–45.

Tveten, U., Brynildsen, L.I., Amundsen, I. and Bergan, T.D. (1998) Economic consequences of the Chernobyl accident in Norway in the decade 1986–1995. *Journal of Environmental Radioactivity*, **41**, 233–255.

UNSCEAR (2000) *Report to the General Assembly: Sources and Effects of Ionizing Radiation* (Volume II, Annex J), pp. 453–551. United Nations, New York (available online at: http://www.unscear.org).

Vakulovsky, S.M., Nikitin, A.I., Chumichev, V.B., Katrich, I.Yu., Voitsekhovitch, O.A., Medinets, V.I., Pisarev, V.V., Bovkum, L.A. and Khersonsky, E.S. (1994) Caesium-137 and Strontium-90 contamination of water bodies in the areas affected by releases from the Chernobyl Nuclear Power Plant accident: an overview. *Journal of Environmental Radioactivity*, **23**, 103–122.

Vandecasteele, C.M., Van Hees, M., Hardeman, F., Voigt, G.M. and Howard, B.J. (2000). The true absorption of iodine and effect of increasing stable iodine in the diet. *Journal of Environmental Radioactivity*, **47**, 301–317.

Voigt, G. (1993) Chemical methods to reduce the radioactive contamination of animals and their products in agricultural ecosystems. *Science of the Total Environment*, **137**, 205–225.

Voigt, G. (ed.) (2001) Special issue: Remediation strategies. *Journal of Environmental Radioactivity*, **56**.

Voitsekhovitch, O.V., Kanivets, V.V. and Shereshevsky, A.I. (1988) The effectiveness of bottom sediment traps created to capture contaminated matter transported by suspended particles. *Proceedings of the Ukrainian Hydrometeorological Institute*, **228**, 60–68 (in Russian).

Voitsekhovitch, O.V., Nasvit, O., Los'y, Y. and Berkovski, V. (1997) Present thoughts on aquatic countermeasures applied to regions of the Dnieper river catchment contaminated by the 1986 Chernobyl accident. In: Desmet, G., Blust, R., Comans, R.N.J.,

Fernandez, J.A., Hilton, J. and de Bettencourt, A. (eds), *Freshwater and Estuarine Radioecology*, pp. 75–86. Elsevier, Amsterdam.

Voitsekhovitch, O.V. (1998) Water quality management of the radioactive contaminated waters (Strategy and Technology). In: Voitsekhovitch, O.V. (ed.), *Radioecology of the Water Bodies at the Areas Affected as a Result of the Chernobyl Accident*, pp. 169–217. Chernobyl Inter-inform, Kiev (in Russian).

Wright, S.M., Smith, J.T., Beresford, N.A. and Scott, W.A. (2003) Prediction of changes in areas in west Cumbria requiring restrictions on the movement and slaughter of sheep following the Chernobyl accident using a Monte-Carlo approach. *Radiation and Environmental Biophysics*, **42**, 41–47.

6

Health consequences

Jacov E. Kenigsberg and Elena E. Buglova

6.1 INTRODUCTION

The Chernobyl accident subjected millions of people in different countries to a wide range of radiation exposures. The highest radiation doses were received by the personnel of the Chernobyl Nuclear Power Plant (NPP), firemen, specialists working to reduce the consequences of the accident and the population of nearby territories. Most of the areas where radioactive contamination exceeded 37 kBq m^{-2} ^{137}Cs were located in the former USSR (Republic of Belarus, Russian Federation, the Ukraine). In the early stages of the accident, radiation exposures were due to short-lived radionuclides, particularly iodine radioisotopes. Several weeks after the accident, as short-lived radionuclides decayed, the long-lived caesium radioisotopes (^{134}Cs, ^{137}Cs) played the most important role in dose formation. Some contribution to the total dose was also made by other long-lived radionuclides such as strontium-90 (^{90}Sr) and transuranium elements. Low-level exposures to the population living on land contaminated by these long-lived radionuclides will continue for decades.

Over the years after Chernobyl, numerous studies have been dedicated to the health effects of the accident. The main results of these studies have been discussed in representative international meetings and summarized in the UNSCEAR (1988, 2000) reports and proceedings of international fora (Karaoglou *et al.*, 1996; Thomas *et al.*, 1999; Chernobyl Conference, 2001). This chapter summarises the large body of work on the radiation-induced health effects of the Chernobyl accident but does not consider health effects from non-radiation factors of the accident (see Chapter 7 for a discussion of social, psychological and economic consequences, and non-radiation impacts on health).

6.2 RADIATION-INDUCED HEALTH EFFECTS

The exposure of organs and tissues of the human body to different radiation doses causes different types of health effects (see Tables 6.1 and 6.2). A very high dose (several grays or more) to the whole body can cause death within days or weeks. A very high dose to a limited area of the body might not prove to be fatal but can cause other early effects (e.g., skin erythema (after exposure of skin) or sterility (after exposure of gonads)). These effects are called deterministic. Deterministic effects occur only if the dose or dose rate is greater than some threshold value, and the effect usually occurs soon after exposure and is more severe as the dose and dose rate increase.

Some types of deterministic effect can occur a long time after exposure. Such effects are not usually fatal, but can be disabling or distressing because the function of some parts of the body may be impaired or other non-malignant changes may arise. The best-known examples are cataract (opacity in the lens of the eye) and skin damage (thinning and ulceration).

If the dose is lower, or is delivered over a longer period of time (chronic exposure), there is a greater opportunity for the body cells to repair, but damage

Table 6.1. The most critical radiation-induced health effects resulting from a radiation exposure.
Adapted from IAEA (2005).

Health effect	Target organ or entity
Deterministic health effects	
Haematopoetic syndrome	Red marrow
Gastrointestinal syndrome	Small intestine for external exposure or colon for internal exposure
Pneumonitis	Lung
Embryo/foetal death	Embryo/foetus in all periods of gestation
Moist desquamation	Skin
Necrosis	Soft tissue
Cataract	Lens of the eye
Acute radiation thyroiditis	Thyroid
Hypothyroidism	Thyroid
Permanently suppressed ovulation	Ovum
Permanently suppressed sperm counts	Testes
Severe mental retardation	Embryo/foetus 8–25 weeks of gestation
Malformation	Embryo/foetus 8–25 weeks of gestation
Growth retardation	Embryo/foetus 8–25 weeks of gestation
Possible verifiable reduction in IQ	Embryo/foetus 8–25 weeks of gestation
Stochastic health effects	
Thyroid cancer	Thyroid
Leukaemia	Red bone marrow
Solid cancers	Different organs

Table 6.2. Examples of stochastic health effects from exposure to radiation.
Adapted from IAEA (2004).

Circumstances of exposure and occurrence	Effects	Sources of information
Any dose or dose rate Risk depends on dose Effect appears years later	Various cancers	Risk factors for human beings estimated by extrapolating the human data for high doses and dose rates.
Any dose or dose rate Risk depends on dose Effect appears in offspring	Hereditary defects	Risk factors for human beings inferred from animal data and the absence of evidence showing hereditary effects in humans.

may still occur. The tissues may have been damaged in such a way that the effects appear only later in life (perhaps decades later), or even in the descendants of the irradiated person. Effects of this type, called stochastic (probabilistic) effects, are not certain to occur. The likelihood of their occurrence increases as the dose increases, but the timing and severity of an effect does not depend on the dose. Cancers and hereditary diseases are examples of stochastic effects (Table 6.2).

6.3 DETERMINISTIC HEALTH EFFECTS AFTER THE CHERNOBYL ACCIDENT

On the night of 26 April, 1986, about 400 workers were on the site of the Chernobyl NPP. As a consequence of the accident, they were subjected to radiation exposure from several sources: external gamma/beta radiation from the radioactive cloud, exposure from fragments of the damaged reactor core scattered over the site, radioactive particles deposited on the skin and inhalation of radioactive gases and particles.

During the first days after the accident more than 500 individuals including personnel of the power plant, firemen and emergency workers were examined in the hospitals of Moscow and Kiev. Initial diagnoses of acute radiation syndrome (ARS) were made for 237 persons. Later the diagnosis of ARS of differing severity was confirmed for 134 of these patients (Table 6.3).

The ARS in emergency workers and firemen was caused by high exposures from different sources: relatively uniform exposure of the whole body from external γ-radiation and exposure of the skin from surface contamination by β-radiation. Doses to the skin were 10–20 times higher than doses to the bone marrow and reached dozens of Gy. High doses of radiation quickly lead to erythema (redness of the skin) with subsequent development of burns of differing severity and, as a result, the skin becomes one of the critical systems determining the survival of

Table 6.3. Emergency workers with ARS.
Table adapted from UNSCEAR (2000).

Degree of ARS	Dose range (Gy)	Number of patients treated	Number of deaths
Mild (I)	0.8–2.1	41	0
Moderate (II)	2.2–4.1	50	1
Severe (III)	4.2–6.4	22	7
Very severe (IV)	6.5–16	21	20
Total	*0.8–16*	*134*	*28*

patients. Damage to the bone marrow and the gastrointestinal system was also of critical importance.

As was mentioned in Section 6.2, cataracts are also potential deterministic effects of high-dose radiation. There is, however, still no commonly accepted consensus on the development of cataracts among the Chernobyl victims. The data of official medical statistics shows an increase of cataract incidence among emergency workers and in the population of contaminated regions. The outcomes of current research projects have not, however, found statistical proof of excess cataract incidence related to radiation exposure (Day *et al.*, 1995; Fedirko, 1999).

6.4 STOCHASTIC HEALTH EFFECTS AFTER THE CHERNOBYL ACCIDENT

Radiation-induced cancer is a stochastic (probabilistic) effect of radiation exposure. Within the exposed population it is determined as the increase of cancer incidence or mortality from cancer in comparison with an unexposed population. This can be studied using epidemiological data. To prove the correlation between increased incidence (or mortality) and exposure it is necessary to show an excessive incidence (or mortality) and to determine the dose dependence of that excessive incidence (or mortality).

6.4.1 Leukaemia

The time since the Chernobyl accident is enough for an increase in leukaemia incidence to have been observed: the latent period of radiation-induced leukaemia (2–3 years following exposure) is much shorter than that of other radiation-induced cancers. The majority of excess leukaemia cases observed in survivors of the Hiroshima and Nagasaki atomic bombs occurred within 15 years of exposure (Pierce *et al.*, 1996). During the period since the accident, numerous studies have been performed in various countries where radioactive fallout occurred. These studies were aimed at detecting increased leukaemia incidence among the exposed population, primarily in children and emergency workers. The studies were conducted within national programmes as well as large-scale international projects. These epidemiological studies did not find evidence of increased leukaemia incidence among children and adults subjected to exposure due to the Chernobyl accident. Since large cohorts of examined subjects were involved in the studies, the statistical power of the epidemiological analysis would have been sufficient to reveal a significant excess incidence. The negative result of the studies conducted among the exposed population can be explained only by the relatively low radiation doses in the majority of people in the affected countries.

The inhabitants of regions located close to the Chernobyl NPP at the early stage of the accident received higher exposures compared to the rest of the population. During the first days and months after the accident more than 116,000 persons were evacuated from the Ukrainian and Belarussian settlements mostly located in the 30-km zone around the Chernobyl NPP. The average effective dose of external exposure to these people was approximately 17 mSv, and individual doses ranged from 0.1–380 mSv. The average dose for the evacuated population of Belarus was estimated as 31 mSv, whilst about 30% of people had effective doses of less than 10 mSv. Approximately 85% received a dose due to external exposure of less than 50 mSv, whilst only 4% received more than 100 mSv. The dose distribution shows that only a small part of the evacuated population received significant (>50 mSv, for example) doses. Considering that the number of children among the evacuated population did not exceed 25%, and in view of the relatively low doses, the probability of observing excess cancer incidence (excluding thyroid cancers discussed below) among children is very low.

Compared to other population groups, it was much more likely that studies would be able to detect excess radiation induced leukaemia among emergency workers since this group received higher doses than the evacuated population. According to the official statistics, more than 600,000 persons were involved in recovery works in the former Soviet Union (fSU) in 1986–1989 (approximately 292,000 people worked in the zone in the period 1986–1987). Information regarding the nature of their work, the duration, radiation exposure and health status is contained in the National Registries of Belarus, Russia, the Ukraine, as well as Lithuania, Latvia and Estonia. The number of emergency workers and their estimated radiation exposures are presented in Table 6.4.

Although the distribution of individual doses of emergency workers has a generally log-normal character, it should be noted that in 1986 the dose limit for emergency workers was fixed at 250 mSv. In 1987, the dose limit was 100 mSv, and in 1988 and 1989 it was 50 mSv. If the dose limit was exceeded, a worker was removed from emergency works. However, because of a lack of comprehensive dosimetry monitoring, some workers are likely to have received doses exceeding these fixed limits.

Considering the large cohorts of emergency workers and the values of doses received during a relatively short period of time, one could hypothesise an increase in leukaemia incidence, as was observed in atomic bomb (A-bomb) survivors at Hiroshima and Nagasaki. An increase in leukaemia incidence has not

Table 6.4. Distribution of external doses to emergency workers as recorded in the registry of emergency workers.

Period of work	Number of emergency workers	Percentage where dose is known	External dose (mSv)			
			Mean	Median	75th percentile	95th percentile
Belarus						
1986	68,000	8	60	53	93	138
1987	17,000	12	28	19	29	54
1988	4,000	20	20	11	31	93
1989	2,000	16	20	15	30	42
1986–1989	91,000	9	46	25	70	125
Russian Federation						
1986	69,000	51	169	194	220	250
1987	53,000	71	92	92	100	208
1988	20,000	83	34	26	45	94
1989	6,000	73	32	30	48	52
1986–1989	148,000	63	107	92	180	240
The Ukraine						
1986	98,000	41	185	190	237	326
1987	43,000	72	112	105	142	236
1988	18,000	79	47	33	50	134
1989	11,000	86	35	28	42	107
1986–1989	170,000	56	126	112	192	293

been recorded among emergency workers of Belarus and the Ukraine. However, according to the data of the Russian Medical–Dosimetry Registry, a two-fold increase was observed in the incidence of non-Chronic Lymphocytic Leukaemia (CLL) between 1986 and 1996 in Russian clean-up workers exposed to an average dose of approximately 110 mGy (Konogorov *et al.*, 2000; Tsyb *et al.*, 2002). Of the 71,217 Russian emergency workers studied, 21 had contracted non-CLL. Approximately 50% of these cases were expected to have been radiation-induced (Tsyb *et al.*, 2002).

6.4.2 Thyroid cancer

Pre-Chernobyl studies

Another stochastic effect that can be detected several years following exposure is thyroid cancer. Thyroid cancer is, fortunately, treatable, usually requiring surgery (total thyroidectomy) and radioiodine therapy for treatment of the distant metastasis. After treatment, patients need regular monitoring and require continuing therapy to replace the hormones which the thyroid normally produces.

Thyroid cancer is one of the rarest forms of malignant tumor. The incidence rate depends on age, sex and ethnicity, amongst other factors. The highest background incidence rate is recorded in Iceland (6.6 and 10.5 cases per 10^5 of the male and female population respectively) and on Hawaii (2.9 and 10.5 cases per 10^5 of males and females respectively). A relatively high incidence rate (exceeding 3 cases per 10^5 for the female population: approximately two times higher than the global average rate) is recorded in Israel, some regions of Canada and the USA. Comparatively low thyroid cancer incidence rates are observed in England (0.7 and 1.5 cases per 10^5 of males and females respectively), Slovenia (1.0 and 1.9 cases per 10^5 of males and females respectively) and India (0.8 and 1.5 cases per 10^5 of males and females respectively) (Parkin et al., 1992). In general, thyroid cancer is more prevalent in females than in males.

Over time, some increase in thyroid cancer incidence has occurred, though the rate of this increase varies in different countries. In Finland, the incidence rate is increasing by 20% every 5 years (Coleman et al., 1993). According to the data of the US Program Surveillance, Epidemiology and Results (SEER), thyroid cancer incidence in the US increased by 7.6% for the period 1987–1991 (Ries et al., 1994). Such increases in background thyroid cancer, observed in Europe and the US, have been seen for some decades and are not linked to Chernobyl radiation (see, e.g., Leenhardt et al., 2004; Chiesa et al., 2004).

Ionising radiation plays a particular role among etiological (i.e., causal) factors of thyroid cancer induction. According to the current view, one of the causes for thyroid cancer increase among populations of highly developed countries is an increase in the number of radiation-induced tumors from the medical application of ionising radiation (Devesa et al., 1987). The first study of a correlation between radiation and thyroid cancer was made by Duffy and Fitzgerald (1950). They analysed data of 28 children with thyroid cancer who had been treated in the Memorial Hospital of New York in 1932–1948. Among them 10 children received radiation therapy at an age of 4–16 months because of thymus hypertrophy. The outcomes of the first study showed evidence for the effect of radiation in the development of thyroid cancer.

During the period 1920–1960, radiation therapy was widely used in the USA for the treatment of various benign diseases: tinea capitis, thymus hypertrophy, cervical lymphoadenopathy, tonsil and lymph node hypertrophy, and skin abnormalities. During treatment of these diseases using ionising radiation, the thyroid gland was exposed. Radiation doses varied from dozens to hundreds of cGy (centi-gray = 0.01 Gy). Data examined in a 1970 study demonstrated that among those subjected to radiation in childhood, 70% of patients had thyroid cancer (Winship and Rosvoll, 1970). Further studies of the side effects observed in the thyroid during radiation therapy of different diseases (Boice et al., 1988; Maxon et al., 1980; Tucker et al., 1991), in parallel with estimates of the radiation effect on A-bomb survivors of Hiroshima and Nagasaki (Nagataki et al., 1994; Wakabayashi et al., 1983), gave evidence for the effect of exposure on the development of thyroid cancer. Numerous radiation epidemiological studies have been focused on the effects of external irradiation of the thyroid at a young age (Pottern et al., 1990; Ron et al.,

Table 6.5. Results of studies of the risk of thyroid cancer development following acute external radiation from atomic bombs and from radiation therapy at an age of <20 years.

Circumstances of exposure	Number of subjects in the cohort		Average dose (cGy)	Number of thyroid cancers		
				Observed		Expected*
	Exposed	Non-exposed		Exposed	Non-Exp.	Exposed
A-bombing, Japan	41,234	38,738	27	132	93	19.2
Thymus hypertrophy	2,475	4,991	136	38	5	2.77
Tinea capitis	10,834	16,226	9	44	16	11.2
Tonsil hypertrophy	2,634	0	59	309	0	125
Lymph node hypertrophy	1,195	1,063	24	10	0	2.2

* The expected number of spontaneous cancers in the exposed population.

1989; Schneider *et al.*, 1993; Shore *et al.*, 1993; Thompson *et al.*, 1994) as shown in Table 6.5.

In all the cohort studies conducted, the number of observed cases of thyroid cancer among the exposed subjects exceeded the number of expected cases (Table 6.5). In some studies, the number of cases among exposed subjects exceeded those among non-exposed subjects by an order of magnitude. Results of case-control studies also showed evidence for the effect of radiation in thyroid cancer induction.

The effect of internal radiation of the thyroid by ^{131}I at a young age has been studied in cohorts subjected to diagnostic medical procedures which use ^{131}I. Studies were carried out on cohorts in the USA and Sweden (Hamilton *et al.*, 1989; Holm *et al.*, 1988), and also in populations exposed during nuclear tests on the Marshall Islands and at the Nevada Test Site, USA (Kerber *et al.*, 1993; Robbins and Adams, 1989) (see Table 6.6).

Pre-Chernobyl data on the efficiency of internal exposures by ^{131}I in induction of thyroid cancer are limited. These data were based on studies with insufficient statistical power because of the low collective dose in the cohort and the number of observed and expected cases of thyroid cancer (Hamilton *et al.*, 1989; Kerber *et al.*, 1993; Hall *et al.*, 1996). In spite of these uncertainties, the data obtained indicated a lower efficiency of internal exposure by ^{131}I in cancer induction as compared to external radiation. Comparison of the efficiency of exposures by external γ-radiation and ^{131}I in the induction of thyroid cancer is also complicated by differences in the exposed cohorts. The majority of those who were exposed internally to ^{131}I were adults who had thyroid diseases which necessitated irradiation for diagnostic or therapeutic purposes. But the majority of subjects exposed to external γ-radiation were healthy children. A statistical assessment of one epidemiological study of patients exposed to ^{131}I for diagnostic purposes, one experimental study of mice exposed by ^{131}I and six epidemiological studies on humans exposed by γ-radiation

Table 6.6. Results of studies of the risk of thyroid cancer development following exposure to ^{131}I at an age of <20 years.

Circumstances of exposure	Number of exposed subjects	Average dose (cGy)	Number of thyroid cancers	
			Observed	Expected
Medical diagnostics (Sweden)	2,408	150	3	1.78
Medical diagnostics (USA)	3,503	80	4	3.7
Nuclear tests, Nevada	2,473	17	8	5.4
Nuclear tests, Marshall Islands	127	1,240	6	1.2*
Combined treatment of juvenile hyperthyroidism	602	8,800	2	0.1

* The expected rate is high for this small population due to a high natural incidence.

implied that the risk of radiation induced thyroid cancer following ^{131}I exposure is 66% of the risk following external γ-radiation. However, wide confidence intervals in these estimates do not allow interpretation of the results in a proper way (Laird, 1987). Evidently, it was necessary to continue studying the effects of ^{131}I exposure in different age groups.

Absolute and relative risks are quantitative expressions of the probability of radiation induced thyroid cancer (see Box 6.1). Each of the studies conducted since 1950 showed values of excess relative risk (ERR) and/or excess absolute risk (EAR) of radiation induced thyroid cancer (Ron et al., 1989; Schneider et al., 1993; Shore et al., 1993; Hamilton et al., 1989; Holm et al., 1988; Kerber et al., 1993; Robbins and Adams, 1989), as shown in Table 6.7.

The risk factor obtained in each individual study reflected to some extent the specifics of the exposed population (medical–demographic features) and conditions of the radiation exposure (type, duration and dose rate). However, to apply the risk coefficient for prognostic (i.e., predictive) purposes one would need to have a value which reflects the dose–response relationship in different populations. Such a value is only obtainable by summarizing all the available data to expand the population under survey and the duration of study (to consider more excess cancers). For the first time accumulated data were summarized by the National Council on Radiation Protection, USA (NCRP, 1985).

The result of the NCRP (1985) review was an acceptance of the role of external γ- and X-ray exposure of the thyroid gland in the development of radiation-induced thyroid cancer. One of the principal parameters used in the assessment of radiation risk is a factor accounting for the efficiency of different types of radiation with respect to cancer induction (the relative biological effectiveness, RBE). It was estimated that for acute external γ- and X-ray exposure, and ^{132}I, ^{133}I, ^{135}I exposure, this coefficient is equal to 1, but for ^{131}I and ^{125}I the coefficient was estimated to be 0.33. In other words, internal exposure from ^{131}I was believed to be three times less damaging (per gray absorbed by the thyroid) than external γ- and

Box 6.1. Excess relative risk (ERR) and excess absolute risk (EAR).

The **ERR** is an estimate of the excess risk of incidence or mortality of cancer (or other diseases) in an exposed population compared with the background incidence in an unexposed population. For example, a population is exposed to 0.5 Gy radiation and subsequently 5 people per 10^5 of that population develop a particular cancer. The background (spontaneous) rate of that cancer in a similar, but unexposed, control population is 1 per 10^5 people. The relative risk of radiation induction of that cancer in the population following an exposure to 0.5 Gy is then:

$$RR = \frac{5/10^5}{1/10^5} = 5$$

The ERR following a 0.5 Gy exposure is the excess risk over and above the background of spontaneous cancers in the population:

$$ERR = \frac{5/10^5 - 1/10^5}{1/10^5} = 4$$

and the ERR per Gray is: 4 per $0.5\,Gy = 8\,Gy^{-1}$.

If the ERR is greater than 0, then there is an increase in incidence of that cancer in the population, though this increase is not necessarily statistically significant (see below).

The **EAR** is an estimate of the rate of radiation-induced cancer incidence (or mortality) per 10^4 people (10^4 is typically used) per year. In the example above, if the particular cancer appears between 5 and 25 years after exposure to radiation, the expected excess cancers for an exposure of 1 Gy is 0.8 per 10^4 people (i.e., ERR per Gy × 1 Gy × the spontaneous incidence rate). The average EAR per Gray during that 20 year period is then:

$$EAR\ per\ Gy = \frac{0.8}{20} per\ Gy = 0.04\ per\ 10^4\ people\ per\ Gray\ of\ exposure\ per\ year$$

$$= 0.04/10^4\ person\text{-}year\text{-}Gy$$

where person-year-Gy is often abbreviated to 'PYGy'. In the simple example above, we have not accounted for any effects of age differences within the population.

The 95% **confidence interval** (CI) is clearly important in interpreting such estimates. If a study determines that the ERR of a particular cancer is 1.2 with CI: −0.6 to 3.1 then that study alone is insufficient to determine (with 95% confidence) that there is an increase in the particular cancer in the population due to radiation (since an ERR of zero is within the CI range). On the other hand, if the ERR is 1.2 with CI: 0.3 to 2.1 then the study has determined (with 95% confidence) that there is an increase in cancer due to radiation.

Table 6.7. Risk of radiation-induced thyroid cancer following radiation exposure at an age of <20 years.

Circumstances of exposure	Average dose (cGy)	Excess risk	
		ERR/Gy (95% CI)	EAR/10^4 person-year-Gy (95% CI)
A-bombing of Japan	27	4.7 (1.7; 10.9)*	2.7 (1.2; 4.6)*
		0.4 (−0.1; 1.2)**	0.4 (−0.1; 1.4)**
Thymus hypertrophy	136	9.1 (3.6; 29.0)	2.6 (1.7; 3.6)
Tinea capitis	9	32.5 (14.0; 57.0)	7.6 (2.7; 13.0)
Tonsils hypertrophy	59	2.5 (0.6; 26.0)	3.0 (0.5; 17.1)
Medical diagnostics (Sweden)	150	0.25 (<0; 2.7)	0.15
Medical diagnostics	80	0.10 (<0; 2.0)	0.05
Nuclear test (Nevada, USA)	17	7.9 (<0; 41.0)	3.3
Nuclear test (Marshall Islands)	1,240	0.32 (0.1; 0.8)	1.5
Combined treatment of juvenile hyperthyroidism	8,800	0.3 (0; 1.0)	0.1

* Individuals exposed at age <15 years; ** individuals exposed at age >15 years.

X-ray exposure (NCRP, 1985). However, there was significant uncertainty in estimates of differences in carcinogenic efficiency of different iodine radionuclides (Archer, 1989) and the range in estimates of RBE for ^{131}I was 0.1–1.0 (NCRP, 1985).

It is known that inclusion in the study of additional groups of exposed people and expansion of the total duration of life under risk allows the risk factor to be more accurately determined. Over the decade following the last NCRP publication, the data of seven studies were reviewed (five cohort studies and two case-control studies). These studies involved 120,000 subjects exposed to medical or accidental radiation (Ron et al., 1995). Average thyroid doses varied from 0.09–12.5 Gy. Calculated values of excess absolute and relative risk were 4.4 per 10^4 person-year-Gy and 7.7 per Gy correspondingly. A higher probability of radiogenic cancer development in women as compared with men was confirmed in this study, and significant decrease of risk of cancer development with age of exposure was observed. A linear dose-dependant relationship was observed in the range 10–2,000 cGy (100 mGy–20 Gy). The duration of the latency period varied from 3–8 years (average of 5 years), time of life under risk was 40 years or more, with the maximum incidence rate at 15 years after the exposure. The efficiency factor of different types of radiation was not different to that of the previous study (NCRP, 1985).

Studies following the Chernobyl accident

Exposures to the thyroid received following the Chernobyl accident included doses of internal radiation from ^{131}I, doses of internal radiation from shorter lived iodine isotopes (^{132}I, ^{133}I, ^{135}I) and tellurium (^{131}Te and ^{132}Te), doses of internal radiation

Table 6.8. Thyroid dose distribution for various age groups in Belarus.

Age at time of accident (years)	Number of subjects in dose group (thyroid dose, Gy)					Total number of people
	0–0.05	0.05–0.1	0.1–0.5	0.5–1	>1	
<1	62,310	45,229	32,611	13,086	4,892	158,128
1–2	58,641	44,683	30,914	13,657	4,320	152,215
2–3	60,838	42,261	29,616	13,716	3,577	150,008
3–4	62,059	47,028	24,129	12,959	3,179	149,354
4–5	63,169	45,569	31,160	3,398	2,693	145,989
5–6	66,001	44,500	29,003	3,201	2,218	144,923
6–7	65,048	46,427	25,292	2,800	1,806	141,373
7–8	63,456	42,894	28,170	3,077	2,196	139,793
8–9	64,192	44,980	24,647	2,706	1,777	138,302
9–10	94,037	15,717	24,294	2,298	1,043	137,389
10–11	102,049	10,878	22,417	2,556	199	138,099
11–12	103,062	10,519	21,560	2,432	49	137,622
12–13	102,706	11,488	19,965	2,309	32	136,500
13–14	104,187	12,730	16,732	2,030	29	135,708
14–15	107,383	10,604	15,256	1,749	17	135,009
15–16	106,064	10,591	14,573	1,279	18	132,525
16–17	106,318	9,833	14,488	906	11	131,556
17–18	213,609	18,155	29,231	1,005	26	262,026
Total children and adolescents	*1,605,129*	*514,086*	*434,058*	*85,164*	*28,082*	*2,666,519*
Adults	*5,597,593*	*502,866*	*727,086*	*46,966*	*596*	*6,875,107*
Total	**7,202,722**	**1,016,952**	**1,161,144**	**132,130**	**28,678**	**9,541,626**

from long-lived radionuclides (^{134}Cs and ^{137}Cs) and exposures from external radiation.

For the majority of people, doses of internal radiation to the thyroid gland from ^{131}I intake with milk and green leafy vegetables were the most significant. These could comprise, according to various estimates, from 90–98% of the total dose to the thyroid. Individual doses to the thyroid varied widely, depending on age at the time of the accident, distance from the Chernobyl NPP, eating habits, and protective measures.

In general, thyroid doses among children were 3–10 or more times higher than among adults living in the same settlement. Children living in the regions of Belarus and the Ukraine located close to the Chernobyl NPP, and also children of some regions of Russia, received maximum doses to the thyroid exceeding 1 Gy (Kenigsberg *et al.*, 2002a; Balonov and Zvonovna, 2002; Likhtarev *et al.*, 1996). As an example, Table 6.8 presents exposures to the thyroid for the population of Belarus. Average doses to the thyroid for exposed people in Croatia, Greece,

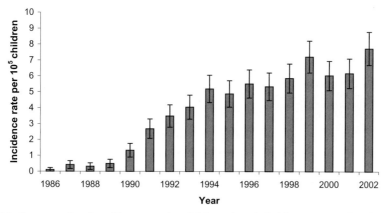

Figure 6.1. Increase in thyroid cancer in children (aged 0–18 years at the time of the Chernobyl accident) in Belarus during the period 1986–2002.

Hungary, Poland and Turkey were much less than those in the Ukraine, Belarus and Russia, being in the range 1.5–15 mGy (UNSCEAR, 2000).

The increase in thyroid cancer incidence among children and adolescents who were exposed to radiation following the Chernobyl accident (age 0–18 years at the time of the accident) may be considered today as a proven fact. Since the accident, the number of patients having thyroid cancer (who were children and adolescents in 1986) in Belarus, Russia and the Ukraine has reached 4,000 (Demidchik *et al.*, 2002; Tronko *et al.*, 2002a; Ivanov *et al.*, 2002). The significant increase of incidence started 4–5 years after the accident and continued at a high level during subsequent years (Figure 6.1). The observed excess incidence of thyroid cancer among exposed individuals aged 0–18 allowed an estimation of the dose-dependant relationship and enabled an analysis of radiation risk as a function of gender and age to be made (Table 6.9). Figure 6.2 shows the excess thyroid cancer risk in regions and cities of the Ukraine, Belarus and Russia as a function of dose to the thyroid (Jacob *et al.*, 1998).

The risk of cancer development decreases with the age at the time of exposure, but appears to be higher for girls than for boys. Similar results were obtained in studies of thyroid cancer in Russia and the Ukraine as well as in studies of joint cohorts (Jacob *et al.*, 1999; Tronko *et al.*, 1999, 2002a,b). Values of excess absolute risk, obtained in all studies conducted after the Chernobyl accident are in broad agreement with the values of EAR described for A-bomb survivors of 2.7 (CI: 1.2–4.6) and a pooled study of A-bomb and medical exposures together gave 4.4 (CI: 1.9–10.1) (Ron *et al.*, 1995). This fact is of principal importance since it demonstrates compatibility between the efficiency of thyroid cancer induction following internal exposure to ^{131}I and external γ- and X-ray radiation. On the basis of a study of thyroid cancer risk in Belarus (in contrast to some pre-Chernobyl studies) Jacob *et al.* (2000) concluded that 'it is not possible to conclude that incorporated ^{131}I is less effective in inducing thyroid cancer than external exposures with high dose rates'. They (Jacob *et al.*, 2000), however, noted

Table 6.9. Risk of thyroid cancer for children and adolescents of Belarus considering gender and age at the time of the accident.

EAR/10^4 PYGy (95% CI)			ERR per Gy (95% CI)		
Male	Female	Both	Male	Female	Both
0–6 years					
1.5	2.6	2.1	86.4	46.2	55.9
(1.2; 1.9)	(1.8; 34)	(1.6; 2.6)	(67.7; 124.0)	(33.8; 63.5)	(43.9; 76.2)
7–14 years					
1.0	2.3	1.7	32.9	21.5	34.2
(0.1; 4.4)	(0.2; 3.4)	(1.2; 2.9)	(0.9; 78.9)	(1.8; 33.5)	(16.9; 47.2)
15–18 years					
0.8	3.9	2.4	18.3	22.5	21.7
(−0.1; 1.8)	(−0.4; 4.9)	(−0.2; 2.9)	(−0.4; 26.9)	(−0.9; 28.4)	(−0.8; 26.7)

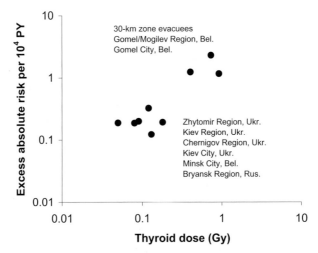

Figure 6.2. Increase in excess thyroid cancer risk in the period 1991–1995 in children born between 1971 and 1986.* The excess absolute risk is plotted against estimated dose to the thyroid. The highest doses and risks were to evacuees of the 30-km Zone and those living in the Gomel and Mogilev Regions of Belarus. Other regions/cities of Belarus, Ukraine and Russia had lower exposures and risks, as indicated.
*From data in Jacob *et al.* (1998).

that there is still significant uncertainty concerning thyroid cancer induction from internal and external exposures.

There is still no agreed opinion concerning excess thyroid cancer incidence among those exposed when adults (19 years and older). In Russia, excess radiation-induced incidence has not been recorded in adults (Ivanov *et al.*, 2003). However, in Belarus six years after the Chernobyl accident (1992) a significant

Table 6.10. Thyroid cancer risk for the exposed adult population of Belarus.

Risk	Male	Female	Both
EAR/10^4 PYGy	0.4	2.5	1.7
(95% CI)	(−0.6; 1.5)	(1.9; 4.7)	(0.3; 3.2)
ERR per Gy	3.9	2.4	3.8
(95% CI)	(−0.9; 5.9)	(0.8; 5.6)	(0.1; 9.8)

increase of the observed incidence over expected was recorded (Kenigsberg et al., 2002b). A dose-dependant relationship was revealed and an initial risk analysis was made (Table 6.10).

Doses to the thyroid of the adult population were much lower than those of children, while spontaneous (background) thyroid incidence is significantly higher. It is therefore more difficult to find an excess radiation-induced incidence in the adult population. But the results obtained in Belarus demonstrate, in principle, the possibility of the development of excess radiation-induced thyroid cancer incidence in adults due to internal exposure to ^{131}I.

6.4.3 Non-thyroid solid cancer

During the years since the Chernobyl accident a large number of epidemiological studies dedicated to cancer incidence among the affected population and emergency workers were conducted. The results of these studies, reviewed in the UNSCEAR (2000) report, showed a lack of evidence for the growth of (non-thyroid) solid cancer incidence among the population and emergency workers. Possibly, the lack of effect could be explained by the relatively low dose level and the chronic nature of the radiation exposure which is believed to decrease the risk of a stochastic effect by 2–10 times. In addition, the latency period of development for radiation-induced solid cancer is very long (not less then 10 years and for some types of cancer 20 or more years). The publications issued since UNSCEAR (2000) similarly did not find evidence of an increase of cancer incidence among the population. There are some doubts whether such an increase can be determined since predictive models estimate that the increase of incidence (mortality) will not exceed a few percent for the lifetime of the exposed populations.

It is more probable that an excess solid cancer incidence (mortality) among emergency workers may be observed. Doses (at least for some categories of emergency workers) were comparable to doses of A-bomb survivors. Intensive studies of emergency workers were conducted on the basis of the Russian National Medical and Dosimetric Registry (Ivanov et al., 2004a,b,c). The results of these studies show a slight – but not significant – increase of cancer incidence during 1996–2001 (after minimum latency period of 10 years): ERR per Gy is 0.19 with 95% CI from −0.66 to 1.27. For emergency workers participating in recovery operations at the Chernobyl NPP the excess risk was not statistically significant: 0.95 per Sv (95% CI from −1.52 to 4.49).

It is likely that the follow-up period is still not sufficient for detecting excess solid cancer incidence. In the future, special attention should be paid to detection of excess incidence of some types of cancer, and first of all breast cancer, since the relative risk of this type of cancer after external radiation at a young age could be higher than the risk of thyroid cancer and leukaemia. The results of ongoing epidemiological studies on this have not been published yet.

6.4.4 Non-cancer diseases

Cardiovascular diseases are the primary cause of death in developed countries. There is evidence of increased mortality from cardiovascular diseases in exposed populations, though the mechanism is not clear. It is difficult to determine any effect of radiation amongst the multiple other factors influencing rates of cardiovascular diseases. Data on the effect of chronic exposure at low doses are insufficient for decision-making concerning this type of radiation damage. However, it is known that extremely high acute doses of external radiation can damage heart and blood vessels (Trivedi and Hannan, 2004). The Chernobyl accident presented a unique opportunity for studying this problem since millions of people of different ages were exposed to radiation at a wide range of doses. A study involving emergency workers from Russia demonstrated that the ERR coefficient for deaths from cardiovascular diseases was 0.54 per Sv (95% CI: 0.18–0.91) (Ivanov et al., 2004c). This ERR value is higher than the value of 0.17 (95 CI 0.08–0.26) for A-bomb survivors (Preston et al., 2003) although the 95% confidence intervals overlap.

6.5 DISCUSSION

As a result of Chernobyl, millions of people of various countries were accidentally exposed to radiation. Personnel of the NPP, firemen, emergency workers and the population of nearby areas received the highest radiation doses. Hundreds of thousands of Belarussian, Russian and Ukrainian people continue to live on land contaminated with long-lived radionuclides and are subject to chronic low-dose exposures.

During the first days after the accident ARS was initially diagnosed for 237 persons and later confirmed for 134 persons. Among them 28 persons died in 1986 due to ARS, and 17 died in 1987–2004 from various causes, not all linked to radiation. No ARS cases have been recorded among the general population.

The studies of long-term health effects after the Chernobyl accident can be divided into four groups. The first group includes radiation epidemiological studies of thyroid cancer (ecological, case control study, cohort study) in which an excess of radiation induced cancer was found, a dose dependant relationship was determined and coefficients of excess absolute and relative risk were estimated. At the present time, the results of the studies for the period of 1986–2002 have been published.

A number of thyroid cancer studies are being continued in order to get more complete data – trying to take account of uncertainties of dose estimates, effects of gender, age, iodine deficiency, screening using updated diagnostic procedures, genetic susceptibility and other factors. However, it is absolutely clear that thyroid exposure to iodine radionuclides in the early stages of the Chernobyl accident (under conditions of inefficient implementation of protective actions) led to an unprecedented increase of thyroid cancer incidence, primarily in children and adolescents of Belarus, Russia, and the Ukraine. Continuing studies of thyroid cancer in people exposed at different ages and in emergency workers will help to improve estimates of the dose-dependant relationship and to design new models of radiation risk.

The second group of studies deals with other stochastic effects of radiation – various types of cancer. Although some experts have suggested only a small probability or even an impossibility of detecting excess radiation-induced incidence or mortality among the affected population, the studies should be continued preferably through methods of analytical epidemiology having strict requirements applied to the design and statistical power of the study. The most promising from this point of view are the long-timescale international cohort studies of emergency workers of Belarus, Russia, the Ukraine and the Baltic countries.

The third group of studies deals with non-cancer diseases – primarily of all cardiovascular diseases. Published results obtained in the studies of Chernobyl emergency workers and A-bomb survivors show possible evidence of a correlation between radiation exposure and increase of mortality. However, the study of the role of the Chernobyl accident in causing cardiovascular diseases is a very complicated task due to the influence of non-radiation effects of the accident on the health of the affected population (see Chapter 7). It is difficult to make quantitative estimates of these factors within a standard epidemiological study.

The fourth group of studies deals with the reaction of cellular and sub-cellular systems in individuals exposed to additional radiation due to the Chernobyl accident. These studies are focused on genetic instability, immune system status, oxidative stress and bystander effect, amongst others. The results of these studies are still difficult to interpret since it is as yet unknown in what way the observed changes correspond to specific diseases.

The Chernobyl accident provided a unique opportunity to study the health effects of low-dose ionising radiation because of the particular features of the dose formation and the involvement of millions of people in the accidental situation. Prior to Chernobyl, the most studied cohort (A-bomb survivors) had fewer persons with doses of less then 200 mSv, and this fact limited the opportunity to study the effects of lower doses. This gap in knowledge of the effects of ionising radiation may be filled in by the results that will be obtained in the course of further studies of health effects in the population affected by the Chernobyl accident.

6.6 REFERENCES

Archer, V.E. (1989) Risk of thyroid cancer after diagnostic doses of radioiodine. *Journal of the National Cancer Institute*, **81**, 713–715.

Balonov, M.I. and Zvonova, I.A. (eds) (2002) *Average Thyroid Doses for People of Different Ages Who Lived in 1986 in the Settlements of the Bryansk, Tula, Orel, and Kaluga Regions Contaminated with Radionuclides due to the Accident at the Chernobyl NPP.* Ministry of Public Health of Russia, Moscow.

Boice, J.D. Jr, Engholm, G., Kleinerman, R.A., Blettner, M., Stovall, M., Lisco, H., Moloney, W.C., Austin, D.F., Bosch, A. and Cookfair, D.L. (1988) Radiation dose and second cancer risk in patients treated for cancer of the cervix. *Radiation Research*, **116**, 3–55.

Chernobyl Conference (2001) Fifteen years after the Chernobyl accident. Lessons learned. *Proceedings of the international conference, Kyiv, Ukraine, 18–20 April, 2001.* Kyiv.

Chiesa, F., Tradati, N., Calabrese, L., Gibelli, B., Giugliano, G., Paganelli, G., De Cicco, C., Grana, C., Tosi, G., DeFiori, E., et al. (2004) Thyroid disease in northern Italian children born around the time of the Chernobyl nuclear accident. *Annals of Oncology*, **15**, 1842–1846.

Coleman, M.P., Esteve, J., Damiecki, P., Arslan, A. and Renard, H. (1993) *Trends in Cancer Incidence and Mortality*, pp. 609–640. IARC Scientific Publication No 121. IARC, Lyon.

Day, R., Gorin, M.B. and Eller, A.W. (1995) Prevalence of lens changes in Ukrainian children residing around Chernobyl. *Health Physics*, **68**, 632–642.

Demidchik, E., Demidchik, Y., Gedrevich, Z., Mrochek, A., Ostapenko, V., Kenigsberg, J. and Buglova, E. (2002) Thyroid cancer in Belarus. In: Yamashita, S., Shibata, Y., Hoshi, M. and Fujimura, K. (eds), *Chernobyl: Message for the 21st Century*, pp. 69–75. International Congress Series 1234. Elsevier, Amsterdam.

Devesa, S.S., Silverman, D.T., Young, J.L., Pollack, E.S., Brown, C.C., Horm, J.W., Percy, C.L., Myers, M.H., McKay, F.W. and Fraumeni, J.F. (1987) Cancer incidence and mortality trends among whites in the United States. *Journal of the National Cancer Institute*, **4**, 701–770.

Duffy, B.J.J. and Fitzgerald, P.J. (1950) Cancer of the thyroid in children: A report of 28 cases. *Journal of Clinical Endocrinology and Metabolism*, **10**, 1296–1308.

Fedirko, P. (1999) Chernobyl catastrophe and the eye: Some results of a prolonged clinical investigation. *Ophthalmological Journal*, **2**, 69–73.

Hall, P., Mattson, A. and Boice, J.D. (1996) Thyroid cancer after diagnostic administration of iodine-131. *Radiation Research*, **145**, 86–92.

Hamilton, P., Chiacchierini, R. and Kaczmarek, R. (1989) *A Follow-up Study of Persons Who Had Iodine-131 and Other Diagnostic Procedures During Childood and Adolescence.* CDRH – Food and Drug Administration, Rockville.

Holm, L.E., Wiklund, K.E., Lundell, G.E., Bergman, N.A., Bjelkengren, G., Cederquist, E.S., Ericsson, U.B., Larsson, L.G., Lidberg, M.E., Lindberg, R.S., et al. (1988) Thyroid cancer after diagnostic doses of iodine-131: A retrospective cohort study. *Journal of the National Cancer Institute*, **80**, 1132–1138.

IAEA (2004) *Radiation, People and the Environment.* IAEA, Vienna.

IAEA (2005) *Development of Extended Framework for Emergency Response Criteria.* TECDOC-1432. IAEA, Vienna.

Ivanov, V., Gorski, A., Tsyb, A., Maksioutov, M., Vlasov, O. and Godko, A. (2002) Risk of radiogenic thyroid cancer in the population of the Bryansk and Oryol regions of Russia after the Chernobyl accident. In: Yamashita, S., Shibata, Y., Hoshi, M. and Fujimura, K. (eds), *Chernobyl: Message for the 21st Century*, pp. 85–93. International Congress Series 1234. Elsevier, Amsterdam.

Ivanov, V., Gorski, A., Maksyutov, M., Vlasov, O., Godko, A., Tirmarche, M., Valenty, M. and Verger, P. (2003) Thyroid cancer incidence among adolescents and adults in Bryansk region of Russia following Chernobyl accident. *Health Physics*, **84**, 46–60.

Ivanov, V.K., Gorski, A.I., Tsyb, A.F., Ivanov, S.I., Naumenko, R.N. and Ivanova, L.V. (2004a) Solid cancer incidence among Chernobyl emergency workers residing in Russia: Estimation of radiation risk. *Radiation and Environmental Biophysics*, **43**, 35–42.

Ivanov, V.K., Ilyin, L.F., Gorski, A.I., Tukov, A. and Naumenko, R. (2004b) Radiation and epidemiological analysis for solid cancer incidence among nuclear workers who participated in recovery operations following the accident at the Chernobyl NPP. *Journal of Radiation Research*, **45**, 37–40.

Ivanov, V.K., Tsyb, A.E., Ivanov, S. and Pokrovsky, V. (eds) (2004c) *Medical Consequences of the Chernobyl Catastrophe in Russia: Estimation of Radiation Risks*. Nauka, St. Petersburg.

Jacob, P., Goulko, G., Heidenreich, W.F., Likhtarev, I., Kairo, I., Tronko, N.D., Bogdanova, T.I., Kenigsberg, J., Buglova, E., Drozdovitch, V., et al. (1998) Thyroid cancer risk to children calculated. *Nature*, **293**, 31–32.

Jacob, P., Kenigsberg, Y., Zvonova, I., Heidenreich, W.F., Buglova, E., Goulko, G., Drozdovovitch, V., Golovneva, A., Demidchik, E.P., Balonov, M. and Paretzke, H.G. (1999) Chernobyl exposure during childhood and thyroid cancer incidence in Belarus and Russia. *British Journal of Cancer*, **80**, 1461–1469.

Jacob, P., Kenigsberg, Y., Goulko, G., Buglova, E., Gering, F., Golovneva, A., Kruk, J. and Demidchik, E.P. (2000) Thyroid cancer risk in Belarus after the Chernobyl accident: Comparison with external exposures. *Radiation and Environmental Biophysics*, **39**, 25–31.

Karaoglou, A., Desmet, G., Kelly, G.N. and Menzel, H.G. (eds) (1996) The radiological consequences of the Chernobyl accident. *Proceedings of the first international conference, Minsk, 18–22 March, 1996*. European Commission, Luxembourg.

Kenigsberg, J., Buglova, E., Kruk, J. and Golovneva, A. (2002a) Thyroid cancer among children and adolescents of Belarus exposed due to the Chernobyl accident: Dose and risk assessment. In: Yamashita, S., Shibata, Y., Hoshi, M. and Fujimura, K. (eds), *Chernobyl: Message for the 21st Century*, pp. 293–300. International Congress Series 1234. Elsevier, Amsterdam.

Kenigsberg, J., Buglova, E., Kruk, J. and Ulanovskaya, E. (2002b) Chernobyl-related thyroid cancer in Belarus: Dose and risk assessment. In: *Proceeding of Symposium on Chernobyl-Related Health Effects*. Radiation Effects Association, Tokyo.

Kerber, R.A., Till, J., Simon, S.L., Lyon, J.L., Thomas, D.C., Preston-Martin, S., Rallison, M.L., Lloyd, R.D. and Stevens, W. (1993) A cohort study of thyroid disease in relation to fallout from nuclear weapons testing. *Journal of the American Medical Association*, **270**, 2076–2082.

Konogorov, A.P., Ivanov, V.K., Chekin, S.Y. and Khait, S.E. (2000) A case-control analysis of leukemia in accident emergency workers of Chernobyl. *Journal of Environmental Pathology, Toxicology and Oncology*, **19**, 143–151.

Laird, N.M. (1987) Thyroid cancer risk from exposure to ionising radiation: A case study in the comparative potency model. *Risk Analysis*, **7**, 299–309.

Leenhardt, L., Grosclaude, P. and Cherie-Challine, L. (2004) Increased incidence of thyroid carcinoma in France: A true epidemic or thyroid nodule management effects? Report from the french thyroid cancer committee. *Thyroid*, **14**, 1056–1060.

Likhtarev, I., Sobolev, B., Kairo, I., Tabachny, L., Jacob, P., Pröhl, G. and Goulko, G. (1996) Results of large-scale thyroid dose reconstruction in Ukraine. In: Karaoglou, A.,

Desmet, G., Kelly, G.N. and Menzel, H.G. (eds), *The Radiological Consequences of the Chernobyl Accident*, pp. 1020–1034. European Commission, Brussels.

Maxon, H.R., Saenger, E.L., Thomas, S.R., Buncher, C.R., Kereiakes, J.G., Shafer, M.L. and McLaughlin, C.A. (1980) Clinically important radiation-associated thyroid disease. *Journal of the American Medical Association*, **244**, 1802–1805.

NCRP (1985) *Induction of Thyroid Cancer by Ionizing Radiation*. NCRP Report N80. National Council on Radiation Protection and Measurements, Bethesda.

Nagataki, S., Shibata, Y., Inoue, S., Yokoyama, N., Izumi, M. and Shimaoka, K. (1994) Thyroid diseases among atomic bomb survivors in Nagasaki. *Journal of the American Medical Association*, **272**, 364–370.

Parkin, D.M., Muir, C.S., Whelan, S.L., Gao, Y.T., Ferlay, J. and Powell, J. (1992) *Cancer Incidence in Five Continents Volume VI*. International Agency for Research on Cancer Scientific Publication No. 120. IARC, Lyon.

Pierce, D.A., Shimizu, Y., Preston, D.L., Vaeth, M. and Mabuchi, K. (1996) Studies of the mortality of atomic bomb survivors. Report 12, Part 1. Cancer: 1950–1990. *Radiation Research*, **146**, 1–27.

Pottern, L.M., Kaplan, M., Larsen, P.R., Silva, J.E., Koenig, R.J., Lubin, J.H., Stovall, M., and Boice, J.D. Jr. (1990) Thyroid nodularity after childhood irradiation for lymphoid hyperplasia: A comparison of questionnaire and clinical findings. *Journal of Clinical Epidemiology*, **43**, 449–460.

Preston, D.L., Shimizu, Y., Pierce, D.A., Suyama, A. and Mabuchi, K. (2003) Studies of mortality of atomic bomb survivors. Report 13: Solid cancer and non-cancer disease mortality: 1950–1997. *Radiation Research*, **160**, 381–407.

Ries, L.A.G., Miller, B.A., Hankey, B.F., Kosary, C.L., Harras, A. and Edwards, B.K. (eds) (1994) *SEER Cancer Statistics Review, 1973–1991: Tables and Graphs*. National Cancer Institute, Bethesda.

Robbins, J. and Adams, W. (1989) Radiation effects in the Marshall Islands. In: Nagataki, S. (ed.), *Radiation and the Thyroid: Proceedings of the 27th Annual Meeting of the Japanese Nuclear Medicine Society*, pp. 11–24. Excerpta Medica, Amsterdam.

Ron, E., Modan, B., Preston, D., Alfandary, E., Stovall, M. and Boice, J.D. Jr. (1989) Thyroid neoplasia following low-dose radiation in childhood. *Radiation Research*, **120**, 516–531.

Ron, E., Lubin, J.H., Shore, R.E., Mabuchi, K., Modan, B., Pottern, L.M., Schneider, A.B., Tucker, M.A. and Boice, J.D. Jr. (1995) Thyroid cancer after exposure to external radiation: a pooled analysis of seven studies. *Radiation Research*, **141**, 259–277.

Schneider, A.B., Ron, E., Lubin, J., Stovall, M. and Gierlowski, T.C. (1993) Dose-response relationships for radiation-induced thyroid cancer and thyroid nodules: Evidence for the prolonged effects of radiation on the thyroid. *Journal of Clinical Endocrinology and Metabolism*, **77**, 362–369.

Shore, R.E., Hildreth, N., Dvoretsky, P., Andresen, E., Moseson, M. and Pasternack, B. (1993) Thyroid cancer among persons given X-ray treatment in infancy for an enlarged thymus gland. *American Journal of Epidemiology*, **137**, 1068–1080.

Thomas, G., Karaoglou, A. and Williams, E.D. (eds) (1999) *Radiation and Thyroid Cancer*. World Scientific, Singapore.

Thompson, D.E., Mabuchi, K., Ron, E., Soda, M., Tokunaga, M., Ochikubo, S., Sugimoto, S., Ikeda, T., Terasaki, M., Izumi, S. and Preston, D.L. (1994) Cancer incidence in atomic bomb survivors. Part II: Solid tumor incidence, 1958–87. *Radiation Research*, **137**, 17–67.

Trivedi, A. and Hannan, M.A. (2004) Radiation and cardiovascular diseases. *Journal of Environmental Pathology Toxicology and Oncology*, **23**, 99–106.

Tronko, M.D., Bogdanova, T.I., Komissarenko, I.V., Epstein, O.V., Oliynyk, V., Kovalenko, A., Likhtarev, I.A., Kairo, I., Peters, S.B. and LiVolsi, V.A. (1999) Thyroid carcinoma in children and adolescents in Ukraine after the Chernobyl nuclear accident: Statistical data and clinico-morphologic characteristics. *Cancer*, **86**, 149–56.

Tronko, N.D., Bogdanova, T.I., Epstein, O.V., Oleynyk, V.A., Komissarenko, I.V., Rybakov, S.I., Kovalenko, A.E., Tereshchenko, V.P., Likhtarev, I.A., Kairo, I.A., *et al.* (2002a). Thyroid cancer in children and adolescents of Ukraine having been exposed as a result of the Chernobyl accident (15-year expertise of investigations). *International Journal of Radiation Medicine*, **4**, 222–232.

Tronko, N.D., Bogdanova, T.I., Likhtarev. I.A., Kairo, I.A. and Shpak, V.I. (2002b) Summary of the 15-year observation of thyroid cancer among Ukrainian children after the Chernobyl accident. In: Yamashita, S., Shibata, Y., Hoshi, M. and Fujimura, K. (eds), *Chernobyl: Message for the 21st Century*, pp. 77–83. International Congress Series 1234.

Tsyb, A.F., Ivanov, V.K., Sokolov, V.A., Gorski, A.I., Maksioutov, M.A., Vlasov, O.K., Khait, S.E. and Godko, A.M. (2002) Radiation risks of Leukemia among Russian emergency workers 1986–87. In: *Radiation and Risk*. Special Issue: Health consequences 15 years after the Chernobyl catastrophe: data of the National Registry, pp. 39–50 (available online at: phys4.harvard.edu/~wilson/radiation/Si2002/TITLE.html).

Tucker, M.A., Morris, J.P.H., Boice, J.D. Jr., Robison, L.L., Stone, B.J., Stovall, M., Jenkin, R.D., Lubin, J.H., Baum, E.S. and Siegel, S.E. (1991) Therapeutic radiation at a young age is linked to secondary thyroid cancer. The Late Effects Study Group. *Cancer Research*, **51**, 2885–2888.

UNSCEAR (1988) *Sources, Effects and Risks of Ionizing Radiation*. UNSCEAR Report to the General Assembly, with Scientific Annexes. United Nations, New York.

UNSCEAR (2000) *Sources, Effects and Risks of Ionizing Radiation*. UNSCEAR Report to the General Assembly, with Scientific Annexes. United Nations, New York.

Wakabayashi, T., Kato, H., Ikeda, T. and Schull, W.J. (1983) Studies of the mortality of A-bomb survivors. Report 7. Part III: Incidence of-cancer in 1959–1978, based on the tumour registry, Nagasaki. *Radiation Research*, **93**, 112–146.

Winship, T. and Rosvoll, R.V. (1970) Thyroid carcinoma in childhood: Final report on a 20-year study. *Clinical Proceedings, Children Hospital*, **26**, 327–349.

7

Social and economic effects

Ingrid A. Bay and Deborah H. Oughton

7.1 SOCIAL AND ECONOMIC EFFECTS AND THEIR INTERACTIONS

The lives of more than seven million people in Belarus, the Ukraine and Russia have been, and continue to be, directly or indirectly affected by the Chernobyl accident. Close to 350,000 people were evacuated or resettled, more than 800,000 were ordered to take part in clean-up operations, 4.5 million are classified as living in areas described as 'contaminated', and almost 150,000 persons are registered as permanent invalids (Table 7.1; UNDP, 2002). The expenditures for countermeasures still constitute a significant proportion of the national budgets in the three countries, and the political upheaval and economic depression following the break-up of the Soviet Union in the early 1990s further aggravated problems faced by relocated families and those living in contaminated areas.

Table 7.1. Estimated number of people affected by the accident in terms of evacuation, resettlement, people living in contaminated areas, liquidators and invalids. From UNDP (2002).

	Ukraine	Belarus	Russia	*Total*
Evacuated people (1986 to 1990)	91,000	24,000	3,400	*118,400*
Resettled people (1991 to 2000)	72,000	135,000	52,400	*231,000*
People living in 'contaminated' areas	1,140,813	1,571,000	1,788,600	*4,500,000*
Liquidators	550,000	108,000	200,000	*858,000*
Invalids*	88,931	9,343	50,000	*148,000*
Total	*3,189,477*	*1,823,153*	*2,091,000*	*7,103,630*

* As will be detailed later in this chapter, it should be noted that the registration of people as 'invalids' does not necessarily imply that radiation exposure was the direct causal factor of the condition.

What happens to people and the society under such circumstances? In what way does the contamination influence their daily lives and the state's economy? What new social and cultural features are being established? The Chernobyl accident has been a subject of much research in many fields, from medicine and natural sciences to sociology and anthropology. During July and August, 2001, an international team conducted a multidisciplinary study into the conditions in which people affected by the Chernobyl accident were living 15 years on. The work was carried out on behalf of the United Nations Development Programme (UNDP) and the United Nations Children's Fund (UNICEF), with the support of the UN Office for the Coordination of Humanitarian Affairs (UN-OCHA) and the World Health Organisation (WHO). Their final report, *The Human Consequences of the Chernobyl Nuclear Accident*, represents one of the most comprehensive sources of information on the impacts of the accident in the former Soviet Union (fSU) (UNDP, 2002).

The UNDP study described how people in Belarus, the Ukraine and Russia were affected by interacting factors at many different levels. For example, when contaminated land became useless for agricultural production and forestry, people lost their livelihoods. This influenced economic activity, both as direct costs on the national budget (e.g., compensation, medical treatment), but also in loss of income, infrastructure and agricultural land. There have been widespread economic problems in all three fSU countries since the break-up of the Soviet Union, and in the affected areas Chernobyl has exacerbated these. The Ukraine, previously known as 'Russia's breadbasket', is today a net importer of agricultural products (UN-OCHA, 1999; UNDP, 2002). The economic breakdown, in turn, led to a variety of social problems, including unemployment, decreases in birth rate, lower living standards and emigration. All of this provoked negative physiological as well as psychosocial health effects, creating a further rebound of social and economic problems. Many of those within affected villages and settlements were portrayed as being trapped in a 'downward spiral' of living conditions (Figure 7.1).

The socio-psychological impact of the accident for the population has been underlined in all international Chernobyl conferences, and experience suggests that it may last a long time (IAEA, 1991, 1996; Rumyantseva *et al.*, 1996; Roche, 1996). As part of the UNDP study, the Institute of Sociology in Kiev identified five factors involved in the interaction of radiation health risk and psychosocial effects (UNDP, 2002):

(1) The psychosocial dimension to the *perception of health risk* involved in radiation and the part that *information policy* plays.
(2) The socio-cultural dimension of *displacement* (through policies of relocation due to heavy contamination) and the consequent social disruption of communities.
(3) The general *pathogenic factor relating to psychological stress* reaction to change in lifestyle, such as dietary habits and the consumption of alcohol and tobacco.
(4) The medical sociological dimension concerning *changes in illness behaviour* of the population and the diagnostic behaviour of doctors.
(5) The socioeconomic dimension relating to the large-scale effects of the Chernobyl accident such as the closure of nuclear plants and the reversion of the sources

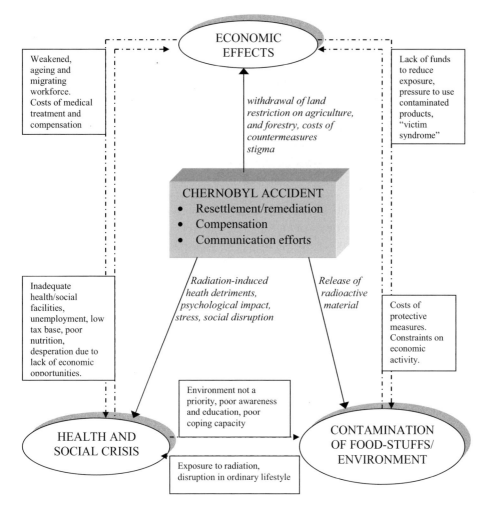

Figure 7.1. Interaction between health, social and economic effects.
Adapted from UNDP (2002).

of energy, as well as the economic transition following the collapse of the USSR.

All factors were evident in the affected communities, making the psychosocial effect the most pervasive underlying cause of ill health and lack of well-being.

These social, psychological and economic problems must be set against a background of immense economic and social difficulties in the fSU countries, particularly during the political changes prevalent in the 1990s, though the accident undoubtedly played a major role in exacerbating these problems. Because of the urgent (and immensely complex) situation following the accident, the authorities made serious mistakes when implementing countermeasures and communicating with the public. In fact, many of their efforts worsened the impact and effects of the accident itself, and led to people losing trust in their authorities.

This interplay between the social, economic and health effects of the Chernobyl accident forms the basis of this chapter. In addition to presenting an overview of what those effects were, we will also attempt to shed some light on some of the factors influencing them. The data on social and economical effects refers mostly to the Belarussian, Russian and Ukrainian populations, and draws heavily from the UNDP report. Three main areas are considered: the medical and psychosocial dimension (i.e., including direct health detriments from radiation exposure and psychological detriments from stress and anxiety following exposure); the socio-economic dimension (i.e., direct and indirect monetary costs of the accident); and the socio-cultural dimension (i.e., the consequences bought about predominately from resettlement and lifestyle changes). The chapter concludes with a brief discussion of factors influencing risk perception, including the way in which understanding these factors can help improve management of radiation risks, in particular relating to information policy.

7.2 HEALTH DETRIMENTS AND ASSOCIATED HARMS DUE TO RADIATION EXPOSURE

A large number of health detriments might be said to arise from radiation exposure, of which alteration in DNA and development of cancer is only one. Broadly speaking, detriments might include those brought about because of the physical nature of ionising radiation and the way it interacts with biological material, as well as a whole suite of psychological effects caused by anxiety and stress. Secondary stress related health effects, such as heart disease and ulcers, can also occur. To put the health effects of Chernobyl in a social context, one needs to address specific issues concerned with perception of the risk (what makes being exposed to radiation so psychologically unpleasant?) and the more general ways in which social disruption and stress can injure health (which health effects arise from psychosocial stress?).

With respect to radiation exposure, two points can illustrate how the social aspects of the Chernobyl risk situation might differ from other hazards. First, when compared to the deaths and harmful injuries caused by other disasters such as an airplane crash or earthquake, the nature of cancer and hereditary effects adds additional layers of psychological and social consequences. Whereas after an earthquake a community can relatively quickly enter a period of recovery and rebuilding, without undue worry that the death toll will rise, the 'lingering threat' from a contaminating accident such as Chernobyl makes recovery much more difficult (Kleinman, 1986). Second, cancer and potential hereditary effects can appear decades after the exposure. Apart from the widely accepted increase in thyroid cancers, epidemiologists have not found a link between radiation exposure and the many ailments and health effects that could be expected to occur in the general population after seventeen years.[1] But one can hardly be surprised if there is a tendency for the people to wonder whether their own, or their child's, particular illness has been bought about by exposure to radiation.

The inherent uncertainties, and the anxiety they cause, mean that not only those individuals that actually go on to develop a radiation-attributable disease or illness are harmed, but to some extent, *all individuals* in the area can be harmed, even people that have not been exposed to radiation. Finally, as for any illness, the social consequences are not only the harm caused to the sufferer. There are numerous knock-on effects on the affected person's family, dependents and loved ones.

The following section describes some of the psychosocial consequences linked to radiation exposure. These include health effects bought about by worry and stress associated with the radiation exposure, as well as economic breakdown and disruption of daily life. The more complex social effects bought about by countermeasures such as relocation and agricultural interventions, which can in turn have their own stress related health effects, are described in Section 7.4.

7.2.1 Radiation exposure of the Chernobyl 'liquidators'

As discussed in Chapters 1 and 6, the most severely irradiated group were those involved in fighting the Chernobyl fire during the days immediately following the accident. A total of 600 men including the plant's fire service, external volunteers, helicopter crews as well as the plant's operating crew, were employed in fire fighting. One hundred and thirty four of them received acute radiation syndrome, due to doses of radiation between 0.8 and 16 Gy (UNSCEAR, 2000).

The reactor fire and radioactive emissions were finally bought under control on 6 May. Around 800,000 men were subsequently involved in clean-up operations that lasted until 1989. Such operations included removing radioactive debris and building the shelter (sarcophagus) over the destroyed reactor. As robots were unable to cope with the extreme radiation levels in and on the reactor building and broke down, Soviet authorities bought in young military conscripts to remove the debris manually. The treatment and ultimate fate of these conscripts is arguably one of the darker sides of the Chernobyl aftermath. Survivors refer to themselves as 'biorobots' – denoting the state's perception of them as disposable biological resources (UN-OCHA, 2000; Petryna, 2002).

Approximately 200,000 liquidators working in and around the reactor in 1986–1987 received an average radiation dose of 100 mSv (Boice, 1997; Cardis *et al.*, 2001). Today many of the liquidators claim to be suffering from damage to their health.

---

[1] The occurrence, or not, of illness (other than thyroid cancer) in the exposed populations remains a matter of some debate. However, most of the 'established sources', such as the International Atomic Energy Agency (IAEA) and WHO suggest that no statistically significant increase can thus far be observed in the general population of affected areas (see Chapter 6).

How many of them have died from the effects of radiation exposure is a controversial question. According to government agencies in the three former Soviet states affected, about 25,000 liquidators have so far died, but to what extent this represents an increase over expected mortality rates, and to what degree any such excess might be related to radiation exposure is a matter of contention. Epidemiological studies so far have not shown significant increases in rates of fatal cancer in the liquidators (see Chapter 6). Similar controversy exists around the association of other ailments and health problems with radiation exposure (Stone, 2001).

7.2.2 Physiological health effects

Initially, many of the investigations on the long-term health consequences of the Chernobyl accident focused on evaluating possible increases in the incidence of cancer and hereditary effects in exposed populations. The scientific community accepts that these illnesses can be bought about by exposure to radiation, but there is some disagreement about whether or not incidence rates of certain cancers have increased in the exposed populations. This applies both to the results of epidemiological studies of the populations at risk, as well as dose-derived estimates of expected cancers calculated from dose–risk coefficients (i.e., contention about the shape of dose–effect relationships).

As detailed in Chapter 6, there has been a clear increase in thyroid cancers in children exposed to radiation from Chernobyl. The appearance of other cancers can be delayed by 15–20 years after exposure, which means that there may still be other effects that have (or will) not be seen, as discussed in Chapter 6. This delayed response can have clear psychological effects both on populations that were exposed to high levels during the early phases, as well as those who continue to live on contaminated land (UN-OCHA, 2000; UN, 2000; Buldakov, 1995; Kossenko, 1994).

Radiation exposure of populations in Belarus and the Ukraine has also been linked to other health effects, including immunological effects, such as depression of the cellular immune system (IAEA, 1991). The population affected by Chernobyl are also reported to have suffered from lung, heart and kidney problems, as well as hyperthyroidism since the accident (UN-OCHA, 2000). There is no evidence that such effects are directly linked to radiation: they may be secondary effects linked to stress and poor living conditions brought about by the accident. Such effects have been associated with other exposure incidents. Studies on people living along the banks of the Techa River area in the South Urals – an area contaminated in the 1950s by releases of radioactive waste from the Mayak plutonium production plant – suggested a variety of health effects in exposed populations including immune system depression, Down's Syndrome and increased stillbirth rate (Buldakov, 1995; Kossenko, 1994). However, both for Chernobyl and Mayak, it is not clear whether these primary effects were caused by radiation exposure or as a result of more complex, psychosocial or psychological factors caused by general poor health of these often very poor rural populations.

7.2.3 Psychological and social effects

In one of the first major international studies on the effects of Chernobyl, five years after the accident, the International Atomic Energy Agency (IAEA) observed that the negative consequences of the accident were not limited to radiation induced health effects (IAEA, 1991). The IAEA claimed that the predominant adverse health effects to people living in or evacuated from contaminated areas were due to secondary psychological effects brought about by stress, worry and social disruptions, and not damages attributable to the primary effects of radiation itself (IAEA, 1991).

Research carried out in the Ukraine indicates that anxiety over the effects of radiation is the most important psychosocial effect of the accident, that such anxiety is widespread and shows no sign of diminishing (UNDP, 2002; Rumyantseva et al., 1996). This stress in itself causes 'illness-behaviour', and may thereby influence reproduction and lifestyle patterns such as consumption of tobacco and alcohol. These responses are known to be persistent, since chronic stress causes depression as well as apathy and fatalism among the affected people. Similar indirect effects have been hypothesised in other populations living in the vicinity of the Chernobyl reactor (IAEA, 1991, 1996), including claims of a multitude of secondary health effects, including anaemia, AIDS and increased tuberculosis (Burns et al., 1994). Doctors in Italy and Denmark reported increased rates of induced abortion after the Chernobyl accident, believed to be a result of anxiety in the pregnant mothers (Knudsen, 1991; Spinelli and Osborn, 1991).

There is consensus that psychological stress following exposure to radiation risks is very real (Weisæth, 1990; WHO, 1990, 1996). As late as five years after the Three Mile Island accident, increased levels of stress hormones were measured in the affected population (Baum, 1987; Baum et al., 1983). Epidemiologists studying the Japanese survivors of Nagasaki and Hiroshima identified a significant excess mortality beyond that from cancer – a large proportion of which could be attributed to heart disease (Shimizu et al., 1992; Preston et al., 2003). Whilst it is possible that radiation exposure could have caused an increase (see Chapter 6), studies have also identified a dose-related increase for a number of known heart disease risk factors including smoking, cholesterol and hypertension, which may reflect indirect causal relationships attributed to the anxiety and stress in survivors (Mendelssohn, 1995).

The origin of these psychosocial effects are complex and relate not only to the accident itself, and the threats that it involved to other aspects of health, but also to the impact of resettlement and disruption of life. In this respect it is interesting to note that control has long been established as an important psychological factor in coping with stress (Lazarus and Folkman, 1984; Folkman, 1984; Baum et al., 1983), and studies have noted the importance of coping in people's attitude to Chernobyl radiation risks (Tønnessen et al., 1995; MacGregor, 1991). Medical literature suggests that lack of individual control and not providing full information to patients can be in itself psychologically *and* physically damaging to health (Egbert et al., 1964; Friedman et al., 1974; Morris et al., 1977; Lazarus, 1990). These factors are not confined to humans: a seminal experiment carried out on rats showed that

Table 7.2. Registered cases of 'Class 16' illnesses in the Ukraine (per 10,000 people), 1982–1992.
From Petryna (2002).

1982	1983	1984	1985	1986	1987	1988	1989	1990	1991	1992
1.3	1.7	1.7	1.9	2.3	2.7	5.9	34.7	108.3	127.4	141.3

lack of control over an otherwise identical source of pain increased the incidence of physical, stress-related symptoms (Weiss, 1968).

Moreover, the health situation in the former Soviet states has changed during the past 17 years. A relevant question to ask is whether the medical phenomena *per se*, or only the medic's *perception* of the phenomena, are subject to changes. According to the Institute of Sociology in Kiev, both changes in illness behaviour of the population and the diagnostic behaviour of doctors influenced the development of the health situation following the Chernobyl accident. Table 7.2 shows the number of registered 'symptoms, signs and ill-defined states' (class 16 in the international classification of diseases) in the Ukraine from 1982–1992. These states include anything from insomnia, fatigue and persistent headaches to personality changes, hallucinations and premature senility. The increased reporting of illnesses may have been linked in part to the fact that compensation was given for (presumed) Chernobyl-related illnesses.

In 1990, the year that laws on Chernobyl social protection were being published by Ukrainian legislators, there was a sharp increase in the clinical registration of illnesses under this category. Not surprisingly, international observers were sceptical and claimed that Ukrainian scientists had failed to prove (or disprove) these claims on the basis of epidemiological criteria and causality. However, one might question whether appeals to scientific criteria and causality are suitable tools for evaluating the complex circumstances occurring within the post-socialist state (Petryna, 2002).

Petryna has highlighted the deep vulnerability of citizens subject to radical changes in structures of governance and democratising processes, and suggests the collective and individual survival strategy of the so-called 'Chernobyl Victim Syndrome' is better described as a kind of 'biological citizenship' (Petryna, 2002). In other words, she underlines that health cannot be reduced to a set of norms of physiological and mental activity, or to a set of cultural differences. In order to determine what undermines and what protects the health of the population, it is necessary to understand the people's knowledge, reason and suffering, and the way it is shaped by local histories and political economies. 'Seen this way, health is a construction as well as a contested way of being and evolving in the world' (Petryna, 2002).

To conclude, physiological and psychological detriments brought about by an accident like Chernobyl, and the knock-on health effects that such stress can promote, are well-accepted medical phenomena. Understanding their social and psychological context can be a great aid to both authorities trying to improve the situations and doctors treating the people whose health is being harmed. In that respect, it seems clear that both information and education, as well as the phenomena of risk perception, is crucial for understanding and combating psychological and mental detriments connected with accidents like Chernobyl. This is further outlined in Section 7.4.6 on 'risk perception'.

7.3 ECONOMIC IMPACT

In 1986, at the time of the Chernobyl accident, the economic and social conditions for the environment and people living in Belarus, Russia and the Ukraine were controlled by the Soviet Union. As the Union broke down in 1991, both Russia and the Ukraine became independent nations and so-called 'economies in transfer'.[2] This represented a huge political change from a welfare system based on the Soviet planned economy, to market-oriented solutions for employment, health care, infrastructure, etc. The region still faces grave economic challenges as a consequence of the political shift, and it is in this context that the social and economic consequences of the accident need to be interpreted and understood (SDC, 2004).

The exact numbers for economic losses due to the nuclear accident are not known, simply because of the scale and complexity of the political situation. The 1990s was a period of great political, social and economical turmoil in the three most affected countries. The collapse of the Soviet Union, including the social tensions, growing economic crisis and hyperinflation, all exacerbated the effects of the Chernobyl accident. It is difficult to predict what the economic situation might have been had these changes not occurred. Nevertheless, even without these exceptional circumstances, an evaluation of the economic impact of the accident needs to consider a wide range of inter-relating parameters (UNDP, 2002).

7.3.1 Expenditures related to countermeasures

Expenditures on remediation have had profound effect on the Belarussian and Ukrainian national budget, as well as the regional Russian economy. The estimates for Belarussian expenditures are set at US$235 billion (thousand million), which corresponds to 32 times the national budget in 1985 (SDC, 2004). All countries implemented so-called 'national programmes' in order to mitigate the social, economic and environmental inter-related consequences of the accident, and to break the economic and social downward spiral. Table 7.3 gives some numbers related to the economic and social challenges the three countries met.

While Belarus and the Ukraine implemented a 'Chernobyl tax' on wages in all non-agricultural firms, Russia borrowed money to cover the extra expenditure. In 1994, the Chernobyl tax stood at a rate of 18%, but was heavily criticised for making Belarussian products uncompetitive. Hence, the tax was eventually reduced, and in 1999 was set at 4% in both Belarus and the Ukraine (Sahm, 1999).

---

[2] Belarus is the last country in Europe to have a centrally planned economy. Executive power has rested with President Alexander Lukashenko since 1994 (SDC, 2004).

Table 7.3. Estimates of total expenditures, proportion of national budgets, numbers of newly built houses, schools and hospitals as a part of remediation and relocation actions in the three most affected countries; Belarus, Russia and the Ukraine.
From UNDP (2002), Chernobyl Interinform Agency (2002), EMERCOM (2001), IEBNAS (2001), Sahm (1999).

	Ukraine	Belarus	Russia	Total
Total expenditure (US$billion)	148	235	3,8	
	(1986–2000)	(1986–2016)	(1992–1998)	
% of national budget 1991	15	16.8	–	
% of national budget in 1996	6	10.9	–	
% of national budget in 2002	5	6	–	
Numbers of new built houses and flats	28,692	64,836	36,779	*130,307*
Numbers of new built schools	48,847	44,072	18,373	*111,292*
Numbers of new built hospitals	4,391	4,160	2,669	*11,220*

The prioritisation of various measures also differed between the three governments. Belarus suffered the highest impact, with 23% of its territory having ^{137}Cs at levels higher than $1\,\mathrm{Ci\,km^{-2}}$ ($37\,\mathrm{kBq\,m^{-2}}$) (UNDP, 2002). This high percentage of land classified as contaminated restricted the movement of people and agricultural production out of the affected areas. Precedence was given to improving conditions for people still living in the contaminated areas, whereas the Russian and Ukrainian governments gave priority to resettlement (Petryna, 2002). Resettlement meant that it was not only necessary to provide houses and flats, but also infrastructure such as roads, energy and water supply to establish new communities. This illustrates the economic challenges met by the three countries.

The eventual closure of the Chernobyl Nuclear Power Plant (NPP) complex and the cancellation of Belarus' nuclear power programme also represented another major economic challenge, since an alternative power strategy had to be developed for both the Ukraine and Belarus. The Ukrainian government also carried the economic burden of closing and decommissioning the nuclear reactors at Chernobyl (UNDP, 2002).

In recent years, money has been reallocated from resettlement to compensation and social protection, both of which now dominate the budgets. This includes social and medical benefits like monthly allowances, free/subsidised medical treatment, free meals for school children and students, and respite or 'health' holidays. In 1998, 524,100 months of such holidays were taken in the Ukraine. Two years later, Ukraine had an economic expenditure totalling some US$235 million (Table 7.4). The annual expenditure has declined in recent years due to economic constraints, dropping from US$939 million in 1997 to US$332 million in 2000. As shown in Table 7.4, more than 80% of this expenditure has gone to 'social protection' which includes cash subsidies, family allowances, free medical care and education and pension benefits for sufferers and the disabled.

The lack of money put heavy constraints on the government's ability to carry out necessary countermeasures, and to pay Chernobyl-related benefits to the victims.

Table 7.4. Chernobyl budget expenditures for the Ukraine (US$ million in 2000).
From UNDP (2002).

Item	Expenditure (year 2000)
Social protection (compensation payments, free medical treatment, respite holidays)	290
Special healthcare to affected population	6.4
Scientific research on environment, health and production of clean food	1.8
Radiation control	2.7
Environmental recovery	0.04
Radio-ecological improvement of settlements and disposal of radioactive waste	0.05
Resettlement of people and improvement of their living conditions	13.7
Actions to mitigate the consequences in the exclusion zone	17.4
Other expenditures	0.4
Total	*332.7*

This has left many projects uncompleted. Resettlement programmes were cancelled leaving huge numbers of people in contaminated areas: some 11,600 people in Belarus and the Ukraine who have agreed to be resettled are still waiting for their homes to be built. Support to farmers to produce less contaminated food has been cut. In some cases, people in less contaminated areas got higher benefits than those living in more severely affected areas, leading to public discontent (UNDP, 2002). The UN system has urged the international community to participate in the ongoing effort to rehabilitate the social and economic situation in the region.

7.3.2 Capital losses

Human health risks and the direct expenditures related to countermeasures and compensation were not the only threat to national economies. 'Capital losses' like loss or damage of buildings, agricultural land, mineral resources, labour, income, production and value of products have created major economic problems.[3] In monetary terms, the largest economic losses were those associated with the removal of agricultural land and forests from use, and the closure of agricultural and industrial facilities. Table 7.5 summarises some of these losses in terms of production area and economic units (i.e., factories, enterprises, etc.).

In the Ukraine alone, more than one million hectares of forest were contaminated, half of which has been removed from timber production. Likewise, almost 800,000 hectares of agricultural land in the three countries has been taken out of production (Table 7.5). In addition to removing agricultural land and forests from

---

[3] This may of course be regarded as two sides of the same case, as the costs in many cases constitute the compensation for the losses.

Table 7.5. Selected losses related to agricultural/forest land, and economic units.
From UNDP (2002).

	Ukraine	Belarus	Russia	Total
Agricultural land removed from use (hectares)	512,000	264,000	171,000	784,320
Forest land removed from use (hectares)	492,000	200,000	2,200	694,200
Agricultural and forest enterprises removed from service	20	54	8	82
Factories, transport and service enterprises removed from service	13	9	0	22
Raw material deposits removed from service	0	0	22	22

use, closure of agricultural and industrial facilities represented major losses for the most affected countries.

7.3.3 Rural breakdown

The agricultural sector has been the part of the economy worst hit by the effects of the accident. Rural areas are particularly vulnerable because of their great economic dependency on agriculture and forestry, and the impacts on many households in Belarus, the Ukraine and Russia were catastrophic (UNDP, 2002). In Belarus alone, 282 rural settlements were closed due to resettlement. On top of this, the economic crisis of the 1990s made the financial situation of rural families extremely precarious. Many of the people lost not only the possibility for forestry and agriculture, but also their secondary source of income, namely hunting, fishing and collection of wild berries and mushrooms. Together with home produce, these foodstuffs represented a considerable fraction of the traditional diet, but were too heavily contaminated for sale and consumption.

Whilst effectively reducing the collective dose, the restrictions on land use undermined the economic activity in the agricultural communities in several ways. The agricultural countermeasures required large additional subsidies to support the proper maintenance of abandoned land and forest. From an economic point of view, only those countermeasures that allowed production at competitive costs were considered cost-effective. In the long run, this was deemed necessary in order to secure economic development.

In addition, imposing radiological controls and implementing agricultural countermeasures like labelling or food limits seriously affected the market for products from affected areas. This stigma led to a reduction in agricultural activity and a decline in certain types of production, particularly in Belarus where radioactive fallout removed some of the best arable land. Lack of understanding on how to rebuild trust between consumers, producers and food standards authorities has further complicated the situation (UNDP, 2002).

As living conditions in the rural areas continue to decline, populations continue to eat contaminated food themselves. There is some evidence that the dose received by 'high risk groups' has increased in recent years (UNDP, 2002), despite radioactive decay and other natural processes. Although clean food is available in the food stores, this is at a higher price than home produced food, which may explain the increased dose to poor people. Furthermore, country people appear to be increasingly likely to ignore restrictions on consuming forest products. While anxiety and fear of radiation persists, this is confounded by mistrust in the authorities and an inability to interpret available data. Hence, people tend to have a pessimistic perception of their own capacity to control radiation risk. Despite authorities providing equipment to enable people to measure radiation levels in self-gathered or produced food, UN delegates were told by a well-educated resident of the Ukraine (UNDP, 2002):

> We're afraid to check the contamination levels of berries and mushrooms that we pick. We don't want to know.

The whole spectrum of additional 'side effect' costs that might be associated with social and environmental costs like loss of access to an amenity or disruption to livelihoods is discussed in more detail in the following section. These parameters are notoriously difficult to measure because they demand that social and environmental values are converted into monetary values – a practice that is controversial in itself.

7.4 SOCIAL COSTS OF COUNTERMEASURE IMPLEMENTATION

Radioactivity released by the accident caused contamination of large areas of Europe. Whereas short-lived isotopes such as ^{131}I represented the main initial health risk, the most serious long-term consequences, both for health and economic constraints arose from the long-lived ^{137}Cs and (to a lesser extent) ^{90}Sr isotopes. A wide variety of countermeasures were instigated by authorities, both in the fSU and other European countries (see Chapter 5). The main aim of countermeasures was to reduce dose exposures to humans, and to restore ecosystems, the livelihood and health of the affected population, and the economic situation in the region.

The different measures, however, can bring about undesirable side effects that in themselves represent a threat to human welfare. Some countermeasures have so many disadvantages that people perhaps would have been better off without them. The UNDP (2002) report provided a good summary of the successes and failures of the policy with respect to the fSU (reproduced in Table 7.6). Research suggested that remediation strategies have significantly improved the scientific understanding, authorities' preparedness, national capacity and effectiveness of different measures, in particular reducing the collective dose (except for rural high-risk groups). However, they suggest that to date the population at large has not experienced much benefit from such gains in technical understanding, as the countermeasures have failed to increase living standards and local capacity to deal with health, economic and environmental challenges. Persistent anxiety and lack of trust gives a strong indication of the discontent and lack of control felt by people in the region.

Table 7.6. Benefits and costs of remediation efforts.
According to UNDP (2002).

Successes	Failures
Reducing collective dose by technical, economic and administrative measures.	A significant number of rural people in high-risk groups are still exposed to substantial and, probably, increasing doses of radiation.
Significantly improving scientific understanding of possible causes, scenarios and consequences of accidents on nuclear power plants.	Environmental contamination still imposes significant economic constraints associated with a variety of protective measures, many of which are not effective in the new economic and political conditions.
Improving preparedness to deal with consequences of nuclear accidents, including understanding of the effectiveness of different protective measures.	Economies and social structures in affected communities are deteriorating, alongside an apparent increase in poverty.
Building the national capacity in Belarus, Russia and the Ukraine to deal with contamination of the environment by radioactive material, including development of expertise, instrumentation and institutions.	The activities undertaken so far have failed to increase trust and reduce anxiety.
	Low local capacity to deal with health, economical and environmental challenges.

In order to examine the reasons for this anxiety and discontent, and to provide more insight to the successes of countermeasures, the following section provides a description of different remediation strategies and their social and economic impact. These have been divided into: (1) evacuation and resettlement, (2) agricultural measures, (3) compensation and (4) information/communication.

7.4.1 Evacuation and resettlement

The 45,000 inhabitants of Pripyat, 4 km from the reactor site, were evacuated in buses on the 27 April. Within 10 days, 116,000 people from the 76 settlements within a 30-km radius of the reactor were evacuated. This area was officially declared as the Chernobyl exclusion zone (or 30-km zone) and still requires special permission to enter. According to official policy in all three nations, protective measures (primarily relocation) should be put in place for people living on territories where contamination would result in doses exceeding 1 mSv per year. In the early 1990s widespread monitoring of the contamination showed the necessity for further evacuation from 'hot spots' outside the exclusion zone. People in less contaminated regions were permitted to remain in the area, with the exception of children who are sent out of the region for several months during summer (UNDP, 2002).

Massive resettlement has resulted in a wide range of social and economic consequences. First of all, the resettlers lost their jobs, social network, and places of particular social or historical value like neighbourhoods, churchyards, etc. As the majority were forced to move, these losses tended to be deemed more unacceptable than if they had been free to choose themselves.

Resettlement changes not only the lives of the people who leave their homes, but also the social structure of the villages or city districts that become their new home. In Belarus, the Gomel region lost about 43% of its population between 1986 and 2000. Demographic parameters such as mortality and birth rate changed dramatically: elderly people in particular didn't want to leave their villages while young people did. Despite the official prohibition on living in the 30-km exclusion zone, at least 800, mostly older, people returned to their former villages (Botsch, 2000; Sahm, 1999).

The emigration of young people, combined with the (actual and perceived) health risks from radiation exposure, impeded the whole social and economic development of the region. A shortage of teachers and doctors is proving detrimental to the education and health care infrastructure, and companies and farms have had to close down, because they could not find skilled workers. This, in turn, led to more families moving away (Botsch, 2000; Sahm, 1999).

The UNDP/UNICEF (2002) report emphasises that when evacuees are relocated to existing villages, tensions often arose between old and new inhabitants. The Chernobyl victims tend to associate mainly with each other, and especially resettlers over 50 years old find it difficult to adapt to their new environment. This is reinforced by the stigma of radiation contamination. In fact, many of the indigenous villagers believe the resettlers to be suffering an infectious disease. This combined with fear for congenital anomalies, makes it difficult for new inhabitants to find partners. Women, who have moved out of the region, try to keep their origin a secret, out of fear that men will not want to marry them (UN-OCHA, 2000; Sahm, 1999; UNDP, 2002).

Resettlement and evacuation, especially immediately after the accident, probably very significantly reduced the collective radiation dose to the population. However, the effectiveness of the measures declines after some time, as the negative impacts start to outweigh the benefits, especially if other potential uses of the resources are being considered. Resettlement appears to have been least successful when implemented inconsistently (e.g., when large numbers of people were left behind in villages designed for evacuation). This has contributed to a lack of trust between the authorities and the population. Research shows that the psychosocial welfare of people who stayed in their homes is better than that of those who were relocated (UNDP, 2002). Hence a number of recent studies have examined how far the present regime of residency restrictions might be relaxed whilst enabling a growing number of people who wish to return to make informed decisions about the risk to their health.

7.4.2 Countermeasures in agricultural food chains

Agricultural countermeasures include both restrictions on land use and measures undertaken at later stages in the food supply chain (see Chapter 5). These may include: (a) actions to reduce levels in food, such as improving pastures by adding fertilisers and liming or use of radiocaesium binders to prevent uptake to animals; or (b) measures that remove contaminated food from the market (e.g., setting standards for radioactive contamination of foodstuffs, systematic monitoring and providing compensation to buy 'clean' food). Because of the economic aspects of agriculture and forestry, maintenance of production can be a primary reason for application of agricultural countermeasure. Hence the main objective, and benefit, of countermeasures is both to significantly reduce the contamination level in the food product, as well as to maintain trust in the market and economic activity in the region.

Whilst effectively reducing the collective dose, experience showed that restrictions on land use in fact undermined the economic activity in the most affected areas. Furthermore, studies on social consequences of countermeasure implementation both in the fSU and Europe have identified a number of factors that can have a knock-on effect on the economic activity of contaminated areas (Hunt and Wynne, 2002).

- Implemented countermeasures were to a large extent too expensive to allow production at competitive costs, which necessitated large additional subsidies from the state to support the proper maintenance of abandoned land and forests.
- Money had to be spent on policing the countermeasure and averting growth of a black market (Gould, 1990).
- The application of countermeasures resulted in stigma from having been identified as an area with significant contamination. This can have profound consequences both for a range of industries and for the local identities of people and groups (Hunt and Wynne, 2002; Flynn et al., 2001).
- Stigma associated with the perceived contamination of products where the countermeasure has been applied, can generate widespread mistrust of the farming/industry in relation to food production. Loss of confidence that produce and derivative products (e.g., cheese) from affected area/industry is 'safe' may result in loss of employment in local 'cottage' industries.
- General disruption to farming/industry and other related activities.

7.4.3 Compensation

A policy for compensation was also introduced in the aftermath of the accident. Belarussian and Russian legislation provides more than 70, and Ukrainian legislation more than 50, different privileges for Chernobyl victims. These depend on factors such as the degree of invalidity and the level of contamination. People with the right to compensation, the so-called 'Chernobyl victims' included the liquidators (people involved in the clean up), people who had been resettled and people who continued to live in the contaminated areas (UNDP, 2002).

Benefits for Chernobyl victims constitute a wide range of measures, including (UNDP, 2002):

- Health care, free medicines, medical check-ups, etc.
- Housing, including provision of heating and gas.
- Travel; health recuperation holidays and summer camps.
- Tax exemptions.
- Access to university education.
- Compensation for property and damage to health.
- Monthly allowances for disabilities linked to Chernobyl.

Unfortunately, the compensation policy has its own set of undesirable side effects, one of the most striking being a pattern of behaviour described as 'Chernobyl Victim Syndrome'. Research carried out by the Institute of Sociology in Kiev indicates that 84% of resettlers expect special medical treatment and 71% claim unemployment allowance (UNDP, 2002). Poverty caused by resettlement, restrictions on land use and the effect of the collapse of the Soviet Union led to more and more people claiming Chernobyl related benefits. The scarce information on radiation related health effects (radiation was treated as secret information in the fSU) resulted in large uncertainty, and an increased pressure to register as a victim for an ever-increasing number of people.

Being a victim became the only means of access to an income and to vital aspects of health provisions, including medicines. This led to a situation where resources were allocated, not on the basis of medical need, but rather on an individual's ability to register as a victim (UNDP, 2002). Many families returned to the contaminated areas in order to claim a higher level of Chernobyl related benefits. The compensation system also created a situation that blocked economic initiatives and investments. For example, tractor mechanics had turned down an opportunity to open their own workshop in fear of losing Chernobyl entitlements. Due to a system that promoted an exaggerated awareness of ill health and a sense of dependency, which has prevented people from taking part in economic and social life, resources available for mainstream provision have been further reduced, both in the affected areas and beyond. The Chernobyl Accident Victim Syndrome has put a focus on the need for changes in the remediation strategies in Belarus, the Ukraine and Russia (UNDP, 2002).

7.4.4 Communication and information

According to both IAEA (1991) and UNDP (2002) one of the most important factors pertaining to psychosocial effects on the population in Belarus, the Ukraine and Russia, was the quality of public information. Also, communication experts and psychologists suggested that many of the experienced conflicts and problems observed following the Chernobyl accident arose because of a lack of information and an inappropriate communication strategy (Gould, 1990;

Otway et al., 1988; Drottz and Sjöberg, 1990; Slovic, 1996). According to Chernobyl expert Astrid Sahm (1999), lack of knowledge seems to have been crucial for the information crisis to develop. She claims that 'Soviet policy in the first two years was clearly based on the assumption that the radio-ecological situation would largely return to normal during this period.'

National radiation protection authorities hold the main responsibility for the assessment and management of radiation risks, and informing the public on both the risks from radiation, and the procedures employed to reduce those risks. This includes publication and dissemination of information materials with specific information on local contamination and advice on reducing exposure to radiation for households, workers, children or other high-risk groups.

In the fSU, most expertise on radiation was regarded as secret information in 1986 and for many Soviet citizens information was withheld for weeks after the accident. The inhabitants of Pripyat, just a few kilometres from the reactor, were not informed about any sort of danger before 36 hours passed after the accident. A teacher took her class down to the reactor to watch the firemen as they struggled to get control over the fires; all of the children developed thyroid carcinomas (UN-OCHA, 2000).

Despite more than a decade of information policy to the affected populations, widespread ignorance still exists on how best to minimise radiation risk. As a 10 year old Belarussian expressed: 'The best way to avoid radiation exposure is to run and hide in the forest' (UNDP, 2002). However, other studies, and particularly those where the information has been produced as a collaboration between both Soviet and non-Soviet scientists, reported greater success in communicating with populations (Hériard Dubreuil et al., 1999). People in the area have a great deal more confidence in the information provided from international sources than facts provided by their own governments. Rehabilitation centres therefore put a lot of effort in trying to inform people about basic risks connected to high-level radiation, and effective measures to minimise this risk (UN-OCHA, 2000).

The ways in which governments act in the immediate aftermath of a large accident have a major influence on *trust* between the public and the authorities, as well as *compliance* with subsequent measures in the following years. Public acceptance of policy is often highlighted as one of the main benchmarks of a successful communication strategy, although there are other criteria than acceptance that should be used to evaluate communication processes, including factors such as openness, empowerment, influence and transparency (Oughton et al., 2002; Oughton, in press). The experience from Chernobyl shows that it is essential that trust is generated, not dissipated, in the time immediately following an accident (Hunt and Wynne, 2002). Evaluation of communication strategies is not simply a matter of how well the public understands the information: perceptions are strongly influenced by factors such as the authorities' openness and the degree to which they, in turn, trust the public to act on the information provided.

Apart from the health detriments caused by lack of personal control, philosophers have argued that withholding information on risk is ethically objectionable. Gewirth (1982) argues that each person has a *right* to informed control over conditions relevant to the possible infliction of injury, because informed control is a component of freedom. Thus, an effective and justifiable communication strategy can be said to rest on two pillars: (1) it reduces hazards, since it is likely that well-informed persons cope with the situation better than uninformed persons; and (2) it fulfils the fundamental right of autonomous citizens.

7.4.5 The European response

Criticism of the information policy has not only been restricted to the Soviet authorities. Many authorities in other parts of Europe were accused of gross incompetence, cover-up, providing false information or not providing full information regarding the risks from Chernobyl and ways of reducing them (Gould, 1990; Samuelsson, 1997). A number of psychologists and sociologists accused authorities of failing to keep the public properly informed after the Chernobyl accident (Wynne, 1989; Drottz and Sjöberg, 1990). On the other hand, scientists have suggested that those with a vested interest in radiation protection exaggerated the risks and stirred-up media attention (Jaworowski, 1999). What most people would agree on was that the majority of authorities initially underestimated the effects of Chernobyl, and the impression given by all but a few government officials was that the radioactive releases would have neither serious nor lasting effects on human health or food production (Gould, 1990). The immediate response of many governments after Chernobyl was that there was no cause for alarm. For example, in France authorities informed the public that 'the cloud had turned at the Alps' (Gould, 1990).

The case in the UK serves as a classic illustration of the situation. When weather reports confirmed that the radioactive cloud would pass over the UK, the Minister for the Environment released a statement that: 'The effects of the cloud have been assessed and none presents a risk to health in the UK' (Hansard, 1986). Less than seven weeks later, however, due to high levels of radiocaesium in lambs grazing mountain pastures in England, Scotland and Wales, the government imposed a 21-day restriction on the movement and slaughter of sheep from certain areas. Authorities were convinced that levels in meat would drop quickly. Unfortunately that optimism proved unfounded. The number of animals under restriction rose quickly to nearly five million. Four years later, 600 farms and over half a million sheep were still under control (Rich, 1991). In 2003, 377 farms and approximately 230,000 sheep were under restriction (RIFE, 2004).

A standard objection to providing full information after an accident is that people will panic and behave in ways that would put themselves at risk (e.g., mass evacuation, overdosing on iodine, abortion), or that they simply will not understand the information. As one spokesman for the Polish government said during a television interview: 'It is not good to give people unnecessary facts ... ordinary people cannot draw conclusions' (Gould, 1990).

But studies by psychologists show that, despite an estimated 40,000 evacuating (by their own choice) the area after Three Mile Island (Alloway and Ayers, 1997), the majority of people behave rationally in emergencies (Rachman, 1978). Whilst the non-radiation risks of evacuation are low (Aumonier and Morrey, 1991) it is likely that uncontrolled evacuation has comparatively high risks of road traffic accidents.

Since communication and information policy can play such a vital role in influencing the social, economic and health effects of an accident, the next section focuses on some of the underlying factors that should be considered when planning such policies. One important area is an understanding of the way public risk perception can be affected by a multitude of factors. Here we consider both the Chernobyl accident as well as knowledge gained from other studies on risk perception.

7.4.6 Risk perception

At the time of the Chernobyl accident, an over simplistic evaluation of risk perception was common within the scientific community. The IAEA was also somewhat simplistic in its evaluation of the social consequences of Chernobyl, often giving the impression that much of the stress was due to an irrational perception of radiation risks in the population rather than focusing on the more complex social and psychological effects bought about by countermeasures themselves (Havenaar et al., 1996). This might be due to the fact that both the nuclear industry and some risk assessors had been campaigning for some time that the public was being irrational in forcing it to waste money on trivial risks – money which, theoretically, might be spent more efficiently saving other lives (Cohen, 1980; Flakus, 1988; Breyer, 1993; Mossman, 1997). For example that:

> ... the waste of monetary and manpower resources due to an irrational phobia, in particular of 'artificial' radiation ... may be seen as one of the many meaningless luxuries which only a few countries are able to afford ... the reasons for this unbalanced perception and these reactions [reflect] the basic psychological problems of less educated persons.
>
> (Becker, 1996)

7.4.7 Factors influencing risk perception

Over the past 20 years, risk assessors, psychologists and anthropologists have conducted numerous surveys, questionnaires and interviews designed to evaluate the public's perception of risk (see reviews in Sjöberg and Drottz-Sjöberg (1994) and Slovic (1996)). These psychometric studies confirmed that experts and the public often disagree in the way they rank risks, but they also showed that the public's perception of risk is complex, multifactorial and by no means unsophisticated. Psychologists now consider it erroneous to label attitudes to radiation risk as irrational or phobic (Drottz-Sjöberg and Persson, 1993). Perhaps one of the most important results of these studies is that varying attitudes to risk cannot always be attributed to a misunderstanding of the probabilities of harm (Slovic et al., 1980, 1990). A variety of factors can influence risk perception (Table 7.7), including choice, control, and consequences for children and future generations. As discussed later, many of the intervention measures introduced to reduce exposure to radionuclides, appear to have aggravated the situation by heightening negative social, psychological

Table 7.7. Factors commonly used to explain the perception of risk.
From Sjöberg and Drottz-Sjöberg (1994), Sjöberg (1996), Slovic (1996).

Condition presumed to bring about a lower risk rating	Condition presumed to bring about a higher risk rating
Hazard related factors	
Voluntary	Imposed
Controllable	Uncontrollable
Well-known, familiar	New, unfamiliar
Visible – known to exposed individual	Invisible – not perceptible
Technological	Natural
Scientific agreement on risks, well-calibrated	Scientific controversy, estimated
Common, un-feared consequences	Dreaded, feared consequences
Delayed harm*	Immediate or sudden harm
Normal distribution	Catastrophic
Reversible harms	Irreversible consequences
Factors related to the social context	
Fairness, equitable distribution	Inequitable distribution of risks and benefits
Clear benefits	Uncertain benefits
Information from trusted experts/authorities	Information from distrusted authorities
Not in the news	Heavy media attention
No special harm to children	Possible harm to children or the foetus
Not harmful to future generations*	Affecting future generations
Causing harm to unidentifiable victims (statistical deaths)	Causing harm to a known or identifiable victim
No viable alternatives	Viable alternatives
Factors related to individual characteristics	
Being male	Being female
High education	Low education
Young	Older
High income	Low income
Laid-back	Anxious
Trained in risk assessment	No training in risk assessment
Factors related to the context of risk judgements	
Risk to self*	Risk to others
Defining risk with emphasis on probabilities	Emphasis on negative consequences

*Some studies suggest the reverse effect.

and ethical factors like feelings of helplessness and confusion, and lack of control and personal autonomy in the exposed populations (Drottz and Sjöberg, 1990; MacGregor, 1991).

Despite the agreement on the social and political relevance of communication, there is still uncertainty on how successful information strategies might be best achieved in practice. Of course, the directive to inform the public often presents considerable practical problems both for experts and the authorities (e.g., whom to inform, when to inform and what to inform about?). Often, all authorities can say is that they have little knowledge of the situation or that they do not know what the outcome will be. Experience of a number of other environmental, public health and food risks (e.g., the BSE outbreak in the UK, Genetically Modified Organisms, outbreaks of foot and mouth disease), shows that the public can perceive official information as propaganda and manipulation (Oughton, in press).

Psychologists have also noted that there can be a problem if the information requirements of the public vary: some wish for a lot of information, some wish only to be reassured (Tønnessen *et al.*, 1995). Hence, it is difficult to develop an information policy that will suit everybody, as attempts to communicate with the public on risk can as easily create unease as reassurance. However, the fact that informing the public is difficult does not undermine the fact that they have a right to be informed. In addition, most psychologists acknowledge that *personal control* is one of the major factors in coping with emergencies (Lazarus and Folkman, 1984).

7.4.8 Control

As Table 7.7 shows, choice, control, familiarity, closeness and numerous other social and psychological factors also play an important role in shaping perceptions towards hazards. It follows that both communication policy and remediation measures that are sensitive to these factors may stand a greater chance of success. For example, for certain countermeasures, reduction in dose need not be the only benefit, or even the main benefit. Some measures will tend to increase personal control or choice regarding the risks (i.e., information on actions that can be taken to reduce exposures), whilst others (i.e., state-controlled interventions) might provoke feelings of helplessness (MacGregor, 1991). Also, actions need not be limited to those that reduce the exposure to radiation (e.g., countermeasures might include better medical attention to reduce all illnesses (Morrey and Allen, 1996)).

As discussed in Chapter 5, a group of French scientists have shown that it is possible to implement intervention procedures that involve the populations themselves, and adhere to the principle of informed personal control over radiation risks (Hériard Dubreuil *et al.*, 1999). The study in question was carried out in Belarussian villages under restrictions due to fallout from the Chernobyl accident. The basic idea was that experts should attempt to teach and advise populations on ways to deal with the situation, rather than adopt the paternalistic role of making decisions on behalf of the public. Thus, scientists instructed the villagers in how to use and interpret radiation-monitoring equipment, and outlined the possible 'self-help' ways of reducing radiation exposures. At least for this study, the conclusion was that this approach not only resulted in reducing exposures with minimal social and psychological side effects, but was also more economically cost effective than the standard bureaucratic, 'top-down' management procedures (Hériard Dubreuil *et al.*, 1999).

7.5 CONCLUSIONS

Scientists from all areas of radiation research agree that there were lessons to be learnt from the Chernobyl accident, and that there was room for improvement both in how authorities acted to reduce risks and in radiation protection policy in general. Chernobyl helped to strengthen the growing awareness that effective management of radiation risks needs to be sensitive to the public's perception of those risks, and that authorities are often faced with moral dilemmas when evaluating actions. The general problems in the aftermath of Chernobyl attributed to a lack of preparedness, knowledge and clarity on distribution of responsibility (Czada, 1990; Marples, 1991; Nohrstedt, 1991).

In this chapter we have tried to highlight some of the economic and social effects imposed on the Belarussian, Ukrainian and Russian societies in the aftermath of the Chernobyl accident. Large-scale deposition of radionuclides, or any contaminant, imposes not only direct risk to people's health, but also to social, economic and cultural conditions in society. The system of Chernobyl related benefits has created expectations of payments and advantages and has undermined the capacity of the individuals and communities concerned to cope with the economic and social problems and search for solutions.

Clearly, the economic rehabilitation of the areas most affected by the accident should be prioritised, as this is crucial for normalising the situation for individuals and communities and for breaking the 'downward spiral' in the region. A remediation strategy therefore should contain development and implementation of countermeasures for economic and social problems, as well as for the associated health effects. Due to the complexity and timescale of such effects, experts in many kinds of disciplines should be involved in the design of such strategies. According to Hunt and Wynne (2002) it is important to monitor and support, rather than presume, effective institutional action, and at least to assess the viability of the existing institutional infrastructures in relation to particular countermeasures.

The further rehabilitation of the Chernobyl region has recently been put on the international agenda. The UNDP has proposed a framework for an international and national response, based on three successive phases, namely: emergency phase (1986–2001), recovery phase (2001–2011) and management phase (long term). The present phase, namely the recovery phase aims at enabling people and societies to take control over their own lives and to acquire the means for self-sufficiency through human and economic development. However, it cannot be assumed that a similar emergency situation will run smoothly simply because people have learned their lesson. First, this learning has never been put into practice; second the pattern from a large-scale nuclear accident is highly unpredictable in terms of contamination deposition as well as its effects (Hunt and Wynne, 2002). As many effects are inter-related, the demands on different official institutions are impossible to predict.

The recommendations offered by the UNDP emphasise the need for a holistic approach to remediation, integrating economic, ecological and health measures, rather than a blinkered preoccupation with 'dose reduction' aspects of radiation protection (UNDP, 2002). This is supported by multidisciplinary research on the long-term management of radioactive contamination (Howard *et al.*, 2002, 2004; Oughton *et al.*, 2004) which also stressed the importance of including the affected population with regard to self-help measures and consensus-driven, decision-making processes. In addition to respecting people's fundamental right to shape their own future and thereby increasing trust and compliance, such an approach may lead to proactive efforts made by people themselves, which also improve factors like cost-effectiveness. In that respect, people's perception of themselves as well as the situation may change to a positive driving force for human and economic development in the region.

7.6 REFERENCES

Alloway, B.J. and Ayres, D.C. (1997) *Chemical Principles of Environmental Pollution*. Blackie Academic and Professional, London.

Aumonier, S. and Morrey, M. (1991) Non-radiological risks of evacuation. *Radiation Protection Dosimetry*, **10**, 287.

Baum, A. (1987) Toxins, technology and natural disasters. In: Vanden Bos, G.R. and Bryant, B.K. (eds), *Cataclysms, Crises and Catastrophes: Psychology in Action* (5th edition, no. 53). American Psychological Association, Washington D.C.

Baum, A., Gatchel, R.J. and Schaeffer, M.A. (1983) Emotional, behavioural, and physiological effects of chronic stress at Three Mile Island. *Journal of Consulting and Clinical Psychology*, **51**, 565–572.

Becker, K. (1996) Perception of natural, medical and 'artificial' radiation exposures. In: *Radiation and Society: Comprehending Radiation Risk*, pp. 191–194. IAEA, Vienna.

Boice, J.D. (1997) Leukaemia, Chernobyl and Epidemiology. *Journal of Radiological Protection*, **17**, 129–133.

Botsch, W. (2000) *Untersuchungen zur Strahlenexposition von Einwohnern Kontaminierter Ortschaften der Nördlichen Ukraine*. University of Hannover, Hannover.

Breyer, S. (1993) *Breaking the Vicious Circle*. Harvard University Press, Cambridge, MA.

Buldakov, L.A. (1995) Medical consequences of radiation accidents. In: *Proceedings of a Symposium, Environmental Impact of Radioactive Releases*, pp. 467–478. IAEA-SM 339, Vienna.

Burns, D.N., Gelbert, G.A. and Crone, R.K. (1994) Tuberculosis in eastern Europe and the former Soviet Union: How concerned should we be? *The Lancet*, **343**, 1445–1446.

Cardis, E., Richardson, D. and Kesminiene, A. (2001) Radiation risk estimates in the beginning of the 21st Century. *Health Physics*, **80**, 349–361.

Chernobyl Interinform Agency (2002). *Interview on 18.04.2002*, Kiev, p. 6.

Cohen, B.L. (1980). Society's valuation of life saving in radiation protection and other contexts. *Health Physics*, **38**, 33–51.

Czada, R.A. (1990) Politics and Administration during a 'nuclear-political' crisis: the Chernobyl disaster and radioactive fallout in Germany. *Contemporary Crises*, **14**, 285–311.

Drottz, B. and Sjöberg, L. (1990) Risk perception and worries after the Chernobyl accident. *Journal of Environmental Psychology*, **10**, 135–149.

Drottz-Sjöberg, B.M. and Persson, L. (1993) Public reaction to radiation: Fear, anxiety, or phobia. *Health Physics*, **63**, 223–226.

Egbert, L., Battit, G., Welch, C. and Bartlett, M. (1964) Reduction of postoperative pain by encouragement and instruction of patients. *New England Journal of Medicine*, **270**, 825–827.

EMERCOM (Ministry of the Russian Federation for Civil Defence Affairs, Emergencies and Elimination of Consequences of Natural Disasters) (2001) *Chernobyl Accident, Results and Problems in Eliminating the Consequences in Russia 1986–2001*, p. 3. Moscow.

Flakus, F.N. (1988) Radiation protection in nuclear energy. *IAEA Bulletin*, **30**, 5–11.

Flynn, J., Slovic, P. and Kunreuther, H. (2001) *Risk, Media and Stigma: Understanding Public Challenges to Science and Technology*. Earthscan, London.

Folkman, S. (1984) Personal control and stress and coping processes: A theoretical analysis. *Journal of Personality and Social Psychology*, **46**, 839–852.

Friedman, C., Greenspan, R. and Mittelman, F. (1974) The decision making process and the outcome of therapeutic abortion. *American Journal of Psychology*, **131**, 1332–1337.

Gewirth, A. (1982). Human rights and the prevention of cancer. In: *Human Rights: Essays on Justification and Applications*, pp. 181–196. University of Chicago Press, Chicago and London.

Gould, P. (1990) *Fire in the Rain: The Democratic Consequences of Chernobyl*. Polity Press, Cambridge.

Hansard (1986). 8th May 1986, House of Commons, U.K.

Havenaar, J.M., Savelkoul, T.J.F., van den Bout, J. and Bootsma, P.A. (1996) Psychosocial consequences of the Chernobyl disaster. In: Karaoglou, A., Desmet, G., Kelly, G.N. and Mezel, H.G. (eds), *The Radiological Consequences of the Chernobyl Accident*, pp. 435–452. European Commission, Brussels.

Hériard Dubreuil, G.F., Lochard, J., Girard, P., Guyonnet, J.F., Le Cardinal, G., Lepicard, S., Livolsi, P., Monroy, M., Ollagon, H., Pena-Vega, A., et al. (1999) Chernobyl post-accident management: The ETHOS project. *Health Physics*, **77**, 361–372.

Howard, B.J., Andersson, K.G., Beresford, N.A., Crout, N.M.J., Gil, J.M., Hunt, J., Liland, A., Nisbet, A., Oughton, D.H. and Voigt, G. (2002) Sustainable restoration and long-term management of contaminated rural, urban and industrial ecosystems. *Radioprotection – colloques*, **37**, 1067–1072.

Howard, B.J., Liland, A., Beresford, N.A., Anderson, K.G., Cox, G., Gil, J.M., Hunt, J., Nisbet, A., Oughton, D.H. and Voigt, G. (2004) A critical evaluation of the Strategy project. *Radiation Protection Dosimetry*, **109**, 63–67.

Hunt, J. and Wynne, B. (2002). *Social Assumptions in Remediation Strategies*. Deliverable 5, STRATEGY project, Lancaster University (available at: www.strategy-ec.org.uk).

IEBNAS (Institute of Economics of the Belarussian National Academy of Sciences) (2001). *Committee on the Problems of the Consequences of the Catastrophe at the Chernobyl NPP 15 Years After the Chernobyl Disaster*. National Academy of Sciences, Minsk.

IAEA (1991) *The International Chernobyl Project: Assessment of Radiological Consequences and Evaluation of Protective Measures*. IAEA, Vienna.

IAEA (1996) *One decade after Chernobyl: Summing up the consequences of the accident, IAEA/WHO/EC International Conference, Vienna*.

Jaworowski, Z. (1999) Radiation risk and ethics. *Physics Today*, September, 24–9.

Kleinman, A. (1986) *Social Origins of Distress and Disease*. Yale University Press, New Haven.

Knudsen, L.B. (1991) Legally-induced abortions in Denmark after Chernobyl. *Biomedicine and Pharmacotherapy*, **45**, 229–231.

Kossenko, M.M (1994) *Analyses of Chronic Radiation Sickness Cases in the Population of the Southern Urals* (AFRI CRR 94-1). Armed Forces Radiobiology Research Institute, Maryland, USA.

Lazarus, R.S. and Folkman, S. (1984) *Stress, Appraisal and Coping*. Springer, New York.

Lazarus, R.S. (1990). Stress, coping and illness. In: Freidman, H. (ed.), *Personality and Disease*, pp. 97–109. Wiley, New York.

MacGregor, D. (1991) Worry over technological activities and life concerns. *Risk Analysis*, **11**, 315–324.

Marples, D.R. (1991) Chernobyl: Five years later. *Soviet Geography*, **32**, 291–313.

Mendelssohn, M.L. (1995) Radiation risks to human health: The perspective from Hiroshima and Nagasaki. In: Sundnes, G. (ed.), *Biomedical and Psychosocial Consequences of Radiation from Man-made Radionuclides in the Biosphere*, pp. 115–128. The Royal Norwegian Society of Sciences and Letters Foundation, Trondheim (IBSN 82-519-1421-3).

Morrey, M., Allen, P. (1996) The role of social and psychological factors in radiation protection after accidents. *Radiation Protection Dosimetry*, **68**, 267–271.

Morris, T., Greer, S. and White, P. (1977) Psychological and social adjustment to mastectomy: A two year follow-up. *Cancer*, **40**, 2381–2387.

Mossman, K.L. (1997) Radiation risks and linearity: Sound Science? In: Walderhaug, T. and Gudlaugsson, E.P. (eds), *Proc. 11th Meeting of Nordic Radiation Protection Society, Reykjavik 26th–29th August 1996*, pp. 39–52. Geislavarnir Rikisins, Reykjavik.

Nohrstedt, S.A. (1991) The information crisis in Sweden after Chernobyl. *Media, Culture and Society*, **13**, 477–497.

Otway, H., Haastrup, P., Connell, W., Gianitsopoulas, G. and Paruccini, M. (1988) Risk communication in Europe after Chernobyl: A media analysis of seven countries. *Industrial Crisis Quarterly*, **2**, 31–35.

Oughton, D.H. (in press) The promises and pitfalls of participation processes. *Global Bioethics*.

Oughton, D.H., Bay, I., Forsberg, E-M., Hunt, J., Kaiser, M. and Littlewood, D. (2002) Social and ethical aspects of countermeasure evaluation and selection – using an ethical matrix in participatory decision-making. Deliverable 4, STRATEGY project, Lancaster University (available at: www.strategy-ec.org.uk).

Oughton, D.H., Bay, I., Forsberg, E-M., Kaiser, M. and Howard, B. (2004) An ethical dimension to sustainable restoration and long-term management of contaminated areas. *Journal of Environmental Radioactivity*, **74**, 171–183.

Petryna, A. (2002) *Life Exposed*. Princeton University Press, Princeton.

Preston, D.L., Shimizu, Y., Pierce, D.A., Suyama, A. and Mabuchi, K. (2003) Studies of mortality of atomic bomb survirors. Report 13: Solid cancer and noncancer disease mortality 1950–1997. *Radiation Research*, **160**, 381–407.

Rachman, S.J. (1978) *Fear and Courage*. Freeman, New York.

Rich, V. (1991) An ill wind from Chernobyl. *New Scientist*, **130**, 26–28.

RIFE (2004) *Radioactivity in Food and the Environment, 2003*. EA, EHS, FSA, SEPA. Environment Agency, Preston.

Roche, A. (1996) *Children of Chernobyl: The Human Costs of the World's Worst Nuclear Disaster*. Harper, London.

Rumyantseva, G.M., Arkhangelskaya, H.V., Zykova, I.A. and Levina, T.M. (1996) Dynamics of social-psychological consequences ten years after Chernobyl. In: Karaoglou, A., Desmet, G., Kelly, G.N. and Menzel, H.G. (eds), *The Radiological Consequences of the Chernobyl Accident*, pp. 529–535. European Commission, Brussels.

Sahm, A. (1999) *Transformation im Shalten von Tschernobyl*, pp. 238–262. LIT Veerlag, Münster.

Samuelsson, C. (1997) Radiation risk information to the public: Principles or common sense. In: Walderhaug, T. and Gudlaugsson, E.P. (eds), *Proc. 11th Meeting of Nordic Radiation Protection Society, Reykjavik 26th–29th August 1996*, pp. 539–544. Geislavarnir Rikisins, Reykjavik.

SDC (2004) International Communications Platform on the Long-term Consequences of the Chernobyl Accident. Website published by the Swiss Agency for Development and Cooperation (SDC) www.chernobyl.info.

Shimizu, Y., Kato, H., Schull, W.J. and Hoel, D.G. (1992) Studies of the mortality of A-bomb survivors. 9. Mortality 1950–1985: Part 3. Non-cancer mortality based on the revised doses (DS86). *Radiation Research*, **130**, 249–266.

Sjöberg, L. (1996) A discussion of the limitations of the psychometric and cultural theory approaches to risk perception. *Rad Prot Dosim*, **86**, 219–225.

Sjöberg, L. and Drottz-Sjöberg, B.M. (1994). Risk perception. In: Lindell, B., Malmfors, T., Lagerlöf, E., Thedéen, T. and Wlinder, G. (eds), *Radiation and Society: Comprehending Radiation Risk* (Volume 1), pp 29–60. IAEA, Vienna.

Slovic, P. (1996) Perception of risk from radiation. *Radiation Protection Dosimetry*, **68**, 165–180.

Slovic, P., Fischhoff, B. and Lichtenstein, S. (1980) Facts and fears: Understanding perceived risk. In: Schwing, R. and Albers, W.A. (eds), *Societal Risk Assessment: How Safe is Safe Enough?*, pp. 181–214. Plenum, New York.

Slovic, P., Kraus, N.N. and Covello, V. (1990) What should we know about making risk comparisons? *Risk Analysis*, **10**, 389–392.

Spinelli, A. and Osborn, J.F. (1991) The effects of the Chernobyl explosion on induced abortion in Italy. *Biomedicine and Pharmacotherapy*, **45**, 243–247.

Stone, R. (2001) Living in the shadow of Chernobyl. *Science*, **292**, 420–424.

Tønnessen, A., Reitan, J., Strand, P., Waldahl, R. and Weisæth, L. (1995) Interpretation of radiation risk by the Norwegian population. In: Sundnes, G. (ed.), *Biomedical and Psychological Consequences of Man-made Radionuclides in the Biosphere*, pp. 215–278. The Royal Norwegian Society of Sciences and Letters Foundation, Trondheim.

UNDP (United Nations Development Programme) (2002). *The Human Consequences of the Chernobyl Nuclear Accident – A Strategy for Recovery* (available at: http://www.undp.org/dpa/publications/chernobyl.pdf).

UN-OCHA (United Nations Office for the Coordination of Humanitarian Affairs) (1999) *United Nations Appeal for International Cooperation on Chernobyl*. Priority Projects Proposed by The Governments of Belarus, Russian Federation and Ukraine, April (available at: http://www.reliefweb.int/ocha_ol/programs/response/cherno/cher.htm)

United Nations Office for the Coordination of Humanitarian Affairs (UN-OCHA) (2000) *Chernobyl – A Continuing Catastrophe*. United Nations, New York and Geneva.

UN (2000) United Nation Office in the Republic of Belarus. *Belarus: Choices for the Future*, p. 67. National Human Development Report, Medium Company, Minsk.

UNSCEAR (2000) *Report to the General Assembly: Sources and effects of ionizing radiation* (Volume II, Annex J), pp. 453–551. United Nations, New York (available at: http://www.unscear.org.)

Weisæth, L. (1990) Reactions in Norway to fallout from the Chernobyl disaster. In: Brustad, T., Landmark, F. and Reitan, J.B. (eds), *Cancer and Radiation Risk*, pp. 149–155. Hemisphere, New York.

Weiss, J.M. (1968) Effects of coping responses on stress. *Journal of Comparative and Physiological Psychology*, **65**, 251–260.

WHO (1990) *Working Group on the Psychological Effects of Nuclear Accidents.* Summary Report EUR/ICP, Geneva.

WHO (1996) *Health Consequences of the Chernobyl Accident. Results of the IPHECA pilot projects and related national programmes.* WHO, Geneva.

Wynne, B. (1989) Sheep farming after Chernobyl: A case study in communicating scientific information. *Environment*, **31**, 33–39.

8

Effects on wildlife

Ivan I. Kryshev, Tatiana G. Sazykina and Nick A. Beresford

In this chapter we discuss observations of effects of irradiation on wildlife living within the Chernobyl exclusion zone. Within this review, the authors have drawn upon many studies by former Soviet Union (fSU) scientists published in Russian language literature.

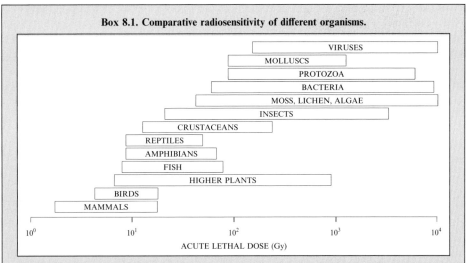

Box 8.1. Comparative radiosensitivity of different organisms.

Summary of data describing the comparative sensitivity of different organisms to radiation (*radiosensitivity*) in terms of **acute lethal dose** (from UNSCEAR, 1996). Within the higher plants radiosensitivity decreases in the order coniferous trees > deciduous trees > shrubs > herbaceous plants. The habits and habitats of different life-stages of some organisms may vary considerably (e.g., bird and egg, larvae and adult forms of insects, seed and plant) and this may lead to very different exposure pathways. Lethal doses to embryonic birds and salmon embryos have both been found to be lower than for adults. Conversely seeds are the least radiosensitive life-stage of plants. An overview of the effects of chronic exposure to ionising radiations on different wildlife groups can be found in Real *et al.* (2004).

Figure 8.1. Area of Red Forest where coniferous trees were killed as a consequence of acute irradiation but deciduous trees continued to grow. Picture taken in the early 1990s.
Photograph by Nikolay Kuchma, supplied by Andrey Arkhipov International Radioecology Laboratory, Slavutych, Ukraine.

8.1 TERRESTRIAL BIOTA

8.1.1 Radiation effects in forests

In 1986 approximately 40% of the 30-km zone surrounding the Chernobyl Nuclear Power Plant (NPP) was forest. Of this 80% was coniferous forest consisting of predominantly pine (*Pinus silvestris*) with spruce (*Picae excelsa*). At the time of the Chernobyl accident the coniferous trees were generally between 30 and 40 years old.

The forests close to the Chernobyl reactor 'filtered' considerable amounts of radioactivity from the contaminated air mass resulting from the accident. The high retention of radioactive material in forest ecosystem resulted in significant radiation damage to the forests (Figure 8.1, above), and in particular coniferous trees (Kozubov *et al.*, 1987, 1988, 1990, 1994; Tikhomirov and Sidorov, 1990; Tikhomirov and Shcheglov, 1994; Abaturov *et al.*, 1996; Sokolov *et al.*, 1994; Ilyin and Gubanov, 2001). Coniferous trees (and consequently forests) are especially sensitive to radioactive contamination (see Box 8.1). Following the Chernobyl accident, coniferous trees were exposed to radioactive contamination during the period of spring growth when the tree is more radiosensitive compared to autumn/winter dormancy periods. This contributed to the significant effects observed in coniferous forests close to the Chernobyl NPP.

Table 8.1. Distribution of radiation damage in forests around the Chernobyl NPP.
From Kozubov *et al.* (1990), Kryshev (1992), Sokolov *et al.* (1994).

Zone and approximate area affected	Degree of radiation damage	Absorbed dose from external γ-radiation (Gy)	Absorbed dose rate (μGy h^{-1}) on 1 October, 1986	Estimated absorbed dose in needles (Gy)
Lethal 4 km^2	Death of pines within several days of exposure Partial damage of deciduous trees	>80–100	>5,000	>100
Sub-lethal 38 km^2	Death of most growth points (90–95% of coniferous trees) Partial die-back of coniferous trees Morphological changes in deciduous trees	10–20	2,000–5,000	50–100
Medium damage 120 km^2	Suppressed reproductive ability Desiccated needles Morphological changes	4–5	500–2,000	20–50
Minor damage	Disturbances in growth and reproduction Morphological disturbances in coniferous trees	0.5–1.2	<200	<10

Consequences in coniferous forests

Along the westward moving plume of high-intensity radiation in the early stages of the release, pines were killed (within several days of exposure) to a distance of 5.5 km around the plant. Within 200–300 m from the plume axis, active growing points of trees were killed or injured up to the crowns height of ~15 m above the ground. Further along the plume axis dead trees were not observed. The height to which the canopy was injured decreased with the distance from the NPP, to about 3 m height at a distance of 8 km. Damage induced reduced sharply laterally from the plume axis. As a consequence a *circa* 4 km^2 area where trees were killed has become known as the Red (or, in Russian, Ginger) Forest. Four zones of damage to forests could be distinguished; these are summarised in Table 8.1 and used for reference throughout this chapter.

Once the radioactive plume passed over the forests, exposure of trees was primarily due to external irradiation from radionuclides deposited on the soil and retained by the above-ground parts of trees. The changes in estimated external irradiation doses from the soil over the first three years after the accident are given in Table 8.2 (Kozubov *et al.*, 1990, 1994). The majority of the external irradiation dose was received in 1986 (N.B. values in Table 8.2 for 1986 do not include estimates of dose received from the contaminated air mass). The highest irradiation levels were recorded in pine stands of 40–50 years age at a distance of about 1.5–2 km to the west of the NPP close to the village Yanov (study plot 9 in Table 8.2). By the autumn of 1986, all the pines here had died, the accumulated external irradiation

Table 8.2. The dynamics of external irradiation dose from soil at study plots with coniferous trees close to the Chernobyl NPP.

Plot number	Bearing from Chernobyl NPP	Distance from the Chernobyl NPP (km)	Absorbed dose to measurement date (Gy) (exposure dose rate at measurement date (μGy h^{-1}))		
			1 October, 1986	1 October, 1987	1 May, 1988
9	260°	2.0	100 (10,000)	126 (2,500)	130 (2,200)
1	260°	5.0	10 (1,000)	12 (250)	12.5 (240)
2	245°	4.0	4 (400)	5.1 (120)	5.5 (120)
4	255°	6.0	2 (200)	2.6 (60)	2.8 (50)
3	165°	3.5	1 (100)	1.2 (30)	1.3 (20)
6	205°	16.0	0.012 (1.2)	0.014 (0.4)	0.015 (0.3)
5	130°	30.0	0.01 (1.0)	0.012 (0.4)	0.013 (0.3)

Note: Absorbed external irradiation doses and exposure dose rates are given with respect to gamma radiation at 1 m above soil surface.

dose from radionuclides in soil being about 100 Gy. Comparison of results for the different study plots demonstrates the heterogeneity in deposition and resultant doses to biota in the region comparatively close to the Chernobyl NPP.

Parts of trees were also exposed to hot particles (see Chapter 2) deposited on surfaces. Microscopic analysis of necrotised shoots revealed tiny particles sticking to wax and resinous matter, determined to be irregular shapes of 2–10 to 30–40 μm by scanning electron microscopy. The local absorbed doses as a consequence of the presence of these hot particles may have been orders of magnitude higher than external irradiation doses. When screening the needles from the pines killed in 1986, hollows could be observed with hot particles lying in the bottom. The hollows had been 'burnt out' by the hot particles of high activity (Kozubov *et al.*, 1990).

Activity contamination in pine needles from some of the same study plots and subsequent estimated internal irradiation doses are available for October 1987 (Table 8.3). Whilst not an estimate of the total internal dose to trees (only needles being considered) a comparison of Tables 8.2 and 8.3 shows that internal dose was a significant component of the total dose rate in autumn 1987.

Whilst the acute exposure of coniferous trees to absorbed doses in the range 80–100 Gy led to the death of above-ground parts of pines, in the autumn of 1986 and spring of 1987 on some seemingly dead trees green shoots emerged which then withered. This suggests that some growing points of pines remained viable even at substantial acute and chronic irradiation. This may be the consequence of deposited radionuclides being located at the top of the soil profile resulting in low exposure to roots allowing nutrients to reach the growing points. In the sub-lethal zone in spring of 1987 the surviving growing points gave rise to robust young shoots with light-green bark and widely-spaced elongated needles.

Table 8.3. Radionuclide activity concentration in pine needles and estimated internal dose rate in October 1987.
From Kozubov et al. (1990) and Kryshev (1992).

Plot number	Activity concentration (kBq kg^{-1} f.w.)						Internal dose rate (µGy h^{-1})
	^{144}Ce	^{106}Ru	^{95}Zr	^{95}Nb	^{134}Cs	^{137}Cs	
9*	13,400	4,100	800	1,500	1,500	4,100	14,200
1	190	100	10	20	20	70	233
2	150	60	8	15	17	72	170
3	2	20	3	5	5	21	21
6	1.5	0.6	0.1	0.17	0.18	0.55	2

*For plot 9, estimates are for dry dead needles. f.w. = fresh weight.

Comparative radiosensitivity of different tree species and parts

Spruce were observed to be more radiosensitive than pine demonstrating disturbances in needle morphogenesis, bud germination and inhibited growth of shoots at absorbed doses as low as 0.7–1.0 Gy. The radiation-induced injuries of spruce in 1986 consisted of inhibited growth of shoots, shortened and elongated needles, death of most growing points and multiple buds. At absorbed doses higher than 3.5–4.0 Gy, needle death was common. In 1987, most spruce trees developed large shoot tips with the needles of 35–40 mm length (cf., a typical length of 9–10 mm) which were twisted, curved or straight.

Deciduous trees in the 30-km zone (predominantly birch (*Betula pendula* Roth.), aspen (*Populus tumula* L.), alder (*Alnus glutinosa* Gaerth.) and oak (*Quercus robur* L.)) were more resistant to radioactive contamination than coniferous trees (Sokolov et al., 1994). Radiation damage to the tree crowns was found only in the immediate vicinity of the reactor at levels of radioactive contamination one order of magnitude higher than in areas of coniferous forest showing similar damage. Most foliage turned yellow and had fallen off by mid-August 1986 in the sub-lethal zone, and in areas of absorbed doses of around 500 Gy young growing shoots of birch trees were necrotised. By the autumn of 1986, necrosis of some large branches of birch trees was observed. In the spring of 1987, surviving birches bloomed abnormally abundantly, and some reproductive parts had an anomalous structure. However, in 1988 birch foliage appeared normal. Figure 8.1 shows the comparative effects of irradiation on deciduous and coniferous trees in an area of forest close to the Chernobyl NPP.

The reproductive parts of pines appeared to be the most sensitive to ionising radiation. Measurements in 1986 showed necrotised microstrobiles (male cones) collected from pines in the sub-lethal zone to have 20 times higher activity concentrations than mature cones; internal doses received by microstrobiles being estimated to be as high as 26 mGy d^{-1} in October 1986 (Kozubov et al., 1987). In 1987, the frequency of meiotic chromosomal aberrations in microsporocytes (male reproductive cells) increased three-fold in comparison with control sites at external irradiation

Table 8.4. Frequency of meiotic chromosomal aberrations in pine microsporocytes. Kozubov et al. (1990).

Estimated absorbed dose (Gy)	Year of study	Number of cells analysed	Number of chromosomal aberrations (%)
0.7–1.1	1987	4,200	22.0
	1988	1,800	14.4
1.7–2.3	1987	6,300	30.2
	1988	2,200	9.6
Control	1987	3,000	5.7
	1988	1,000	5.8

doses in the range 0.7–1.1 Gy and five-fold at doses of 1.7–2.3 Gy (Table 8.4) (Kozubov et al., 1990). In 1988, the occurrence of chromosomal abnormalities in pine decreased. Pine trees, which received absorbed doses of 3.5–4.7 Gy showed a reduction in pollen viability from 76% (control) to 48% in 1987. At absorbed doses of 1.8–2.6 Gy, about 50% of seeds were necrotised, and at 3.8–5.1 Gy about 75% (Kozubov et al., 1990). Whilst second-year cones retained viability in the sub-lethal zone (some producing germinating seeds) many first-year cones were unviable in all zones other than that of minor damage.

Recovery of pine forest

Experimental plots were established at forest sites across a range of estimated dose rates by Arkhipov et al. (1994) and observations made over the period 1986–1991 (Table 8.5). Apparently normal growth was established in plots which had received a dose of below 10 Gy by 1991. The authors reported a rapid growth in the population of insect pest in plots with injured trees. In areas of high pine mortality there was a natural succession by deciduous species including birch and ash.

8.1.2 Radiation effects in herbaceous vegetation

The seed germination rate and chlorophyll mutation frequency of *Dactylis glomerata* (cocksfoot or orchard grass) within the 30-km zone were studied by Shershunova and Zainullin (1995). In 1986 in a plot where the dose rate was $2-2.5\,\text{mGy}\,\text{h}^{-1}$ no seeds germinated. In 1989 a sharp increase in chlorophyll mutation frequency was observed in plots with dose rates in excess of $0.4\,\text{mGy}\,\text{h}^{-1}$. However, studies suggested that the radiation had not caused significant changes in key parameters of seed viability such as mass, germination rate, germination energy and growth strength of seeds in the majority of the 30 km zone. Exceptions were relatively small plots with comparatively high radioactive contamination where enhanced genetic variations in herbaceous plants were also found.

Taskaev et al. (1988) grew the radiosensitive plant spiderwort (*Tradescantia* spp.) on study plots with different dose rates. Mutation alterations in spiderwort were recorded in stamen filament hair, with cells changing colour from blue to pink.

Table 8.5. Temporal dynamics of conditions in study pine stands.
From Arkhipov et al. (1994).

Stand injury (dose, Gy)	Year					
	1986	1987	1988	1989	1990	1991
No injury (<0.1)	Normal	Normal	Normal	Normal	Normal	Normal
Low (0.1–1)	Growth depressed	Occasional morphosis	Normal	Normal	Normal	Normal
Medium (1–10)	Strong growth depression Morphosis Occasional tree death	Partial restoration Morphosis No flowering	Rehabilitation of timber output Morphosis	Normal	Normal	Normal
High (10–60)	Interrupted timber output Browning of needles Death of individual stands	Rehabilitation of individual stands	Rehabilitation of timber output Numerous foliar morphoses	Growth of foliage Undergrowth of grass	Growth of foliage	Undergrowth of grass
Acute (>60)	Total forest destruction	Needles fall-off Bark splintering	Bark fall-off Creation of foliage and undergrowth of grass	Collapsing of stems Creation of new plant community	Collapsing of stems Creation of new plant community	Creation of new plant community

A linear relationship ($p < 0.001$; $R^2 = 0.97$) was established between frequency of mutations and increase of external irradiation up to a dose rate of $2\,\text{mGy}\,\text{h}^{-1}$ (Table 8.6).

The mutation burden of natural populations of the common annual plant *Arabidopsis thaliana* within the 30-km zone did not exceed control levels (8%) in the first two years after the accident at sites with an initial dose rate of up to $0.1\,\text{mGy}\,\text{h}^{-1}$. The mutation burden increased to 13% in the subsequent generation indicating a tendency of accumulating the genetic burden (Abramov et al., 1995). In *Arabidopsis* populations at sites of high initial dose rates (2–$2.4\,\text{mGy}\,\text{h}^{-1}$) the level of mutations was as high as 40–80% in the first years after the accident but later decreased. At sites of sufficiently high exposure, however, the mutation burden level remained elevated (up to 30–50%).

Genetic effects observed in herbaceous species growing on the most contaminated areas of the Chernobyl zone include some previously not routinely observed in acute experimental gamma-radiation studies. In particular, multiple chromosome aberrations were detected in dandelion plants (*Taraxacum officinale*).

Table 8.6. Frequency of mutation in the stamen filament hair of spiderwort at different dose rates.

Year of observation	External gamma dose rate ($\mu Gy\,h^{-1}$)	Frequency of mutations (%)
1986*	3.5	0.22 ± 0.02
	50	0.56 ± 0.05
	150	0.92 ± 0.04
1987[†]	200	1.90 ± 0.08
	1,000	8.40 ± 0.26
	2,000	14.95 ± 0.32
	2,500	15.13 ± 0.55
Control	0.1	0.22 ± 0.03

* 180,000 samples; [†] 53,000 samples.

8.1.3 Radiation effects in soil faunal communities and other insects

Invertebrates are amongst the least radiosensitive of organisms (see Box 8.1). However, as the bulk of radionuclides from the Chernobyl fallout were concentrated in the upper soil layer, soil fauna were subjected to a more intensive radiation exposure as compared with other animals. Because of the shielding effect of soil, soil-dwelling fauna were less affected than those living in (less dense) pine litter (Meyers-Schöne and Talmage, 2003).

A 30-fold reduction in the number of soil-dwelling mites and sexually immature invertebrates was observed in pine forest soils 3 km from the Chernobyl NPP compared with forests 30 and 70 km to the south (Krivolutsky, 1994; Krivolutsky et al., 1990, 1999; Viktorov, 1999) in July 1986. The larvae and nymphs of many species were absent from pine forest soils within 3 km of the Chernobyl NNP; changes in vegetative communities may have impacted on invertebrate populations. In agriculturally cultivated soils (in July 1986) the abundance of young earthworms appeared to be lower at a site 3–7 km distant from the Chernobyl NNP compared with a site 80 km distant (Krivolutsky et al., 1990, 1999). However, the effects of radiation on soil invertebrates in agricultural soils appeared to be much lower than in forest soils. Whilst soil invertebrate populations have subsequently increased, the biodiversity at highly contaminated sites was reported to be significantly reduced in 1995 (Krivolutsky, 1996). Beresford et al. (2005) make reference to a study demonstrating reduced soil faunal biological activity in the early 2000s.

No significant changes have been recorded in the structure and dynamics of the above-ground insects. An exception were aphid populations, which showed a reduction in the number of species; on birch trees where 12–14 aphid species may have been expected only two species were observed. Increases were observed in the morphological variability of Colorado (*Leptinotarsa decemlineata*) and other leaf beetles, and the level of wing asymmetry for a number of dragonfly species (Sokolov et al., 1994). As a consequence of floristic succession on abandoned agricultural lands the abundance of grasshoppers and crickets increased significantly.

8.1.4 Radiation effects in mammal populations

Small 'mouse-type' rodents (i.e., *Microtus* spp., *Apodemis* spp., *Clethrionomus* spp.) are of particular importance for ecological monitoring as they belong to a particularly radiosensitive group (i.e., mammals) and have comparatively high population densities and breeding rates.

Studies of mouse populations within the Chernobyl 30-km zone showed a significant reduction in the number of animals in 1986 in the most contaminated areas. Soon after the accident, estimated monthly absorbed doses reached 22 Gy for gamma irradiation and 860 Gy of beta irradiation at study sites a few kilometres to the west of the reactor (Testov and Taskaev, 1990). High doses appeared to have caused an increase in death rate of animals by the autumn of 1986; subsequently rodent numbers increased probably due to migration from adjacent uncontaminated territories. The evacuation of the human population and subsequent plentiful food supplies was also reported to have contributed to population increases in 1987 (Testov and Taskaev, 1990).

Considerable embryonic mortality was observed with only 67% of the potential numbers of animals being born at highly contaminated study sites, compared to 94% for controls in 1986. In the spring of 1987, the number of live embryos was lower at study sites with the highest radioactive contamination levels. In the mid-1990s estimated doses to small rodents in the Red Forest and other highly contaminated areas of the zone were still in excess of those known to impair reproduction (Chesser *et al.*, 2000).

Small rodents from the experimental plots close to the Chernobyl NPP demonstrated abnormalities in blood and liver resembling radiation induced symptoms. Some of the animals had liver cirrhosis (Matery, 1990) and biochemical changes such as increased peroxide oxidation of tissue lipids were observed (Shishkina *et al.*, 1990). Enlarged spleens were observed by Baker *et al.* (1996) in a few individuals amongst their sampling of a range of small rodent species from within 1 km of the Chernobyl NPP in the mid-1990s.

Liver samples from mice (*Apodemus agrarius* and *Apodemus sylvaticus*) and voles (*Clethrionomys glareolus*) collected from close to the Chernobyl NPP in 1996 were analysed for oxidative stress enzyme activity (Holloman *et al.*, 2000); decreases in levels of enzymes such as superoxide dismutase and catalase having been previously observed in animals exposed to radiation during laboratory studies. Of the three species collected from the Chernobyl zone, only *A. agrarius* demonstrated significantly reduced oxidative stress enzyme activity compared with animals collected from less contaminated areas. The authors make reference to work of Il'enko and Krapivko (Il'enko and Krapivko, 1994; Krapivko and Il'enko, 1989) which demonstrates radioresistance having developed in *C. glareolus* and *A. sylvaticus* within the Chernobyl zone. They therefore speculate that these two species may have developed a higher radioresistance than *A. agrarius* leading to their observations of decreased oxidative stress enzyme activity in the latter species only.

Increased genetic diversity was observed in *C. glareolus* populations collected from the Red Forest and Glubokoye Lake (around 4 km north of the Red Forest) compared with a control population to the south of the Chernobyl exclusion zone (Matson et al., 2000). Genetic diversity was greater in the Red Forest population than that from Glubokoye Lake where radiation doses were also estimated to be lower. The authors acknowledged that whilst this increase in genetic diversity may be as a consequence of exposure to radiation it may also reflect the history of populations which had migrated into the two sites following the accident.

Few (if any) studies have considered radiation effects in large wild mammals inhabiting the exclusion zone. However, radiation effects in domesticated animals left in the exclusion zone in 1986 have been reported (Ilyin and Gubanov, 2001; Suvorova, 1993; Spirin et al., 1990; Sazykina et al., 2003). Most domestic animals were evacuated from the zone shortly after the accident. But some cattle and horses left on an island 6 km from Chernobyl received thyroid doses from ^{131}I in contaminated fodder of 150–200 Sv (IAEA, 1991). The IAEA (1991) report summarises the effects on these animals: 'All the horses died and the first cattle died after 5 months. Necropsy revealed the absence of thyroid tissue. Surviving cattle were all hypothyroid and of stunted growth. The second cattle generation seems to be normal.'

In the years after the accident, a UN Food and Agriculture Organisation team investigated evidence of birth anomalies in domestic and wild animals in the affected areas. The team concluded that 'none of the abnormalities reported were other than could be seen anywhere else in the world ... a large scale survey found no [excess] effects that could be attributed to the accident' (IAEA, 1991).

Considerable ^{137}Cs and ^{90}Sr data (predominantly from the 1990s) are available for larger mammals (see Gaschak et al., 2003) Highest activity concentrations were found in wild boar and roe deer; ^{137}Cs activity concentrations in muscle ranging up to 270 kBq kg^{-1} (f.w.) and ^{90}Sr activity concentrations in bone being as high as 730 kBq kg^{-1} (f.w.).

8.1.5 Radiation effects in bird populations

There are relatively few studies of radiation effects in bird species inhabiting the Chernobyl zone compared to other organisms. Pathological effects were observed in the liver of a number of bird species including crows and pigeons (Suvorova et al., 1993; Ilyin and Gubanov, 2001). Lebedeva et al. (1996) reported greater variation in the eggs of great tits (*Parus major*) between 1989 and 1992 at a field site 3 km from the Chernobyl NNP; the mean ^{90}Sr activity concentration in the shell of eggs was 65 kBq kg^{-1} (d.w.).

An increased frequency of partial albinism (up to 15% in 1991) among adult and nesting barn swallows (*Hirado rustica*) close to the Chernobyl NPP was noted compared with those from uncontaminated areas (<2%) (Ellegren et al., 1997). Partial albinism had previously been suggested to be detrimental presumably as a consequence of greater predation and reduced mating success. Significant declines in the population of barn swallows in the Chernobyl exclusion zone were observed between 1986 and 1996. As with many observations made in the Chernobyl zone, it is difficult to say whether these observations, which could be an effect of radioactivity, actually are or whether they are an effect of some other stressor (e.g., changes in community composition and habitat). Depressed immunoglobulin levels were also observed for this species (Camplani et al., 1999).

The database described by Gaschak et al. (2003) contains values of the ^{137}Cs and ^{90}Sr activity concentrations in a number of bird species including raptors measured in the mid to late 1990s; ^{137}Cs activity concentrations in muscle of 220 Bq kg^{-1} (f.w.) (juvenile white-tailed eagle; *Haliaeetus albicilla*), 1,200 Bq kg^{-1} (f.w.) (adult goshawk; *Accipiter gentiles*) and 8,500 Bq kg^{-1} (f.w.) (juvenile harrier; *Circus* spp.). Strontium-90 activity concentrations in the bone of juvenile white-tailed eagle and harriers were 320 Bq kg^{-1} and 144,000 Bq kg^{-1} (f.w.) respectively.

8.2 FRESHWATER BIOTA

8.2.1 Exposure of aquatic biota

Aquatic organisms can be exposed to external irradiation predominantly from radionuclides in water (all organisms) and/or sediments (bottom-dwelling fish and benthic organisms). In the Chernobyl Cooling Pond, the highest estimated external irradiation doses from water were about 2–3 mGy d^{-1} in the first days after the accident; short-lived radionuclides (including ^{131}I, ^{132}I, ^{132}Te, ^{140}Ba and ^{140}La) largely predominating. By June 1986, the estimated external irradiation from water had decreased by two orders of magnitude due to settling of radionuclides to the bottom and radioactive decay. By 1988, this had further decreased by more than an order of magnitude although still five-times higher than the background (Table 8.7) (Kryshev, 1992; Sokolov et al., 1994; Kryshev and Sazykina, 1995).

The maximum levels of external irradiation from bottom sediments also occurred in the first days after the accident and were estimated to be within the range 100–200 mGy d^{-1}. By June 1986, the dose rates from bottom sediments had decreased to 20–50 mGy d^{-1} as a result of radioactive decay of short-lived radionuclides. By autumn of 1986, the radioactive contamination of the bottom was principally determined by long-lived radionuclides and further reductions in the external dose from this source were consequently slow.

Other organisms living in the littoral zone may also be exposed to external radiation close to the surfaces of aquatic plants. In the first days after the accident, external irradiation doses near the surface of aquatic plants were estimated to have been above 10 mGy d^{-1}.

The estimated dose rates due to internal irradiation for plankton in the Chernobyl Cooling Pond were 5–10 mGy d^{-1} in the days following the accident, reducing to 0.3 mGy d^{-1} by September 1986. For molluscs, estimated internal dose rates were about 5 mGy d^{-1} in December 1986, reducing to 0.1–0.2 mGy d^{-1} in July–August of 1987 (Kryshev, 1992; Sokolov et al., 1994). The maximum internal dose rates for aquatic plants were estimated to be approximately 100 mGy d^{-1} soon after the accident, decreasing to 5–10 mGy d^{-1} in late summer 1986 and 1–2 mGy d^{-1} in autumn 1987.

Table 8.7. Effects of chronic radiation exposure on reproduction and offspring of silver carp (*Hypophthalmichthys molitrix*) in the Chernobyl Cooling Pond.
From Belova *et al.* (1993) and Makeeva *et al.* (1994).

General conditions of study	Exposure regime	Description of effects
Adult fish from a fish farm that survived the accident. Some studies were made on free-swimming fish from the Cooling Pond.	The highest dose rates after the Chernobyl accident (1986) were about 8–9 mGy d^{-1}; in 1989–1992 dose rates were 0.4 mGy d^{-1}. Accumulated dose to adult fish: 4 Gy (by 1989); ~4.5 Gy (by 1992). In 1989–1992 the main contributor to dose rate was ^{137}Cs.	**1989:** There were 5.7% sterile specimens, and 8.6% of specimens with gonad asymmetry. 25% of males had anomalies of sexual cells. In the control population less than 0.25% of specimens were sterile. **1990:** There were 12.5% sterile specimens and 16.7% of specimens with gonad asymmetry; 47.1% of fish had anomalies of sexual cells. **1991:** There were 23.1% of specimens with gonad asymmetry. No sterile specimens were detected. Cytological analysis: 68.8% of fish had anomalies of sexual cells. **1992:** 42.9% of males had a deformed shape of gonads; 100% of males and 33.3% of females had some anomalies of sexual cells. Freely living fish from the Cooling Pond: 15.4% of males were partially sterile, and 9.1% of females had gonad asymmetry; 89.5% of fish had anomalies of sexual cells.
Offspring of the above, housed in fish farm enclosures in the Cooling Pond.	See above for doses to parents.	**Offspring born 1989.** In 1992, 28.7% of young fish had anomalies, including 2.8% sterile bisexual specimens, 11.1% anomalies in gonad shape, 8.3% anomalies of body shape, 3.7% anomalies of the swim bladder, and 2.8% other anomalies. **Offspring born 1990.** In 1992, 12.1% of fish had anomalies, including 3.2% with anomalies of gonad shape and 8.9% anomalies of body shape. No sterile specimens were observed in this generation.

Note: for details of dose reconstruction see Kryshev *et al.* (1996) and Kryshev (1998).

Exposure of fish in the Chernobyl Cooling Pond

Radiation doses to fish were determined by their ecology and feeding habits. For non-predatory fish species, such as members of the carp family (*Cyprinus* spp., *Abramis* spp. and *Alburnus* spp.), the highest estimated internal irradiation doses were 2–3 mGy d^{-1} in 1986 (Kryshev, 1992; Sokolov *et al.*, 1994; Kryshev and Sazykina, 1995). By 1988 this had reduced to 0.2–0.4 mGy d^{-1}. Temporal changes in estimated dose for predatory species differ to those for non-predatory fish (see Chapter 4). Estimates of the internal dose increased in 1987 and 1988 compared with 1986 for some species (e.g., *Lucioperca lucioperca*) as a consequence of the accumulation of radiocaesium at higher tropic levels (Rowan and Rasmussen, 1994; Smith *et al.*, 2000).

For bottom-dwelling fish, dose rates of external irradiation provided a significant contribution to the total absorbed doses. Consequently, the highest estimated total doses were for benthophagous fish (e.g., *Cyprinus carpio*, *Abramis brama* and *Blicca bjoorkna*). The total absorbed doses accumulated by fish born in 1986 and 1987 were estimated to be greater than 10 Gy by 1990–1995.

8.2.2 Radiation effects in aquatic biota

Radiation-induced damage was studied in the gonads of silver carp (*Hypophthalmichthys molitrix*) surviving in the Chernobyl Cooling Pond after the accident in 1986, and in subsequent generations. Silver carp were commercially farmed in enclosures in the Cooling Pond prior to the accident. At the time of the accident the fish studied were 1–2 years old. Estimated accumulated doses to the fish by 1989 were of the order 4.5 Gy; whole-body dose rates decreased from about 8–9 mGy d^{-1} soon after the accident to about 0.4 mGy d^{-1} in 1989–1992 (Belova *et al.*, 1993; Kryshev *et al.*, 1996; Kryshev, 1998; Kryshev *et al.*, 1999). Of the 70 silver carp examined over the period 1989–1992, 7.1% were sterile, and 35% of females and 48% of males showed gonad abnormalities (Belova *et al.*, 1993; Makeeva *et al.*, 1994; UNSCEAR, 1996). In wild populations of silver carp, sterility is usually less than 0.25%.

Exposed silver carp from the Cooling Pond were used in 1989–1990 to produce offspring which were reared in the Cooling Pond and subsequently examined in 1992 when 3 years old (Makeeva *et al.*, 1994). Approximately 14% of specimens had significant gonad deformities, including 2.8% sterile bisexual specimens with abnormalities of the reproduction system being common. In addition, about 15% of the specimens had other morphological abnormalities (abnormal body shape, etc.). Details of this work are summarised in Table 8.7.

Measurements of chromosomal aberrations in carp (*Cyprinus carpio*) corneal epithelium and developing silver bream (*Blicca bjorkna*) embryos showed no increase over expected ranges. An increased frequency of chromosome aberrations was observed in chironomids (non-biting midge larvae) in the Chernobyl NPP Cooling Pond, and nearby ponds at Yanov (Petrova, 1990). In frogs, the frequency of aberrant cells was higher than in controls by 4–7-fold in 1987, 2–5-fold in 1988 and 2–3-fold in 1989 (Krysanov and Krysanova, 1990). Reductions in the mollusc population of the Chernobyl NPP Cooling Pond were also reported (Sokolov et al., 1990).

A higher rate of DNA strand breaks was observed in catfish (*Ictalurus punctatus*) collected from the Chernobyl NPP Cooling Ponds compared to control fish (Sugg et al., 1996). DNA strand breaks were correlated with the radiocaesium activity concentration in muscle. Variations in cellular DNA content were also observed in a range of fish species collected from within 10 km of the Chernobyl NPP (Dallas et al., 1998).

8.3 THE CHERNOBYL EXCLUSION ZONE – A NATURE RESERVE?

The effects of radiation released by the Chernobyl accident on wildlife in the surrounding area were evident within a few days with the death of pine trees. Subsequently many other effects including reductions in rodent populations close to the reactor and reproductive changes in fish living in the cooling ponds were observed.

However, perhaps the largest impact of the accident on the ecology of the Chernobyl exclusion zone was brought about by the removal of the human population. Consequently, activities such as agricultural production and the associated usages of herbicides, pesticides and fertilisers ceased. As a result, floral and faunal biodiversity and abundance has increased considerably (Baker and Chesser, 2000; Sokolov et al., 1990, 1994). In spring 1988, numbers of wild boar (*Sus scrofa*) were estimated to be 8-fold higher than prior to the accident. Other species including wolf (*Canis lupus*), moose (*Alces alces*), roe deer (*Capreol capreolus*), black stork (*Ciconia nigra*), raptors (including eagle species) and game birds are also abundant. Baker and Chesser (2000) report that when trapping small mammals they achieved a higher success rate in the exclusion zone compared to uncontaminated areas. The floral diversity of the exclusion zone is similar to that observed in protected habitats outside of the zone (Baker and Chesser, 2000). Abundance of wildlife is not restricted to the countryside: the town of Pripyat is being reclaimed by nature (see Figures 8.2 and 8.3).

Box 8.2. Recommended dose limits for biota.

International organisations have recommended guideline dose limits for biota below which significant effects are unlikely to be observed. The NCRP (1991) concluded that a dose limit of $10\,\text{mGy}\,\text{d}^{-1}$ would ensure protection at the level of population. The IAEA (1992) proposed dose rates below which no effects are likely to be observed at the level of population of $10\,\text{mGy}\,\text{d}^{-1}$ for aquatic animals, $1\,\text{mGy}\,\text{d}^{-1}$ for terrestrial animals and $10\,\text{mGy}\,\text{d}^{-1}$ for terrestrial plants.

The net positive effect of removing humans from the exclusion zone therefore appears to exceed the negative impacts of radiation. However, dose rates in small

Sec. 8.3] The Chernobyl exclusion zone – a nature reserve? 281

Figure 8.2. Pripyat sports stadium (September 2002).

Figure 8.3. Kestrels nesting on the roof of a tower block, Pripyat.
Photo: Sergei Gaschak.

areas of the exclusion zone remain in excess of international guidelines (see Box 8.2) and there continue to be examples of, what could be, the effects of chronic radiation exposure as determined by modern molecular biological techniques (i.e., damage to genetic material in natural biota).

It should be noted, however, that there are a number of possible confounding factors in considering observations of the longer term effects of radiation in the Chernobyl exclusion zone on biota. It is often difficult to separate the consequences of residual acute exposure in the immediate aftermath of the accident from what may be responses to chronic lower level exposure. In addition, some of the studies discussed above, investigating potential responses to radiation, have reported responses in some of the parameters investigated but not others. Where a number of parameters are studied for effects, it increases the likelihood that a correlation with radiation levels will be potentially observed by chance in one or two of those parameters. Finally, possible radiation effects are often difficult to separate from effects potentially brought about by the significant changes in habitats and population compositions which have occurred in the exclusion zone since the accident. If we are to fully understand the ecological impact of the Chernobyl accident, long-term investigations of genetic diversity and damage, life-expectancy, fertility and fitness, and adaptation of natural biota to radioactivity need to be conducted within the exclusion zone.

8.4 REFERENCES

Abaturov, Yu.D., Abaturov, A.V., Bykov, A.V. et al. (1996) *Effects of Ionizing Radiation on Forests in the Near Zone of the Chernobyl NPP*. Nauka, Moscow (in Russian).

Abramov, V.I., Dineva, S.B., Rubanovich, A.V. and Shevchenko, V.A. (1995) Genetic consequences of radioactive contamination of the Arabidopsis thaliana populations growing in 30 km of Chernobyl NPP Radiation Biology. *Radioecology*, **35**, 676–689 (in Russian).

Arkhipov, N.P., Kuchma, N.D., Askbrant, S., Pasternak, P.S. and Musica, V.V. (1994) Acute and long-term effects of irradiation on pine (Pinus-silvestris) stands post-Chernobyl. *Science of the Total Environment*, **157**, 383–386.

Baker, R.J., Hamilton, M.J., van den Bussche, R.A., Wiggins, L.E., Sugg, D.W., Smith, M.H., Lomakin, M.D., Gaschak, S.P., Bundova, E.G., Rudenskaya, G.A., et al. (1996) Small mammals from the most radioactive sites near the Chornobyl nuclear power plant. *Journal of Mammalogy*, **77**, 155–170.

Baker, R.J. and Chesser, R.K. (2000) The Chornobyl nuclear disaster and subsequent creation of a wildlife preserve. *Environmental Toxicology and Chemistry*, **19**, 1231–1232.

Belova, N.V., Verigin, B.V., Yemelianova, N.G., Makeeva, A.P. and Ryabov, I.N. (1993) Radiobiological analysis of silver carp *Hypophthalmichthys molitrix* in the cooling pond of the Chernobyl NPP in the post-accidental period. I: The condition of the reproductive system of fish which survived the accident. *Voprosy ihtiologii (Problems of Ichthyology)*, **33**, 814–828 (in Russian).

Beresford, N.A., Wright, S.M., Barnett, C.L., Hingston, J.L., Vives, I., Batlle, J., Copplestone, D., Kryshev, I.I., Sazykina, T.G., Proh, I.G., Arkhipov, A., et al. (2005) A case study in the Chernobyl zone. Part 2: Predicting radiation induced effects in biota. *Radioprotection – colloques*, **40**, S299–S305.

Berger, M.J. (1971) Distribution of absorbed dose around point sources of electrons and β-particles in water and other media. *Journal of Nuclear Medicine*, **12**, 5–23.

Blaylock, B.G., Frank, M.L. and O'Neal, B.R. (1993) *Methodology for Estimating Radiation Dose Rates to Freshwater Biota Exposed to Radionuclides in the Environment*. ES/ER/TM-78. Oak Ridge National Laboratory, Oak Ridge, Tennessee.

Brownell, G.L., Ellett, W.H., Reddy, A.R. (1968) Absorbed fractions for photon dosimetry, MIRD Pamphlet N.3. *Journal of Nuclear Medicine*, **9**, 27–39.

Camplani, A., Saino, N. and Møller, A.P. (1999) Carotenoids, sexual signals and immune function in barn swallows from Chernobyl. *Proceedings of the Royal Society*, **266**, 1111–1116.

Chesser, R.K., Sugg, D.W., Lomakin, M.D., van den Bussche, R.A., DeWoody, J.A., Jagoe, C.H., Dallas, C.E., Whicker, F.W., Smith, M.H., Gaschak, S.P., et al. (2000) Concentrations and dose rate estimates of ^{134,137}cesium and ^{90}strontium in small mammals at Chernobyl, Ukraine. *Environmental Toxicology and Chemistry*, **19**, 305–312.

Dallas, C.E., Lingenfelser, S.F., Lingenfelser, J.T., Holloman, K., Jagoe, C.H., Kind, J.A., Chesser, R.K. and Smith, M.H. (1998) Flow cytometric analysis of erythrocyte and leukocyte DNA in fish from Chernobyl-contaminated ponds in the Ukraine. *Exotoxicology*, **7**, 211–219.

Ellegren, H., Lindgram, G., Primmer, C.R. and Møller, A.P. (1997) Fitness loss and germline mutations in barn swallows breeding in Chernobyl. *Nature*, **389**, 593–596.

Gaschak, S., Chizhevsky, I., Arkhipov, A., Beresford, N.A. and Barnett, C.L. (2003) The transfer of Cs-137 and Sr-90 to wild animals within the Chernobyl exclusion zone. In: *International Conference on the Protection of the Environment from the effects of Ionizing Radiation*, pp. 200–202. IAEA-CN-109, Stockholm.

Holloway, K.A., Dallas, C.E., Jagoe, C.H., Tackett, R., Kind, J.A. and Rollor, E.A. (2000) Interspecies differences in oxidative stress response and radiocaesium concentrations in rodents inhabiting areas highly contaminated by the Chernobyl nuclear disaster. *Environmental Toxicology & Chemistry*, **19**, 2830–2834.

IAEA (1991) *The International Chernobyl Project Technical Report*, 640 pp. IAEA, Vienna.

IAEA (1992) *Effects of Ionising Radiation on Plants and Animals at Levels Implied by Current Radiation Protection Standard*. IAEA TRS 332. IAEA, Vienna.

Il'enko, A.I. and Krapivko, T.P. (1994) Radioresistance of populations of bank voles (*Clethrionomys glareolus*) in radionuclide-contaminated areas. *Doklady Biological Sciences*, **336**, 262–266.

Ilyin, L.A. and Gubanov, V.A. (eds) (2001) *The Heavy Radiation Accidents: Consequences and Counter-measures*, 752 pp. Izdat, Moscow (in Russian).

Krapivko, T.P. and Il'enko, A.I. (1989) First features of radioadaptation in a population of red-backed voles (*Clethrionomys glareolus*) in a radiation biogeocenosis. *Doklady Akademy Nauk USSR*, **302**, 656–659 (in Russian).

Kozubov, G.M., Taskaev, A.I., Ladanova, N.V., Kusivanova, S.V. and Artemov, V.A. (1987) *Radioecological Studies of Coniferous Trees in the Area Exposed to the Chernobyl Contamination*. Syktyvkar, USSR Academy of Sciences, Moscow (in Russian).

Kozubov, G.M., Bannikova, V.P., Taskaev, A.I., Artemov, V.A., Ostapenko, E.K. and Sytnik, K.M. (1988) *Investigation of Reproductive Capacity of Pines in the Area*

Exposed to the Chernobyl Contamination. A series of preprints 'Scientific reports' – Institute of Botany, Academy of Sciences of the Ukraine, Kiev (in Russian).

Kozubov, G.M., Taskaev, A.I., Ignatenko, E.I., Artemov, V.A., Ostapenko, E.K., Ladanova, N.V., Kusivanova, S.V., Kozlov, V.A. and Larin, V.B. (1990) *Radiation Exposure of Coniferous Forest in the Area Exposed to the Chernobyl Contamination.* Syktyvkar, Academy of Sciences of Russia, Moscow (in Russian).

Kozubov, G.M. and Taskaev, A.I. (1994) *Radiobiological and Radioecological Investigations of Wood Plants.* Nauka, St. Petersburg (in Russian).

Krivolutsky, D.A. and Pokarzhevsky, A.D. (1990) Changes in populations of soil fauna caused by accident on the Chernobyl NPP. In: Ryabov, I.N. and Ryabtsev, I.A. (eds), *Biological and Radioecological Aspects of the Consequences of the Chernobyl Accident.* Abstracts of the Ist International Conference ('Zeleny Mys' 10–18 September 1990). USSR Academy of Sciences, Moscow (in Russian).

Krivolutsky, D.A. (1994) *Soil Fauna in Ecological Control,* 269pp. Nauka, St. Petersburg (in Russian).

Krivolutsky, D.A., Pokarzhevsky, A.D., Usacheov, V.L., Shein, G.N., Nadvorny, V.G. and Viktorov, A.G. (1990) Effect of radioactive contamination of environment on soil fauna in the area of Chernobyl NPP. *Ecology,* **6**, 32–42.

Krivolutsky, D.A. (1996) Dynamics of biodiversity in ecosystems under the conditions of radioactive contamination. *Transactions of the Russian Academy of Sciences,* **347**, 567–570 (in Russian).

Krivolutsky, D.A., Martushov, V.Z. and Ryabtsev, I.A. (1999) Effect of radioactive contamination on animal kingdom in the area of Chernobyl NPP in first period after accident (1986-1988). In: Ryabov, I.N. and Ryabtsev, I.A. (eds), *Bioindication of Radioactive Pollutants,* pp. 106–122. Nauka, St. Petersburg.

Krysanov, E.Yu. and Krysanova, I.A. (1990) Cytogenetic monitoring in the zone of accidental contamination of the Chernobyl NPP. In: Ryabov, I.N. and Ryabtsev, I.A. (eds), *Biological and Radioecological Aspects of the Consequences of the Chernobyl Accident.* Abstracts of the 1st International conference ('Zeleny Mys', 10–18 September, 1990). USSR Academy of Sciences, Moscow (in Russian).

Kryshev, A.I. (1998) Modelling the accidental radioactive contamination and assessment of doses to biota in the Chernobyl NPP's cooling pond. In: *Proceedings of the topical meeting of International Union of Radioecologists, Mol, 1–5 June, 1998, pp. 32–38.* Balen, Belgium.

Kryshev, I.I. (ed.) (1992) *Radioecological Consequences of the Chernobyl Accident,* 142 pp. Nuclear Society, Moscow.

Kryshev, I.I. and Sazykina, T.G. (1995) Assessment of Radiation Doses to Aquatic Organisms in the Chernobyl Contaminated Area. *Journal of Environmental Radioactivity,* **28**, 91–103.

Kryshev, I.I., Sazykina, T.G., Ryabov, I.N., Chumak, V.K. and Zarubin, O.L. (1996) Model testing using Chernobyl data. II: Assessment of the consequences of the radioactive contamination of the Chernobyl Nuclear Power Plant Cooling Pond. *Health Physics,* **70**, 13–17.

Kryshev, I.I., Sazykina, T.G., Hoffman, F.O., Thiessen, K.M., Blaylock, B.G., Feng, Y., Galeriu, D., Heling, R., Kryshev, A.I., Kononovich, A.L. *et al.* (1999) Assessment of the consequences of the radioactive contamination of aquatic media and biota for the Chernobyl NPP cooling pond: Model testing using Chernobyl data. *Journal of Environmental Radioactivity,* **42**, 143–156.

Lebedeva, N.V., Ryabtsev, I.A. and Beloglazov, M.V. (1996) Population radioecology of birds. *Uspekhi Sovremennoi Biologii (Achivements of Modern Biology)*, **4**, 432–446 (In Russian).

Makeeva, A.P., Yemelianova, N.G., Belova, N.V. and Ryabov, I.N. (1994) Radiobiological analysis of silver carp *Hypophthalmichthys molitrix* in the cooling pond of the Chernobyl NPP in the post-accidental period. II: Development of the reproductive system of the fish in the 1st generation. *Voprosi ihtiologii (Problems of Ichthyology)*, **34**, 681–696 (in Russian).

Matery, L.D. (1990) Morphological disturbances in the systems of blood and liver of mouse-type rodents in the zone of the Chernobyl NPP. In: Ryabov, I.N. and Ryabtsev, I.A. (eds), *Biological and Radioecological Aspects of the Consequences of the Chernobyl Accident*, p. 217. Abstracts of the Ist International Conference ('Zeleny Mys', 10–18 September, 1990). USSR Academy of Sciences, Moscow (in Russian).

Matson, C.W., Rodgers, B.E., Chesser, R.K. and Baker, R.J. (2000) Genetic diversity of *Clethrionomys glareolus* populations from highly contaminated sites in the Chornobyl region, Ukraine. *Environmental Toxicology and Chemistry*, **19**, 2130–2135.

Meyers-Schöne, L. and Talmage, S.S. (2003) Contaminant sources and effects: Nuclear and themal. In: Hoffman, D.J., Rattner, B.A., Burton, G.A. and Cairns, J. (eds), *Handbook of Ecotoxicology* (2nd edition), pp. 615–643. CRC Press, Boca Raton.

NCRP (1991) *Effects of Ionizing Radiation on Aquatic Organisms*. National Council on Radiation Protection and Measurements, Report No. 109, Bethesda, Maryland.

Pentreath, R.J. and Woodhead, D.S. (2000) A system for environmental protection: reference dose models for fauna and flora. In: *Proceedings of the International Conference 'Harmonization of Radiation, Human Life and the Ecosystem'*, Hiroshima, May 14–19, 2000, Japan.

Petrova, N.A. (1990) Changes in frequency of structural reconstruction of chromosomes in chironomids from the Chernobyl zone in 1987–1988. In: Ryabov, I.N. and Ryabtsev, I.A. (eds), *Biological and Radioecological Aspects of the Consequences of the Chernobyl Accident*, p. 159. Abstracts of the Ist International conference ('Zeleny Mys', 10–18 September, 1990). USSR Academy of Sciences, Moscow (in Russian).

Real, A., Sundell-Bergman, S., Knowles, J.F., Woodhead, D.S. and Zinger, I. (2004) Effects of ionising radiation exposure on plants, fish and mammals: Relevant data for environmental radiation protection. *Journal of Radiological Protection*, **24**, 123–137.

Rowan, D.J. and Rasmussen, J.B. (1994) Bioaccumulation of radiocesium by fish: The influence of physicochemical factors and trophic structure. *Canadian Journal of Fisheries and Aquatic Science*, **51**, 2388–2410.

Sazykina, T.G., Jaworska, A. and Brown, J.E. (eds) (2003) *Dose-effects Relationships for Reference (or Related) Arctic Biota*. Deliverable Report 5 for the EC project EPIC (Contract no. ICA2-CT-200-10032). Norwegian Radiation Protection Authority, Østerås.

Shershunova, V.I. and Zainullin, V.G. (1995). Monitoring of natural populations of *Dactylis glomerata* L. growing within zone of the Chernobyl NPP. Radiation Biology. *Radioecology*, **35**, 690–695 (in Russian).

Shishkina, L.N., Kudryashova, A.G., Zagorskaya, N.G., Matery, L.D. and Taskaev, A.I. (1990) Regulation system of peroxide oxidation of lipids and disturbances in liver of rodents at areas of Chernobyl NPP accident. In: Ryabov, I.N. and Ryabtsev, I.A. (eds), *Biological and Radioecological Aspects of the Consequences of the Chernobyl Accident*, p. 231. Abstracts of the Ist International Conference ('Zeleny Mys', 10–18 September, 1990). USSR Academy of Sciences, Moscow (in Russian).

Smith, J.T., Kudelsky, A.V., Ryabov, I.N., Daire, S.E., Boyer, L., Blust, R.J., Fernandez, J.A., Hadderingh, R.H. and Voitsekhovitch, O.V. (2002) Uptake and elimination of radiocaesium in fish and the 'size effect'. *Journal of Environmental Radioactivity*, **62**, 145–164.

Spirin, D.A., Martyushev, V.Z., Smirnov, V.G. and Tarasov, O.V. (1990). Impact of radiation contamination on fauna in the 30-km zone of the Chernobyl NPP. In: Ryabov, I.N. and Ryabtsev, I.A. (eds), *Biological and Radioecological Aspects of the Consequences of the Chernobyl Accident*, p. 71. Abstracts of the Ist International Conference ('Zeleny Mys', 10–18 September, 1990). USSR Academy of Sciences, Moscow (in Russian).

Sugg, D.W., Bickham, J.W., Brooks, J.A., Lomakin, M.D., Jagoe, C.H., Dallas, C.E., Smith, M.H., Baker, R.J. and Chesser, R.K. (1996) DNA damage and radiocaesium in channel catfish from Chernobyl. *Environmental Toxicology and Chemistry*, **15**, 1057–1063.

Suvorova, L.I., Spirin, D.A., Martyushev, V.Z., Smirnov, E.G., Tarasov, O.V. and Shein, G.P. (1993) Assessment of biological and ecological consequences of radioactive contamination of biogeocenoses. In: Izrael, Yu.A. (ed.), *Radiation Aspects of the Chernobyl Accident* (Volume 2), pp. 321–325. Gidrometeoizdat, St. Petersburg (in Russian).

Taskaev, A.I., Shevchenko, V.A., Popova, O.N., Abramov, V.I., Frolova, N.P., Sergeeva, S.A., Semov, A.B., Shershunova, V.I., Niliva I.N., Shevchenko V.V., *et al.* (1988) Eco-genetic Consequences of the Chernobyl Accident for Flora. A series of preprints 'Scientific reports'. Syktyvkar: Komi Scientific Center of the Academy of Sciences of Russia (in Russian).

Testov, B.V. and Taskaev, A.I. (1990) Dynamics of mouse-type rodent populations in the zone of the Chernobyl NPP. In: Ryabov, I.N. and Ryabtsev, I.A. (eds), *Biological and Radioecological Aspects of the Consequences of the Chernobyl Accident*, p. 86. Abstracts of the 1st International Conference ('Zeleny Mys', 10–18 September, 1990). USSR Academy of Sciences, Moscow (in Russian).

Tikhomirov, F.A. and Sidorov, V.P. (1990) Radiation damage of forest in the zone of the Chernobyl NPP. In: Ryabov, I.N. and Ryabtsev, I.A. (eds), *Biological and Radioecological Aspects of the Consequences of the Chernobyl Accident*, p. 18. Abstracts of the Ist International Conference ('Zeleny Mys', 10–18 September, 1990), p. 18. USSR Academy of Sciences, Moscow (in Russian).

Tikhomirov, F.A. and Shcheglov, A.I. (1994) Main results of investigations on the forest radioecology in the Kyshtym and Chernobyl accident zones. *Science of Total Environment*, **157**, 45–47.

Sokolov, V.E., Ryabov, I.I., Ryabtsev, I.A., Tikhomirov, F.A., Shevchenko, V.A. and Taskaev, A.I. (1990) Ecological and genetic consequences of the Chernobyl accident. *The first international working group on severe accidents and their consequences* (Dagomys, 30 October–3 November 1989), pp. 173–183. Nauka, Moscow.

Sokolov, V.E., Ryabov, I.N., Ryabtsev, I.A., Kulikov, A.O., Tikhomirov, F.A., Shcheglov, A.I., Shevchenko, V.A., Kryshev, I.I., Sidorov, V.P., Taskaev, A.I., *et al.* (1994) Effects of Radioactive Contamination on flora and fauna in the vicinity of the Chernobyl Nuclear Power Plant. In: Turpaev, T.M. (ed.), *Soviet Scientific Reviews Supplement Series Section F Physiology and General Biology Reviews* (8). Harwood Academic Publishers, New York.

Viktorov, A.G. (1999) Radiosensivity and radiopathology of earthworms, their using in bioindication of radioactive contaminations. In: Krivolutsky, D.A. (ed.), *Bioindication of Radioactive Pollutants*. pp. 213–217. Nauka, Moscow (in Russian).

UNSCEAR (1996) *Effects of Radiation on the Environment*, Annex to *Sources and Effects of Ionizing Radiation (1996 Report to the General Assembly, with one Annex)*. United Nations Scientific Committee on the Effects of Atomic Radiation, UN, New York.

U.S. Department of Energy (2000) *A Graded Approach for Evaluating Radiation Doses to Aquatic and Terrestrial Biota*. U.S. Department of Energy Interim Technical Standard, Washington, D.C., U.S.

9

Conclusions

Jim T. Smith and Nick A. Beresford

In this concluding chapter, we summarise the current radiation exposures to the most affected populations living in and around the evacuated areas and predict future exposures arising from environmental contamination after Chernobyl. Health, social and economic effects of the accident are summarised and conclusions are made concerning the impact of the accident on the ecosystem. Finally, we briefly consider the implications of the accident for the future of nuclear energy worldwide.

9.1 CONTAMINATION OF THE ENVIRONMENT

9.1.1 Current radiation exposures in the Chernobyl affected areas

Radiation exposures to people living in and around the most affected areas of the Ukraine, Belarus and Russia remain higher than the natural background exposure in this region. Exposures are dominated by ^{137}Cs, though (to a much lesser extent) ^{90}Sr and other isotopes also form part of the dose. For most of the population living around the evacuated areas, average doses are currently less than 1 mSv per year. In Chapter 3, however, it was shown that populations utilising contaminated forests and consuming milk from private cows and wild foodstuffs (mushrooms, berries, game animals, freshwater fish) tended to have higher radiation exposures.

In 1996, populations in villages near forest areas of the Bryansk region of Russia had average exposures of around 2–3 mSv yr^{-1} and the 'critical group' of forestry workers had average exposures of 5–6 mSv yr^{-1} (Fesenko *et al.*, 2000). The mean contamination of the area of Bryansk studied by Fesenko *et al.* (2000) was approximately 750 kBq m^{-2}. This area and these exposures can be considered to be reasonably representative of the highest doses encountered by other rural populations living around the exclusion zones of the Ukraine, Belarus and Russia and utilising forests. These exposures are lower than those which could arise if people returned to the most highly contaminated areas of the 30-km zone.

There are people living illegally in the 30-km zone and in other abandoned areas, but few (if any) are living within the most contaminated parts of the zone. Because of the absence of countermeasures in the exclusion zone, transfers of radiocaesium and radiostrontium to food products are likely to be higher than for other areas of the fSU (there is some evidence for this in Beresford and Wright, 2005).

During 2001, 412 people inhabited 15 settlements within the Ukrainian sector of the 30-km zone. Of these 15 settlements, seven had fewer than 20 residents. A study carried out in 1997–1999 (A.N. Arkhipov, Chernobyl Scientific and Technical Centre, pers. commun.) measured external dose rates to settlers in this area in the range 0.5–$1.1\,\mathrm{mGy\,yr^{-1}}$ (≈ 0.5–$1.1\,\mathrm{mSv\,yr^{-1}}$). The same study reports ingestion doses in 12 settlements during 2002 which ranged between 0.18–$0.8\,\mathrm{mSv\,yr^{-1}}$ from ^{90}Sr and 0.43–$4.6\,\mathrm{mSv\,yr^{-1}}$ from ^{137}Cs. It should be noted that there are other settlements in more contaminated areas than these, so doses may be higher to some people than the ones indicated here.

Exposures to people living in and around the abandoned areas of the Ukraine, Belarus and Russia are therefore still significantly elevated over average natural background doses in this region. As shown in Chapter 1, however, there is considerable variation worldwide in natural background doses due to enhanced occupational exposures to natural radiation and to differences in geological and environmental conditions both within and between countries. Looked at in the context of variation in natural radiation doses, current dose rates to the most exposed populations in the Chernobyl affected areas represent an equivalent annual exposure to that experienced by long-haul airline crew, some miners or people living in areas of high exposure to natural radiation. There are, for example, now relatively few areas (even within the 30-km zone) where external dose rates exceed 10–$12\,\mathrm{\mu Sv\,hr^{-1}}$, the dose rate from cosmic rays which was typically observed on Concorde, the highest altitude passenger airliner (though passengers and air crew were obviously only exposed at this rate during flights).

9.1.2 Future environmental contamination by Chernobyl

On the basis of information on fallout and environmental transfers of radionuclides, we can make tentative predictions of the very long term (hundreds to thousands of years) environmental consequences of the Chernobyl accident. In general, areas outside the 30-km zone are contaminated primarily by ^{137}Cs. There are, however, some areas within the 30-km zone which are also significantly contaminated by ^{90}Sr, ^{241}Am and very long-lived Pu isotopes. Based on measurements at a forest site, Kopachi, near to the reactor, we have estimated fallout of various radionuclides in the soil (Table 9.1). The fallout at the Kopachi site (approximately $2{,}000\,\mathrm{kBq\,m^{-2}}$ of ^{137}Cs) illustrates the contamination levels in the more contaminated western and northern parts of the 30-km zone (see Figure 1.9). Many parts of the 30-km zone are significantly less contaminated than this, but some areas, such as the Red Forest and an area around Glubokoye Lake, are approximately 10–15 times

Table 9.1. Estimated activity concentrations of long-lived radionuclides at Kopachi, 6 km south-east of Chernobyl, from measurements in Murumatsu et al. (2000). Estimates are also shown of future activity concentrations assuming declines by physical decay only (thus leading to a likely over-estimate) 50, 150, 500 and 1000 years after the accident. For illustrative purposes, estimates are compared with the UK Generalised Derived Limit (GDL) of radionuclides in soil (NRPB, 1998).[†]

Isotope	UK GDL for well-mixed soil ($kBq\,kg^{-1}$)	Activity concentration (area deposition) $kBq\,kg^{-1}$ ($kBq\,m^{-2}$)				
		1986	2036	2136	2486	2986
^{90}Sr	0.4	**98.8**[1] (938)[1]	**29.7** (282)	**2.7** (25.4)	$\mathbf{6 \times 10^{-4}}$ (6×10^{-3})	**0**
^{137}Cs	1	**203**[2] (2040)[3]	**64.4** (648)	**6.5** (65.2)	$\mathbf{2 \times 10^{-3}}$ (0.02)	**0**
^{238}Pu	5	**0.60**[4] (6.2)[4]	**0.40** (4.2)	**0.18** (1.9)	**0.01** (0.12)	$\mathbf{2 \times 10^{-4}}$ (2×10^{-3})
^{239}Pu	5	**0.51**[2] (5.3)[3]	**0.51** (5.3)	**0.51** (5.3)	**0.50** (5.2)	**0.50** (5.1)
^{240}Pu	5	**0.77**[2] (8.0)[3]	**0.77** (8.0)	**0.76** (7.9)	**0.73** (7.6)	**0.69** (7.2)
^{241}Pu	200*	**102**[5] (1063)[5]	**9.2** (96.0)	**0.08** (0.78)	**0**	**0**
^{241}Am	5*	**0.07**[6] (0.76)[6]	**3.0** (31.2)	**2.8** (29.3)	**1.6** (16.8)	**0.7** (7.5)

[†] The GDL is the activity concentration of a particular radionuclide which could give rise to a dose of $1\,mSv\,yr^{-1}$ (using generally conservative assumptions). * Note that in the longer term the vast majority of ^{241}Am arises from ingrowth from ^{241}Pu and that doses from ingrown ^{241}Am are accounted for in the ^{241}Pu GDL. [1] Estimated from ^{137}Cs data assuming ^{90}Sr:^{137}Cs ratio at Kopachi of 0.46, from data in Lux et al. (1995) and Belli and Tikhomirov (1996); [2] L/Of/Oh horizon; [3] L/Of/Oh + Ah/B horizons; [4] Estimated from ^{239}Pu data assuming ^{238}Pu:^{239}Pu ratio of 1.17, from data in UNSCEAR (2000); [5] Estimated from ^{239}Pu data assuming ^{241}Pu:^{239}Pu ratio of 200, from data in UNSCEAR (2000); [6] Estimated from ^{239}Pu data using ^{241}Am:^{239}Pu ratio from initial fallout (Ivanov et al., 1994; UNSCEAR, 2000) plus ingrowth from ^{241}Pu.

more contaminated than at Kopachi. Some small areas within a 2-km² 'hot spot' in the Red Forest area may have even higher radioactivity concentrations (Kashparov et al., 2003).

As discussed in Chapters 2 and 3, there are a number of processes leading to declines in concentrations of radioactivity in surface soils and in foodstuffs. Most important of these are: physical decay of the radionuclide, erosion from soils, migration to deeper soil layers and increased binding ('fixation') of the radionuclide to the soil. For very long-lived radionuclides, rates of transport and migration are difficult to predict over the very long timescales required: soil and radionuclide erosion rates may change significantly if there is significant climate change, for example. It is, however, valuable to assess future contamination using an illustrative 'worst case' scenario in which erosion and 'fixation' processes are relatively unimportant, so the decline in environmental contamination is determined primarily by physical decay. Future activity concentrations at the Kopachi site are estimated under this assumption (Table 9.1). Values in Table 9.1 can be multiplied by a factor of 10–15 to illustrate typical activity concentrations in the most highly contaminated parts of the 30-km zone.

It is seen (Tables 9.1 and 9.2) that in highly contaminated areas, radiation doses will remain significant for many more decades. However, due primarily to falling external exposures from ^{137}Cs, much of the currently abandoned area will gradually

Table 9.2. Tentative prediction of future contamination by Chernobyl.*

Time after accident	Likely scenario assuming little removal of radioactivity from the surface layer of soils (hence declines in contamination due primarily to physical decay).
50 yr	Exposures still dominated by ^{137}Cs and ^{90}Sr. A large part of the 30-km zone would give rise to doses lower than intervention limits but restrictions on consumption of wild foodstuffs would remain both within and outside the 30-km zone. External exposures in currently abandoned areas outside the 30-km zone would be relatively low.
150 yr	Exposures still dominated by ^{137}Cs and ^{90}Sr. Outside 30-km zone unlikely to be any significant contamination, though ^{137}Cs activity concentrations could be up to hundreds – thousands of Bq kg^{-1} in some wild foodstuffs. Inside 30-km zone, some areas could still give significant doses (order several mSv yr^{-1}) to (hypothetical) critical groups.
500 yr	^{90}Sr and ^{137}Cs have decayed away. Exposures are dominated by ^{241}Am and $^{239+240}$Pu. No significant doses found outside the 30-km zone. The majority of the 30-km zone is now 'uncontaminated' and external doses are unlikely to be significant, but some areas (Red Forest, Glubokoye Lake – an area of order of tens of km^2 or a few percent of the current evacuated area) could possibly give rise to low doses (of order 1 mSv yr^{-1}) to critical groups.
1,000 yr	Situation may be relatively unchanged, though concentrations in foodstuffs are likely to have declined significantly by transfers from surface soils to deeper soil layers and by runoff to rivers over these long time periods.
10,000 yr	Exposures dominated by $^{239+240}$Pu. Unlikely to be any areas giving rise to significant doses (i.e., doses much less than 1 mSv yr^{-1}).
100,000 yr	$^{239+240}$Pu has decayed away. No significant doses from environmental contamination. Low levels of very long-lived ^{99}Tc, ^{129}I, ^{242}Pu, ^{237}Np and U isotopes may be measureable in some areas, though it is likely that erosion processes would have transported the vast majority from the surface soils.

* Estimates for times up to 150 years from present are likely to be accurate. Scenarios for 500 years or more are more speculative, and (conservatively) assume little long-term reduction in doses by erosion of radioactivity from surface soils.

become habitable. However, as shown in Chapters 2–4, ^{137}Cs activity concentrations in some foodstuffs, particularly mushrooms, berries, game and freshwater fish are declining with an effective half-life of the order of 15–30 years, approaching the physical decay half-life of the radionuclide. Hence, restrictions are likely to remain on consumption of some foodstuffs for many decades (Beresford *et al.*, 2001).

Over the next few hundred years, ^{90}Sr and ^{137}Cs isotopes will decay to insignificant levels and exposures will be dominated by ^{241}Am and Pu isotopes (Table 9.1). These doses are likely to be very low in all but the most contaminated 'hot spots'. Though some areas would still be described as 'contaminated', exposures to a hypothetical population would be well within typical variations in natural background doses worldwide. In the very long term (several tens of thousands of years and more), some ^{99}Tc, ^{129}I, ^{237}Np, ^{242}Pu and isotopes of uranium could remain, but levels of these radionuclides in surface soils will not be significant.

It can be concluded with reasonable confidence that low-level environmental contamination by Chernobyl will remain in some areas of the 30-km zone for two hundred years or so. On longer timescales, environmental contamination is unlikely to result in doses greater than $1\,\text{mSv}\,\text{yr}^{-1}$. Note, however, that we have not here considered the destroyed reactor or the many waste disposal sites within the 30-km zone. The long-term disposal of nuclear wastes at Chernobyl is clearly an important problem to be solved in the coming decades.

9.1.3 Countermeasures and emergency response

The initial response of the Soviet authorities to the accident was one of denial and delay which resulted in radiation exposures which could have been avoided. The implementation of measures to reduce exposure to ^{131}I, for example, should have been much more rapid and effective. The first key mistake was the failure to warn the local population to stay indoors and thus reduce external exposures and inhalation of airborne radioactivity. The second failure was the inadequate distribution of potassium iodide tablets, and the failure to prevent people from consuming milk contaminated by ^{131}I during the weeks after the accident. This led to unnecessarily high radiation doses to the thyroid. More effective countermeasures at this early stage could have substantially reduced the number of thyroid cancer cases after the accident.

In many cases, mistakes made in the response to the accident could be attributed not to a lack of knowledge in the scientific literature, but (for many complex reasons) to an ineffective distribution of this knowledge both within the scientific community itself and to decision makers and the general public. For example, restrictions on consumption of milk contaminated by ^{131}I were implemented after the 1957 Windscale accident in the UK, and this action led to a major reduction in the potential dose to the population (Jackson and Jones, 1991). Once the full extent of the Chernobyl accident was acknowledged, however, the evacuation of the 30-km zone was rapidly and efficiently conducted. The huge effort by the many Chernobyl 'liquidators' brought the accident under control and significantly reduced its potential consequences.

The accident led to significant developments in countermeasures to reduce radioactivity in terrestrial foodstuffs. As outlined in Chapter 5, soil management practices and use of fertilisers on agricultural land significantly reduced uptake of radionuclides to plants. A number of methods were developed to administer caesium-binding agents to grazing animals which successfully reduced transfers of radiocaesium to milk and meat.

Many countermeasures were implemented to protect aquatic systems. In general, these measures were ineffective and expensive and in some cases led to relatively high exposures to workers implementing them. The most effective countermeasure was the early restriction of drinking water abstraction and changing to alternative supplies. Restrictions on consumption of freshwater fish have also proved effective in some areas, but have not always been adhered to. It is unlikely that any future countermeasures to protect surface waters contaminated by Chernobyl will be justifiable in terms of economic cost per unit of dose reduction.

There has been increasing appreciation that countermeasure application is not only a 'radiological problem'. For the implementation of robust and effective strategies in contaminated areas, wider social and ethical aspects, environmental considerations and quality of life must be taken into account. The potential effectiveness of holistic and inclusive approaches to coping with contaminated environments has been demonstrated in the fSU in the work of, for example, Lepicard and Hériard Dubreuil (2001). In some western European countries stakeholder groups have been established to consider acceptable countermeasures and their implementation should they be needed in the future (Nisbet et al., 2005).

9.2 CONSEQUENCES OF THE ACCIDENT

9.2.1 Damage to the ecosystem

It is clear that the extremely high levels of radioactivity deposited during and shortly after the accident damaged the ecosystem in parts of the 30-km exclusion zone. As discussed in Chapter 8, in an area of approximately $4\,km^2$, pine trees were killed shortly after the accident and serious damage to trees was observed over a much larger area (several tens of km^2). The damage to pine trees was exacerbated because the exposure occurred during the period of spring growth when the tree is most sensitive to radiation damage. Deciduous trees were much more resistant to radiation than coniferous, though in areas of very high radiation doses, damage to deciduous trees was also observed.

Several months after the accident, and in 1987, damaged trees began to recover from the radiation exposure, although developmental aberrations could be observed in newly grown leaves and reduced pollen and seed viability was observed in pines. Abnormalities in development and reduction of seed viability were also observed in herbaceous vegetation. Increases in genetic mutation rates in trees and various other vegetation types were observed in the short term after the accident, and there was evidence of increased microsatellite mutation rates in wheat several years after the accident. Deciduous trees are replacing pine in the areas which received highest exposure.

Although the direct evidence of fatalities in animals after the accident is sparse, it is likely that in the most highly contaminated areas fatalities occurred both through the direct effects of radiation and through damage to habitats. Reductions in rodent populations were observed in the most contaminated areas of the 30-km zone during 1986, and increases in embryonic mortality were observed in 1986 and 1987. During 1987, however, rodent populations increased due in part to inward migration from less contaminated areas. Overall increases in populations of rodents in the exclusion zone during the years after the accident were influenced by the removal of the human population.

In the first few years after the accident, effects of radiation were observed on the liver of small mammals and some individuals had enlarged spleens. Dose rates to small mammals in the most contaminated areas were above those expected to cause reproductive damage. There is also evidence of 'radioresistance' developing in some species of small mammals in the Red Forest. In general, histological examinations of small rodents from the contaminated sites do not show significant levels of abnormalities (Baker et al., 1999; Jackson et al., 2004).

Invertebrates are amongst the least radiosensitive of organisms (see Box 8.1). However, as the bulk of radionuclides from the Chernobyl fallout were concentrated in the upper soil layer, soil fauna were subjected to a more intensive radiation exposure compared with other animals. Because of the shielding effect of soil, soil-dwelling fauna were less exposed than those living in (less dense) pine litter. However, soil communities will also respond to loss of habitat and changes in land use which have occurred in the exclusion zone since the accident. Evidence of increased morphological variability (e.g., level of wing asymmetry in dragonfly species) has also been observed.

Due largely to the shielding effect of water, radiation doses to aquatic organisms were generally lower than to terrestrial organisms. Exposure of aquatic organisms is unlikely to have had lethal effects, although reproductive, morphological and chromosomal aberrations have been observed in fish and other aquatic organisms.

In assessing the ecological consequences of the Chernobyl accident, the negative impact of radiation on the environment must be weighed against the positive impact the removal of humans from the area has had on wildlife habitats. Nearly 20 years after the accident there is some (often contradictory) evidence of continuing radiation damage to organisms, but this appears to be relatively minor (although also poorly understood). On a macro-ecological (i.e., large-) scale, however, there has been a dramatic increase in populations of wild mammals and bird species living in the abandoned lands (Williams, 1995; Baker and Chesser, 2000). Wildlife in the vast majority of the Chernobyl zone has not only recovered, but is now more abundant and diverse than it was before the accident. The area is now home to large populations of wild boar, wolves and many bird species. Although radiation levels in the 30-km zone are in many areas much higher than those considered safe for human habitation, wildlife has benefited from the absence of human disturbance and damage through, for example, agriculture, hunting and fishing.

Radiation is considered to be a risk to humans when there is a small, but significant, probability of cancer induction in later life. Though cancer induction in animals is possible, a small additional cancer risk does not affect wild populations as a whole. Even at the level of the individual, animals in the wild are less prone to cancer than human populations: wild animals are unlikely to live long enough to be subject to cancer which is in general a disease of old age. Wild animals are most likely to be killed by natural predators or starvation before they reach an age at which cancer risk increases.

Although we cannot, of course, ignore the terrible human consequences of the accident, some observers have argued that in purely ecological terms the accident has been a net *benefit* to the environment. Though this is a somewhat controversial view, it is clear that the consequences of the accident to the human population have far outweighed any ecological consequences. What is also clear is that, although radiation can damage organisms at extremely high levels (such as those seen in the Red Forest shortly after the accident), animal and plant populations are resistant to levels of radiation which are orders of magnitude higher than limits set for human protection. With the removal of human impacts in the exclusion zone, wildlife has thrived in an area considered unfit for human habitation.

9.2.2 Direct health effects of the accident

In the early phase after the accident, 28 people died of radiation sickness. Amongst the 134 confirmed radiation sickness cases, a further 11 people died in the period 1987–1998 from various causes including myelodysplastic syndrome (disorder of the bone marrow which has been linked to radiation), heart disease and cirrhosis of the liver (UNSCEAR, 2000). Approximately 4,000 cases of thyroid cancer were observed in children and adolescents during the period to 2003. These were overwhelmingly due to exposures to ^{131}I during the first weeks after the accident. Studies (Jacob et al., 1998; Jacob et al., 2000; Ivanov et al., 2004; Kenigsberg et al., 2002) have shown an increasing thyroid cancer risk with increasing radiation dose from short-lived ^{131}I. It is expected that thyroid cancer incidence will continue to be elevated in these groups as a result of their past exposure to ^{131}I. In a study in Belarus, Jacob et al. (2000) observed 569 excess cases for the period 1991–1996 and predicted 12,000 excess cases for the period 1997–2036. These workers, however, noted that there was a large uncertainty in this prediction, the range in estimates being 4,000–40,000 cases during the 40-year period.

Although very serious, thyroid cancer is, fortunately, treatable, usually requiring surgery and radioiodine therapy for treatment of the distant metastasis (Demidchik, 2004; Reiners et al., 2002). During the period 1986–2002 in Belarus 1,152 thyroid cancer patients (children and adolescents) underwent surgery: 14 patients (1.2%) have died. Of these, 8 cases of death were caused by progression of the thyroid cancer. Actuarial cause-specific survival at 10 years is 99% (Demidchik, 2004). After treatment, patients need regular monitoring and require continuing therapy to replace the hormones which the thyroid normally produces.

Radiation exposure also potentially leads to an increased incidence of leukaemia. The majority of excess leukaemia cases observed in survivors of the Hiroshima and Nagasaki atomic bombs occurred within 15 years of exposure (Pierce et al., 1996). No clear evidence of increased incidence of leukaemia has yet been observed in populations exposed to radiation from Chernobyl (Boice, 1997; UNSCEAR, 2000). No increased incidence of leukaemia has so far been observed in Ukrainian and Belarussian emergency workers. One study of Russian emergency workers (Konogorov et al., 2000; Tsyb et al., 2002), however, observed an increase in the incidence of non-Chronic Lymphocytic Leukaemia (CLL) between 1986 and 1996. Of the 71,217 Russian emergency workers studied, 21 had contracted non-CLL. Approximately 50% of these cases were expected to have been radiation-induced (Tsyb et al., 2002).

The UNSCEAR (2000) report has summarised the effects of Chernobyl on pregnancy outcomes as follows: 'Several studies on adverse pregnancy outcomes related to the Chernobyl accident have been performed in the areas closest to the accident and in more distant regions. So far, no increase in birth defects, congenital malformations, stillbirths or premature births could be linked to radiation exposures caused by the accident'.

In addition to thyroid cancer, radiation exposures can lead to increases in other 'solid' cancers (solid cancers are cancers other than cancers of the blood, bone marrow or lymphatic system, such as lung and breast cancer). Increases in solid cancers were observed in survivors of the Hiroshima and Nagasaki atomic bombs (see Chapter 1). Thus far, no increase in solid cancers (other than thyroid cancer) has been observed in people exposed to radiation after the Chernobyl accident. A future increase in the number of solid cancers cannot, however, be ruled out. Such cancers typically have a long latency period and the cancer incidence increases with time after exposure (cancer incidence is greater in older people). Therefore, if there is an increase in solid cancer incidence from Chernobyl exposures, most cases will be seen in the coming decades.

However, there is good reason to believe that no increase in solid cancers will be observed. The average dose to the liquidators working in the 30-km zone during 1986–1987 was 100 mSv. This dose leads to a predicted increase in fatal radiation-induced cancer incidence in a lifetime of less than 1%. Set against the 'background' fatal cancer incidence in industrialised countries of around 25% (around one in four people die of cancer) the potential increase due to radiation exposure is relatively small. This potential increase may not be observable, given the difficulties in carrying out the large-scale cancer incidence and dosimetric studies required to detect it.

Most of those affected by Chernobyl received relatively low radiation exposures in comparison with many of the survivors of the Hiroshima and Nagasaki atomic bombs (on which cancer incidence estimates are primarily based). Even in the Chernobyl emergency worker ('liquidator') group, only 4% received doses in excess of 250 mSv (Boice, 1997). In addition, the exposures from the Hiroshima and Nagasaki atomic bombs occurred at an extremely high dose rate – approximately 6 Sv per second (Tubiana, 2000), compared to typical exposures of 0.1 Sv or less per year after Chernobyl. There is evidence (e.g., NRPB, 1995) that cancer incidence (for the same total dose) is higher after acute (high dose rate) exposures compared with chronic (low dose rate) exposures. Further, because of the difficulty in detecting small increases in cancer incidence, there is little empirical evidence of a radiation link to cancer in children or adults at doses of less than about 100 mSv (even after acute exposure).

The uncertainties surrounding the potential carcinogenic effects of chronic low-dose radiation mean that attempts to estimate the total number of deaths which have resulted and will result from Chernobyl are somewhat speculative and can be misleading. Apart from less than 100 deaths which have been linked directly to radiation, the actual number of deaths attributable to Chernobyl is not known accurately, and probably never will be. Cardis *et al.* (2001) predicted excess cancer deaths among liquidators and the residents of contaminated territories of 2,000 and

2,500, respectively. These figures assume that the effects of chronic, low-dose radiation can be extrapolated from the acute, high-dose Japanese bomb survivor data. These workers (Cardis *et al.*, 2001) noted that 'these increases ... would be difficult to detect epidemiologically against the expected background number [in the two groups] of 41,500 and 433,000, respectively'. It should also be noted that there is significant debate on the extrapolation of the acute, high-dose Japanese data to chronic exposures of the order of 100–200 mSv or less (e.g., Jaworowski, 1999; Tiubana, 2000; Gofman, 1990).

The number of excess deaths resulting from the Chernobyl accident is not precisely known. It is clear, however, that the media reports (discussed in *Nature*, 1996) stating that 125,000 people had died within the first 10 years after the accident were misleading. In fact, this figure was the number of deaths from all causes in the affected regions (at the rate expected in unexposed populations) and had no link to radiation exposure (Voice, 1996).

9.2.3 Social and economic consequences

As the information presented in this book has shown, some of the most important consequences of the Chernobyl accident are not due to the direct effects of radiation on human health. The social and economic consequences of the accident were at least as serious as the direct health and environmental effects. As described in Chapter 7, the social consequences of moving over three hundred thousand people from their homes were disastrous. Within a context of general economic decline after the break up of the Soviet Union, the additional losses in production, and remediation costs due to Chernobyl, placed a terrible economic burden on the Ukraine, Belarus and the Bryansk region of Russia.

The psychological and social damage caused by the radiation exposures is difficult to quantify, but was (and still is) huge. As noted in Chapter 7, victims of radiation exposure have to live with the (to them, often unknown) future health risks of that exposure for the rest of their lives. Whilst to the scientific community these risks may be deemed low or insignificant compared with other risks encountered in daily life, the exposure may do enormous psychological (and social) damage which in turn can have real health effects.

Populations in the contaminated areas have been shown to have higher levels of stress, worse perceived health and greater use of medical facilities (e.g., number of doctor visits) than similar unaffected populations. There have been media reports (e.g., Bate, 2004) of large numbers of abortions carried out because of mothers' fears of radiation damage to their unborn child. An increased number of abortions has been reported in the most contaminated regions of Belarus and Russia (UNSCEAR, 2000). There was no evidence of increases in induced abortion rates in Sweden or Austria after Chernobyl (Odlind and Ericson, 1991; Haeusler *et al.*, 1992), but studies in Italy and Denmark reported increased rates as a result of anxiety in pregnant mothers (Knudsen, 1991; Spinelli and Osborn, 1991). Any (wholly understandable) fears for the health of unborn children were unfounded. As discussed above, the scientific evidence has not shown any increase in adverse pregnancy outcomes due to radiation (UNSCEAR, 2000), though in the most affected areas there was an enhanced risk of (treatable) thyroid cancer.

Though many people showed great kindness in supporting Chernobyl 'refugees', in some cases relocated people were shunned because they were considered 'contaminated'. There were also cases of people concealing their status as Chernobyl victims for fear that they would not find partners among the unexposed population (UNDP, 2002).

The economic and social decline of the affected areas was, within a context of general economic decline in the Ukraine, Belarus and Russia, clearly linked to the radioactive contamination of the environment. The perception that towns and villages around the evacuated areas were (and still are) 'contaminated' has led to severe social and economic consequences. Populations in towns and villages in these areas have declined since the accident and there has been a significant demographic shift as young people have moved out of the area. Evidence suggests that it is particularly difficult to attract key workers such as teachers and doctors to towns in the affected areas.

The economic difficulties have in turn affected radiation exposures to the population. Consumption of contaminated wild foodstuffs continues because it is an important source of free food for very poor people. When asked about his consumption of contaminated fish, a fisherman at Kozhany, Russia answered 'the scientists keep telling me that I shouldn't eat the fish – but what else is there to eat?' The economic situation has also affected countermeasure implementation: Chapter 5 reports the decline in use of potassium-based fertilisers in recent years, and the consequent increases in radiocaesium accumulation in some crops.

The social and psychological consequences of Chernobyl were exacerbated by the tendency of the Soviet government to conceal or downplay the seriousness of the accident. This reaction was, to a lesser extent, also seen in many western European countries. Once lost, public confidence in authorities is difficult to regain, and there is widespread mistrust in the Ukraine, Belarus and Russia (and in western Europe) of the pronouncements of government and scientists on radiation issues.

As early as 1991, an International Atomic Energy Authority (IAEA, 1991) study found the psychological effects of Chernobyl to be 'wholly disproportionate to the biological significance of the radiation'. This study identified as a priority the provision of accurate information on the real health risks of radiation to the affected population. Fourteen years after this report, the provision of accurate public information by trusted sources is still urgently needed.

Some scientists have argued that the permanent evacuation of people from the contaminated areas was a serious mistake. Average additional radiation doses to people living in the majority of the evacuated areas would have been from 0.6–6 mSv per year for the period 1986–1996 (Jaworowski, 1999). As discussed above, these doses were in many cases lower than the occupational exposure of many people to natural radiation (e.g., air crew and some miners) and are lower than natural background radiation in many parts of the world. In the UK, for example, measures must be taken to reduce radon gas in houses when levels exceed 200 Bq m^{-3}, a predicted dose rate of approximately 5 mSv yr^{-1} (Denman et al., 1999). In view of the relatively low dose rates in most of the areas affected by Chernobyl, Tiubana (2000) argued that 'the forced evacuation of so many people from their – presumably – contaminated homes calls for ethical scrutiny. A comparison ought to be made between the psychological and medical burden of this measure (anxiety, psychosomatic diseases, depression and suicides) and the harm that may have been prevented'. The International Commission on Radiation Protection (ICRP) recommend that intervention actions would almost always be justified in circumstances where total annual individual doses are above $100\,\text{mSv}\,\text{yr}^{-1}$ and would normally not be justified below $10\,\text{mSv}\,\text{yr}^{-1}$ (ICRP, 1999).

The lack of accurate information and misperception of radiation risk is believed also to have led to changes in behaviour of some affected people. There is evidence to suggest that some people take greater risks with their own health because they believe that their previous exposure to radiation makes other health risks insignificant. One Chernobyl liquidator said that 'I don't worry about the dangers of smoking because I've been exposed to radiation at Chernobyl'. In fact, the average dose to liquidators working in the zone in 1986–1987 was approximately 100 mSv, leading to a predicted potential risk of radiation induced fatal cancer in later life of 0.5%. A lifetime of smoking leads to a 50% risk of death from a smoking-related illness.

9.2.4 Future management of the affected areas

Nearly 20 years after the accident, environmental contamination from Chernobyl is not primarily a problem of dose reduction – it is a social and economic problem. Economic recovery in the most affected areas is crucial. With appropriate countermeasures many (possibly all) of the abandoned areas outside the 30-km zone could be brought back into production given appropriate crop choice, countermeasures and a market for the products. In Belarus this is being done to a certain extent, though it is not clear whether this measure is accepted by the population. As discussed above, the provision of accurate, trusted information to the public is a key component of any remediation measures.

Studies since the Chernobyl accident have highlighted the importance of self-help strategies in reducing radiation doses and enhancing quality of life in the affected areas. Such strategies allow people to make their own (informed) decisions concerning their diet and intake of radioactivity. By encouraging countermeasure implementation by the people themselves, they restore a sense of control over the difficult situation people have to live with. The social and psychological benefits of such measures can far outweigh the benefits in radiation dose reduction. Such measures, however, have thus far only been applied in a few communities as relatively small-scale pilot projects in the fSU.

The management of the destroyed reactor and the waste sites within the 30-km zone will need to continue until a final storage or disposal option has been implemented. In 1999, work was completed to strengthen the sarcophagus roof and stabilise the ventilation stack, which was in danger of collapse (EBRD, 2000). Recently a 20,000-tonne steel shelter has been proposed as a protective cover for the sarcophagus. The last reactor at Chernobyl, Unit 3, was closed in the year 2000 and decommissioning activities are now being carried out at the site. These include construction of facilities for spent fuel storage and treatment and processing of low- and medium-level liquid and solid waste.

9.2.5 Chernobyl and the Nuclear Power Programme

Civilian nuclear energy programmes are built on the premise that nuclear power is so safe that the risk of a major nuclear accident is negligible. Obviously, accidents at nuclear power stations seriously damage public and political confidence in nuclear power. In the USA, no new nuclear power stations have been ordered during the 25 years after the 1979 accident at Three Mile Island. The Chernobyl accident effectively put a stop to nuclear power programmes in both western Europe and the fSU. Public opposition to nuclear power is to a large extent grounded on understandable fears of 'another Chernobyl'. In recent years, this has been exacerbated by the fear of terrorist attacks on nuclear installations.

Recently, however, nuclear power has returned to the political agenda primarily because of concerns over increasing Global energy demand (Figure 9.1), declining fossil fuel reserves and Global climate change. Nuclear energy generation emits virtually no greenhouse gasses in comparison to coal, oil and gas. Whilst the Chernobyl accident had tragic consequences for the human population, even this most terrible of accidents did relatively little long-term environmental damage. This may help to justify a more level consideration of the environmental costs and benefits of nuclear power vs. other power sources. A leading environmentalist, James Lovelock, has recently concluded (contrary to the views of most environmentalists) that 'there is no sensible alternative to nuclear power if we are to sustain civilization'.

Nuclear power generation is increasing worldwide and currently accounts for 17% of global electricity generation (Figure 9.2). The fastest increases in capacity are in the developing economies of Asia, with China in particular rapidly expanding its nuclear programme. At the end of 2002, there were no nuclear power plants under

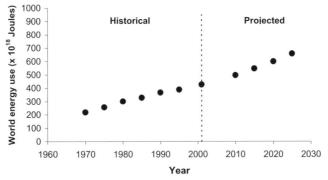

Figure 9.1. Rise in world primary energy consumption from 1970–2025.
EIA (2004).

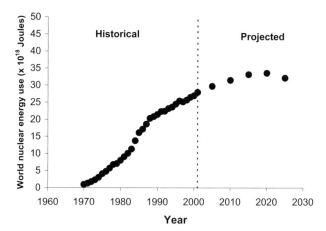

Figure 9.2. World consumption of nuclear energy from 1970 and projected future use. EIA (2004).

construction in western Europe (IAEA, 2003), though one is now planned in Finland (WNA, 2004). There are five reactors currently under construction in Russia, and the Ukraine has completed construction of one of two reactors which will replace lost output from Chernobyl (WNA, 2004). In the US, the current administration supports development of a new generation of nuclear power stations as a key component of national energy policy.

It seems clear that, for the coming decades at least, nuclear power will form an important (and probably increasing) role in global electricity generation. With hundreds of reactors in operation worldwide, reactor design and operation clearly needs to avoid past mistakes. Should a nuclear accident occur, the response of authorities must not only consider the direct protection of the population from radiation exposure, but also the huge social and psychological consequences of radioactive pollution. If 'another Chernobyl' is to be avoided, governments, scientists and the nuclear industry must not forget the lessons learnt in the years since 1986.

9.3 REFERENCES

Baker, R.J., Dewoody, J.A., Wright, A.J. and Chesser, R.K. (1999) On the utility of heteroplasmy in genotoxicity studies: An example from Chornobyl. *Ecotoxicology*, **8**, 301–309.

Baker, R.J. and Chesser, R.K. (2000) The Chernobyl nuclear disaster and subsequent creation of a wildlife preserve. *Environmental Toxicology and Chemistry*, **19**, 1231–1232.

Bate, R. (2004) Chernobyl comes of age. *National Review Online*, September 23, 2004. http://www.nationalreview.com/comment/bate200409230840.asp.

Belli, M. and Tikhomirov, F. (eds) (1996) *Behaviour of Radionuclides in Natural and Seminatural Environments*. Report EUR16531, European Commission, Luxembourg.

Beresford, N.A., Voigt, G., Wright, S.M., Howard, B.J., Barnett, C.L., Prister, B., Balonov, M., Ratnikov, A., Travnikova, I., Gillett, A.G., *et al.* (2001) Self-help countermeasure strategies for populations living within contaminated areas of Belarus, Russia and the Ukraine. *Journal of Environmental Radioacitivty*, **56**, 215–239.

Beresford, N.A. and Wright, S.M. (2005) Non-linearity in radiocaesium soil to plant transfer: Fact or fiction? *Radioprotection – colloques*, **40**, 67–72.

Boice, J.D. (1997) Leukaemia, Chernobyl and Epidemiology. *Journal of Radiological Protection*, **17**, 129–133.

Cardis, E., Richardson, D. and Kesminiene, A. (2001) Radiation risk estimates in the beginning of the 21st Century. *Health Physics*, **80**, 349–361.

Demidchik, Yu.E. (2004) The results of treatment for children and adolescents with thyroid carcinoma. *Proceedings of the 2nd International and Practical Conference 'Mitigation of the Consequences of the Catastrophe at the Chernobyl NPP: State and Perspective'* pp. 30–32. Gomel, Belarus.

Denman, A.R., Barker, S.P., Parkinson, S., Marley, F. and Phillips, P.S. (1999) Do the UK workplace Radon Action Levels reflect the radiation dose received by the occupants? *Journal of Radiological Protection*, **19**, 37–43.

EBRD (2000) Shelter implementation plan. European Bank for Reconstruction and Development, London, 24 pp. (Available at: http://www.iaea.or.at/worldatom/Press/Focus/Chernobyl-15/shelter-fund.pdf)

EIA (2004) *International Energy Outlook*. DOE/EIA-0484(2004). Energy Information Administration, Washington (available at: http://www.eia.doe.gov/oiaf/ieo/).

Fesenko, S.V., Voigt, G., Spiridonov S.I., Sanzharova N.I., Gontarenko I.A., Belli, M. and Sansone U. (2000) Analysis of the contribution of forest pathways to the radiation exposure of different population groups in the Bryansk region of Russia. *Radiation and Environmental Biophysics*, **39**, 291–300.

Gofman, J.W. (1990) *Radiation-induced Cancer from Low-dose Exposure*. Committee for Nuclear Responsibility Book Division, San Francisco.

Haeusler, M.C.H., Berghold, A., Schoell, W., Hofer, P. and Schaffer, M. (1992) The influence of the post-Chernobyl fallout on birth-defects and abortion rates in Austria. *American Journal of Obstetrics and Gynecology*, **167**, 1025–1031.

IAEA (1991) *The International Chernobyl Project Technical Report*, 640 pp. IAEA, Vienna.

IAEA (2003) *Energy, Electricity and Nuclear Power: Estimates for the Period up to 2030*. Reference Data Series No. 1 (RDS-1). IAEA, Vienna.

ICRP (1999) *Protection of the Public in Situations of Prolonged Exposure*. Annals of the ICRP No. 29. Elsevier, Amsterdam.

Ivanov, E.A., Ramzina, T.V. Khamyanov, L.P., Vasilchenko, V.N., Korotkov, V.T., Nosovskii, A.V. and Oskolkov, B.Y. (1994) Radioactive contamination of the environment with ^{241}Am as a result of the Chernobyl accident. *Atomic Energy*, **77**, 629–633.

Ivanov, V.K., Tsyb, A.E., Ivanov, S. and Pokrovsky, V. (eds) (2004) *Medical Consequences of the Chernobyl Catastrophe in Russia: Estimation of Radiation Risks*. Nauka, St. Petersburg.

Jackson, D. and Jones S.R., (1991) Reappraisal of environmental countermeasures to protect members of the public following the Windscale nuclear reactor accident 1957. In: *Comparative Assessment of the Environmental Impact of Radionuclides Released During Three Major Nuclear Accidents: Kyshtym, Windscale and Chernobyl*, pp. 1015–1055. EUR 13574. CEC, Luxembourg.

Jackson, D., Copplestone, D. and Stone, D.M. (2004) Effects of chronic radiation exposure on small mammals in the Chernobyl Exclusion Zone. *Nuclear Energy*, **43**, 281–287.

Jacob, P., Goulko, G., Heidenreich, W.F., Likhtarev, I., Kairo, I., Tronko, N.D., Bogdanova, T.I., Kenigsberg, J., Buglova, E., Drozdovitch, V., *et al.* (1998) Thyroid cancer risk to children calculated. *Nature*, **293**, 31–32.

Jacob, P., Kenigsberg, Y., Goulko, G., Buglova, E., Gering, F., Golovneva, A., Kruk, J. and Demidchik, E.P. (2000) Thyroid cancer risk in Belarus after the Chernobyl accident: Comparison with external exposures. *Radiation and Environmental Biophysics*, **39**, 25–31.

Jaworowski, Z. (1999) Radiation risk and ethics. *Physics Today*, **52**, 24–29.

Kashparov, V.A., Lundin, S.M., Zvarych, S.I., Yoshchenko, V.I., Levchuk, S.E., Khomutinin, Y.V., Maloshtan, I.M. and Protsak, V.P. (2003) Territory contamination with the radionuclides representing the fuel component of Chernobyl fallout. *Science of the Total Environment*, **317**, 105–119.

Kenigsberg, J., Buglova, E., Kruk, J. and Golovneva, A. (2002) Thyroid cancer among children and adolescents of Belarus exposed due to the Chernobyl accident: Dose and risk assessment. In: Yamashita, S., Shibata, Y., Hoshi, M. and Fujimura, K. (eds). *Chernobyl: Message for the 21st Century*, pp. 293–300. International Congress Series 1234. Elsevier, Amsterdam.

Knudsen, L.B. (1991) Legally-induced abortions in Denmark after Chernobyl. *Biomedicine and Pharmacotherapy*, **45**, 229–231.

Konogorov, A.P., Ivanov, V.K., Chekin, S.Y. and Khait, S.E. (2000) A case-control analysis of leukemia in accident emergency workers of Chernobyl. *Journal of Environmental Pathology, Toxicology and Oncology*, **19**, 143–151.

Lepicard, S. and Hériard Dubreuil, G. (2001) Practical improvement of the radiological quality of milk produced by peasant farmers in the territories of Belarus contaminated by the Chernobyl accident. The ETHOS Project. *Journal of Environmental Radioactivity*, **56**, 241–251.

Lux, D., Kammerer, L., Rühm, W. and Wirth, E. (1995) Cycling of Pu, Sr, Cs, and other long-living radionuclides in forest ecosystems of the 30 km zone around Chernobyl. *Science of the Total Environment*, **173/4**, 375–384.

Muramatsu, Y., Rühm, W., Yoshida, S., Tagami, K., Uchida, S. and Wirth, E. (2000) Concentrations of ^{239}Pu and ^{240}Pu and their isotopic ratios determined by ICP-MS in soils collected from the Chernobyl 30 km zone. *Environmental Science and Technology*, **34**, 2913–2917.

Nature (1996) Editorial: Chernobyl's legacy to science. *Nature*, **380**, 653.

Nisbet, A.F., Howard, B., Beresford, N. and Voigt, G. (eds) (2005) Workshop to extend the involvement of stakeholders in decisions on restoration management (WISDOM). *Journal of Environmental Radioactivity*. (Available online at time of writing.)

NRPB (1995) *Risk of Radiation-induced Cancer at Low Doses and Low Dose Rates for Radiation Protection Purposes*. Documents of the NRPB, vol. 6, no. 1. National Radiological Protection Board, Chilton.

NRPB (1998) *Revised Generalised Derived Limits for Radioisotopes of Strontium, Ruthenium, Iodine, Caesium, Plutonium, Americium and Curium*. Documents of the NRPB, vol. 9 no. 1. National Radiological Protection Board, Chilton.

Odlind, V. and Ericson, A. (1991) Incidence of legal-abortion in Sweden after the Chernobyl accident. *Biomedicine and Pharmacotherapy*, **45**, 225–228.

Pierce, D.A., Shimizu, Y., Preston, D.L., Vaeth, M. and Mabuchi, K. (1996) Studies of the mortality of atomic bomb survivors. Report 12, Part 1. Cancer: 1950–1990. *Radiation Research*, **146**, 1–27.

Reiners, Ch., Biko, J., Demidchik, E., Demidchik, Yu. and Drozd, V. (2002) Results of radioactive iodine treatment in children from Belarus with advanced stages of thyroid cancer after the Chernobyl accident. In: Yamashita, S., Shibata, Y., Hoshi, M. and Fujimura, K. (eds), *Chernobyl: Message for the 21st Century*, pp. 205–214. International Congress Series 1234. Elsevier, Amsterdam.

Spinelli, A. and Osborn, J.F. (1991) The effects of the Chernobyl explosion on induced abortion in Italy. *Biomedicine and Pharmacotherapy*, **45**, 243–247.

Tsyb, A.F., Ivanov, V.K., Sokolov, V.A., Gorski, A.I., Maksioutov, M.A., Vlasov, O.K., Khait, S.E. and Godko, A.M. (2002) Radiation risks of Leukemia among Russian emergency workers 1986–1987. *Radiation and Risk* (Special Issue: Health consequences 15 years after the Chernobyl catastrophe: data of the National Registry), pp. 39–50 (available at: phys4.harvard.edu/~wilson/radiation/Si2002/TITLE.html).

Tubiana, M. (2000) Radiation risks in perspective: radiation-induced cancer among cancer risks. *Radiation and Environmental Biophysics*, **39**, 3–16.

UNDP (United Nations Development Programme) and UNICEF (2002). *The Human Consequences of the Chernobyl Nuclear Accident – A Strategy for Recovery* (available at: http://www.undp.org/dpa/publications/chernobyl.pdf).

UNSCEAR (2000) *Report to the General Assembly: Sources and Effects of Ionizing Radiation* (Volume II, Annex J), pp. 453–551. United Nations, New York (available at: http://www.unscear.org).

Voice, E. (1998) Chernobyl legacy. *Nature*, **381**, 642.

Williams N. (1995) Chernobyl: Life abounds without people. *Science*, **269**, 304.

WNA (World Nuclear Association) (2004) *World Nuclear Power Reactors 2003–2004 and Uranium Requirements* (available at: http://www.world-nuclear.org/info/reactors.htm).

Index

Abandoned areas 24, 123
Abortion, induced 298
Absorption from gastrointestinal tract 102
Accident sequence 2
Activated charcoal 197
Acute radiation syndrome (emergency workers) 5, 219, 296
AFCF: ammonium-ferric(III)-cyanoferrate(II) 193–194, 206
Aggregated transfer factor
   For foodstuffs 54, 82, 84, 127
Agricultural products 104
   Exposure via 121
Alpha-particle 5
Americium 71, 107, 290
Aquatic biota
   Exposure of 277
   Radiation effects 278–280
   Uptake of radionuclides 168–181
Aquatic plants 174
Atmospheric nuclear weapons tests 27

Baltic Sea 177, 181
Beta-particle 5
Becquerel, definition of 8
Bentonite 193
Berries (forest) 84, 110, 114–115, 124
Biological half-life 55, 105
Birds 276

Black Sea 160, 175–178
Book of Revelations 4
Bryansk Region (Russia) 200

Calcium, influence on strontium transfer to milk 95
Caesium 69
   Influence of soil properties 61, 87
   In lakes 154–162
   In milk 98, 100, 103
   In rivers 147–154
   Isotopic ratio in Chernobyl fallout 40
   Migration in soil 41–46
   Transfer to animals 93–98
   Uptake to plants 86, 93
Cardiovascular disease 232–233
Cataracts 220
Chernobyl Cooling Pond 162, 166, 170, 174, 277
Chernobyl reactor closure 10, 301
Chernobyl Victim Syndrome 255
Chlorophyll mutation frequency 272
Clean feeding 195–196, 206
Communication 256
Compensation 254
Concentration factor 55, 169
Concentration ratio 54, 82
Coniferous trees 268

Index

Countermeasures 191–203, 293–294
   Effect on economy 247–249, 254
   Implementation in the fSU 204–208
   Social costs 251-254
Cosmic rays 28
Crops, transfer of radionuclides to 86–91

Dairy products 103
Death of pine trees 269, 294
Decontamination 10, 200
Deposition
   Maps 15–17
   Mechanisms 35
Dietary advice 202
Distribution coefficient (K_d) 45, 89, 141–143, 159
DNA strand breaks 280
Dnieper River 139, 147, 149, 170, 174, 177
Dose
   Internal *see* Ingestion dose
   Limiting lifetime dose 23
   To emergency workers 222
   To populations in the fSU 19–21
   To thyroid 21, 228
Drinking water 146, 197–198

Ecological half-life 56, 61–64, 103, 105, 126, 147–148, 152, 163, 165
Economic breakdown 250
Ecosystem damage 294–296
Elbe River 143, 148
Emergency workers *see* Liquidators
Evacuation of populations 6, 239, 252
   Doses to 221
   Effect on wildlife 280–282
External dose 22, 46
   reduction (fSU experience) 200–201

Ferrocyn 206
Finland 149, 152
Fish
   Caesium 168–172
   Consumption in the fSU 141
   Effects of radiation 277–281
   Iodine 173
   Marine 172, 178
   Size effect 171
   Strontium 173

Fixation 57, 59–63, 87
Flooding 153, 198
Foliar absorption 86
Food preparation 203
Forestry workers, doses to 289
France 257
Frayed edge sites 57, 62, 152
Fresh wt./dry wt. ratio 83
Fuel particles *see* 'Hot' particles
Fungi 84, 110–114, 124, 207

Gamma-ray 5
Game animals 116
Genetic diversity 275
Germany 81
Global climate change 301
Grass, transfer of radionuclides to 90
Gray, definition of 8
Groundwater, radionuclides in 179

Half-life
   Physical 8, 12, 55
   Biological 55
   Ecological 56
   Effective ecological 56
Health effects
   Deterministic 16, 218
   Emergency workers 5, 219
   Radiation induced 218
   Stochastic 17, 218
Hiroshima 17, 25, 220, 223
'Hot' particles 17, 25, 220, 223
   Cancer risk 39
   Exposure of trees to 270

Inhalation dose 22, 51
Iodine 65, 82
   In fish 173
   In lakes 161
   In milk 100, 104
   Uptake by thyroid 96
Insects 274
Interception (by plants) 85
   Tree canopies 108, 268
Ingestion dose
   Contributions of milk and meat 204
   Key foodstuffs 207–208

Intervention limits 125
  CFILs 120
  In the fSU 121
  Exceedance in the fSU 206
  Exceedance in western Europe 193
Irrigation water 180
Italy 81

K_d see Distribution coefficient
Kiev Reservoir 139, 160, 162, 171, 181
  countermeasures 197

Lakes, radioactivity in 154–165
  I-131 161
  Sr-90 160
  Water residence time 157
Lake Constance 160, 162, 166
Lake Zurich 154
Leukaemia 220–222, 296
Lime
  Addition to lakes 198
  Effect on radiostrontium uptake 193
Liquidators 11, 239, 296
Live monitoring 196

Marine systems 175–179
Marshall Islands 224
Mayak accident 25, 244
Meat, transfer of radionuclides to 94
Medical radiation 27
Milk 208
  Consumption banned 204
  Transfer of radionuclides to 94, 99–101

Nagasaki 17, 25, 220, 223
Natural radioactivity 27–31
Nevada Test Site 224
Norway 181, 194, 202
Nuclear energy 301
Nuclear weapons fallout 26, 63, 106, 166

Pine forest recovery 272
Physical half-life 55
Physiological health effects 244
Ploughing, effect on radionuclide uptake 105, 192
Plutonium 71, 107, 115, 290

Potassium fertiliser
  Addition to lakes 199
  Effect on radiocaesium uptake 171, 193
Potassium iodide tablets 192, 293
Pripyat 256, 280
  Evacuation 6, 252
  River 139, 144–149, 180, 198
'Private' farms (home food production in fSU) 81
Prussian Blue see AFCF and Ferrocyn

Radical improvement 204
Radioactive decay 55
Radioactive waste 11
'Radioresistance' 295
Radiosensitivity (of different organisms) 267, 295
  Comparative (different tree species) 271
RBMK-1000 reactor 3
'Red Forest' 179, 268, 275, 295
Reindeer 196–197, 202
Relative biological effectiveness 5, 225
Relative risk 226
Release estimate 12
Resuspension 48–53, 86
Rhine 147
Risk perception 258–260
Rivers, radionuclide transport by 53, 143–154, 177
Rodents 275–276
Root uptake 86

Sami 202
Sapropell 204
Sarcophagus 7, 179, 300
Scandinavia 81, 194
Seed germination rate 272
Sediments, radionuclides in 165–168
Selective feeding regime 205
Self-help countermeasures 202, 207–208, 260
Sheep 196
Sievert, definition of 8
Silver-110m 99
Socio-psychological impact 240, 245
Soil
  Radionuclide migration in 41–46, 111
  Influence on radionuclide uptake 87

Soil fauna 274
Solid cancer 231
Stress 245
Strontium 64, 67
   In fish 173, 176
   In game animals 118
   In milk 96–100
   Soil migration 42–43
   Uptake to crops 91–94
Sweden 196, 198, 203

Three Mile Island 25, 257
Thirty-kilometre zone
   Doses to people living in 290
   Establishment of 6
   Groundwater contamination 200
   Inhabited settlements 25
   Lake Glubokoye 168, 290
   Livestock 204

Thyroid cancer 19, 222, 229, 233, 296
   Adult population 231
   Radiation therapy 223
   Surgery for 296
Thyroid dose (Belarus) 228
   Cattle and horses 276
Transfer coefficient 94
   Effect of Ca on transfer to milk 195
Trees 109, 118, 268–272

United Kingdom 82, 145, 180, 194, 257
Uranium 13

Vegetables, radionuclide uptake of 91–94

Weathering half-life 86
Windscale accident 25

Zeolite 197–198
Zones of control 24

Printing: Mercedes-Druck, Berlin
Binding: Stein+Lehmann, Berlin